Roland Trimen, James Henry Bowker

South-African Butterflies - A Monograph of the Extra-Tropical

Species

Vol. II - Erycinidæ and Lycænidæ

Roland Trimen, James Henry Bowker

South-African Butterflies - A Monograph of the Extra-Tropical Species
Vol. II - Erycinidæ and Lycænidæ

ISBN/EAN: 9783744756372

Printed in Europe, USA, Canada, Australia, Japan

Cover: Foto ©berggeist007 / pixelio.de

More available books at **www.hansebooks.com**

SOUTH-AFRICAN BUTTERFLIES:

A MONOGRAPH

OF THE

EXTRA-TROPICAL SPECIES.

BY

ROLAND TRIMEN, F.R.S., F.L.S., F.Z.S., F. Ent. S., &c.

CURATOR OF THE SOUTH-AFRICAN MUSEUM, CAPE TOWN;

ASSISTED BY

JAMES HENRY BOWKER, F.Z.S., F.R.G.S.

COLONEL (RETIRED) IN THE CAPE SERVICE,
LATE COMMANDANT OF FRONTIER ARMED AND MOUNTED POLICE,
GOVERNOR'S AGENT IN BASUTOLAND,
AND CHIEF COMMISSIONER AT THE DIAMOND FIELDS OF GRIQUALAND WEST.

VOL. II.

ERYCINIDÆ AND LYCÆNIDÆ.

LONDON:
TRÜBNER & CO., LUDGATE HILL.
1887.

CONTENTS.

RHOPALOCERA.
    PAGE
    FAMILY II.—ERYCINIDÆ .   1
    Sub-Family—LIBYTHEINÆ   2
    FAMILY III.—LYCÆNIDÆ   7

SYSTEMATIC INDEX   239
LIST OF SPECIES FIGURED IN THE PLATES   241

RHOPALOCERA.

FAMILY II.—ERYCINIDÆ.

Erycinidæ, Swains., "Phil. Mag., Ser. II. vol. i. p. 187 (1827)."
Erycinides and *Libythides*, Boisd., Sp. Gen. Lep., i. pp. 164 and 167 (1836).
Erycinidæ (excl. *Theclinæ*, &c.), Swains., Hist. and Nat. Arrangem. Ins., p. 94 (1840).
Erycinidæ, Westw., Intr. Mod. Class. Ins., ii. p. 357 (1840).
Erycinidæ and *Libytheidæ*, Westw., Gen. D. Lep., ii. pp. 412, 415 (1851).
Erycinidæ, Bates, Journ. Ent., 1861, p. 220; 1864, p. 176.
Lemoniidæ, Kirby, Syn. Cat. Diurn. Lep., p. 282 (1871).

IMAGO.—First pair of legs small and slender; in the ♂ much aborted, the tarsus being without articulations or terminal claws; in the ♀ longer, the tarsus fully developed and with terminal claws.

LARVA.—Of ordinary elongate form, or rather short and subonisciform, usually more or less pubescent; second segment sometimes bearing dorsally two erect spines.

PUPA.—Suspended vertically or obliquely by the tail only, or horizontally by the tail and a silken girdle.

The only constant characters apparently prevailing throughout this extensive Family (containing 69 genera and about 900 species) are those afforded by the fore-legs, which differ so remarkably in the sexes. These organs are in the female, besides the complete development of the tarsi, sometimes twice as large as in the male. Other features characteristic of these butterflies are the usually very small and slender palpi—often scarcely noticeable from above; the three-branched subcostal nervure of the fore-wings; the slenderness of the body; the smoothness of the middle and hind legs, and the small size of their terminal claws; and the thin and fragile structure of the wings;—but all these characters, as Mr. H. W. Bates remarks in his *Catalogue of the Erycinidæ* (*Journ. Linn. Soc. Lond., IX. Zool.*, p. 367, 1868), are liable to many exceptions. Though amazingly varied in form, colouring, and pattern, the insects of this group are all of small size,—the largest of them (*Stalachtis* and *Sospita*) being less than 2½ inches in expanse of wings, the great majority of much smaller stature, and many (such as *Mesene*, *Calydna*, *Parnes*, *Anteros*) among the smallest known butterflies.

Mr. Bates, who paid special attention to them in the Amazons Valley, describes (*loc. cit.*) the habits of the *Erycinidæ* as very varied.

Their flight is of short duration, but in some is very slow and lazy, while in others it is excessively rapid. A large number of the genera have the custom (found also in many *Hesperidæ*) of settling with expanded wings on the under side of leaves near the ground; while many others hold the wings vertically in repose, and a few perch on the upper surface of leaves with the wings only half elevated. Very few species were noticed to frequent flowers.

The Family is essentially Neo-Tropical, and especially abundant in the Equatorial zone. With the exception of the *Libytheinæ*—a small but cosmopolitan Sub-Family—the only members found in the Old World belong to the *Nemeobiinæ*, and it is remarkable that one of these, the well-known *Nemeobius Lucina*, prevails widely in Europe, and is not uncommon in England. More than two-thirds of the known species belong to the *Erycininæ*, a Sub-Family which, unlike the *Eurygoninæ*, has a few representatives to the north of Mexico.

Very few species of *Erycinidæ* have been found in the Ethiopian Region, viz., three *Libytheinæ* and three *Nemeobiinæ*. The former are *Libythea Labdaca*, Westw., of Western Africa; *L. Laius*, Trim., of the Eastern and South-Eastern Coast; and *L. Cinyras*, Trim., of Madagascar and Mauritius; the latter, all of a single genus likewise, are *Abisara Gerontes* (Fab.), and *A. Tantalus*, Hewits., of Western Africa, and *A. Tepahi* (Boisd.), of Madagascar. No representative of the latter group has hitherto been met with in South Africa, and the solitary Erycinide known to inhabit this wide territory is the *Libythea Laius* just mentioned.

Sub-Family—LIBYTHÆINÆ.

Libythides, Boisd., Sp. Gen. Lep., i. p. 167 (1836).
Libytheidæ, Westw., Gen. Diurn. Lep., ii. p. 412 (1851).
Libytheinæ, Bates, Journ. Ent., 1861, p. 220; 1864, p. 176.

IMAGO.—*Head* rather wide, densely hairy above and frontally; *eyes* smooth; *palpi* extremely long, deep at base, closely approximated, projecting horizontally, very densely clothed with scales and hair throughout, but especially on middle joint,—basal joint very small,—middle joint of moderate length, rather swollen and rounded,—terminal joint very much elongated, slender; *antennæ* short and thick, with a gradually or very gradually formed subcylindrical club, blunt at the tip.

Thorax rather robust, clothed superiorly with scales and hairs—the latter long posteriorly,—inferiorly with short dense hair; *pterygodes* longer than usual, hairy. *Fore-wings:* with costa slightly arched; apex squarely acute; hind-margin slightly dentated, more or less strongly angulated on lower radial nervule, below which it is deeply emarginate, but is usually prominent again about extremity of first median nervule; inner margin nearly straight; costal nervure short, ending about middle of costa; first subcostal nervule originating at some distance, second one very little, before extremity of discoidal cell,—third one short, originating far beyond cell,—fourth near third, also short, terminating at apex; upper disco-cellular extremely short—in some cases hardly present,—middle one slender, of moderate length, slightly curved,—lower one rather longer, continuous of curve of third, joining third median nervule at a considerable distance beyond origin of latter; submedian nervure curved upward near base and then more strongly downward; internal nervure slender, almost straight, its extremity anastomosing with submedian nervure just beyond commencement of latter's downward curve. *Hind-wings*: rather large; *costa* prominent at base, and usually more or less so just before apex (in the European *L. Celtis* exceedingly so in both places); hind-margin more or less sinuate-dentate, sometimes with a more prominent dentation at extremity of first median nervule, or at extremity of submedian nervure; inner-margins moderately convex, forming an incomplete, shallow groove; costal nervure running close to and following curves of costa, ending at apex; subcostal nervure branching at a long distance from base; upper disco-cellular nervule slender, transverse, of moderate length, united to second subcostal nervule at some distance from origin of latter,—lower one extremely slender, or almost obsolete, longer and more oblique than the upper, joining third median nervule just beyond origin of latter; internal nervure slender, much curved, very short, terminating on inner margin at some distance before middle. *Fore-legs* of ♂ very small,—femur and tibia about equal in length, the latter very densely hairy,—tarsus rather shorter, also densely hairy, not articulated, without terminal claws; of ♀ much larger and longer, not hairy (except on under side of femur) but scaly,—femur much thicker,—tarsus fully articulated, spiny beneath, and with well-developed terminal claws. *Middle* and *hind legs* rather thick, densely clothed with scales,—tibia rather shorter than femur, its terminal spurs rather short,—tarsus longer than tibia, strongly spinose beneath, with strongly curved terminal claws. *Abdomen* slender, short.

LARVA.—Elongate, cylindrical, shortly pubescent; much resembling that of the *Pierinæ*.

PUPA.—Rather stout, subangulated; median dorsal line ridged, and dorso-thoracic prominence elevated and acute; head with a single pointed projection curving inferiorly; wing-covers projecting beyond line of abdomen. Suspended vertically by the tail only.

(These characters of larva and pupa are given from figures of those of the European *Libythea Celtis*, published by Hübner, Duponchel, and Boisduval respectively.)

The extraordinary length of the palpi, which form a most conspicuous beak-like process in front of the head, readily distinguish the *Libythæinæ* from the rest of the *Erycinidæ*, and, indeed, from all other butterflies. In robustness of structure, angulated fore-wings, and general neuration they approach *Eurytela* and allied genera among the

Nymphalinæ, but do not present the swollen nervures characteristic of the latter; while the perfect tarsi of the fore-legs of the female altogether separate them from that group of butterflies. It is very remarkable, too, to find them sharing with the *Danainæ* a slender but distinct internal nervure of the fore-wings anastomosing with the submedian nervure; and it was probably this character which led Dr. Felder[1] to place them between the *Danainæ* and *Erycinidæ*. The larva, again, is quite unlike that of any group of the *Nymphalidæ*, and is not like those known among the other *Erycinidæ*, but very closely resembles that of the totally distinct *Pierinæ*; while the pupa, on the contrary, does not differ widely from the Nymphalide type.

The twelve or thirteen species of this Sub-Family belonging to but one genus—*Libythea*—are singularly scattered over all the warmer parts of the globe, except, I believe, the continent of Australia and Polynesia. The type of the genus, *L. Celtis*, Fuessly, inhabits Southern Europe and Asia Minor; the Ethiopian Region has three species; India and the Indo-Malayan Islands three; the Austro-Malayan and Australasian Islands two or three; two are natives of the United States and the West Indies; and one is found in Surinam and Brazil. It does not seem improbable that these few and widely-scattered congeners are but the surviving representatives of what was at some former period a numerous and generally-prevalent group.

Genus LIBYTHEA.

Libythea, Fab., "Illiger's Mag., vi. p. 284 (1807);" Latreille, Encyc. Meth., ix. p. 10 (1819); Westw., Gen. Diurn. Lep., ii. p. 412 (1851).

Characters those of the Sub-Family.

There is considerable variation in the different species as regards the length of the palpi, which (as Felder has pointed out) attains its maximum in the American species; and *L. Celtis* is the only member of the genus that I have examined which has the antennæ so thick and so very gradually incrassate from the base. The form of the hind-wings is also variable, none of the species rivalling *Celtis* in the striking sinuosity of the costa; while the hind-margin is in some (the Indian *Myrrha*, Godt., and allies) rounded, without special prominence of any particular dentation,—in the European and American species has a moderate projection at the anal angle, and in the African and some other species presents a very decided process at the extremity of the first median nervule.

The *Libythea* are rather below the middle size, and their colouring is mostly rather dull, consisting of a few fulvous or ochreous-yellow and white spots on a dark-brown ground, except in the case of the males of *L. Geoffroyi* and *Antipoda*, which have the upper side violet or violet-blue.

---
[1] *Diagnoses Lepidopterologicæ*, No. VI. p. 10 (Wien. Entom. Monatschr., 1862).

I have not seen any of these butterflies alive, but from the notes of various collectors they are evidently very active insects, resembling in their flight and habits the smaller *Nymphalinæ*. The Cingalese *L. Rama*, Moore, is noted by Messrs. Hutchison and Mackwood (Moore's *Lepidoptera of Ceylon*, p. 68) as frequenting forest lands and the vicinity of jungle, and darting about and settling on the ground; while the Natalian *L. Laius*, Trim., seems only to settle on the stems and twigs of trees.

The larva of the European species feeds on the "Nettle Tree" (*Celtis australis*), but I am not aware that the food-plant of any other *Libythea* has been recorded.

118. (1.) Libythea Laius, Trimen.

PLATE VII. fig. 3 (♀).

Libythea Laius, Trim., Trans. Ent. Soc. Lond., 1879, p. 337.

Exp. al., 1 in. 10½ lin.—2 in. 2 lin.

♂ *Dark-brown, glossed with bronzy-yellow reflections; with pale fulvous-ochreous bars and spots.* Fore-wing: a longitudinal bar from base occupies lower half of discoidal cell, widening gradually to its abrupt extremity just above origin of first median nervule; immediately beyond bar (sometimes touching or even merged in it) a large, roughly subquadrate spot, which is widest interiorly, and anteriorly reaches to end of discoidal cell; a little beyond and below this spot, a larger paler subovate spot, the upper part of which is traversed by the second median nervule; on outer portion of inner margin a very faint pale-ochreous cloud; an irregular, oblique, subapical row of three white spots, of which the first is nearest middle, largest, exteriorly elongated and crossed in its upper part by the subcostal nervure, and the others are together apart, of about equal size, one above and the other below the second radial nervule. *Hind-wing*: costal margin rather broadly dull-whitish from base as far as a quadrate, pale, whitish-ochreous spot about middle, lying between costal nervure and first subcostal nervule; below and beyond this spot, just above second subcostal nervule, a small ochreous spot; beyond middle, a straight transverse bar formed of four contiguous spots (of which the first is largest and the fourth smallest and less distinct than the rest), extending from just below second subcostal nervule, not far from apex, to first median nervule, not far from its origin. UNDER SIDE.—Hind-wing and apex of fore-wing very variable in tint and clouding, the prevalent ground-colour being *glossy-grey irrorated and hatched with black and fuscous*. *Fore-wing*: paler than on upper side, the bars and spots larger; discoidal cell filled by much enlarged and united bar and spot; a small whitish spot immediately beyond upper part of extremity of cell; second and third spots of subapical row united in one crescent-shaped marking. *Hind-wing*: in two specimens the upper side costal spot and discal bar represented by two irregular whitish transverse rays, interiorly broadly bordered and in parts intruded on by dark-brown clouding,—a similar patch of the clouding being present on hind-margin about extremity of radial nervule; in a third example, only the whitish ray near base is indistinctly represented, while the whole discal region and lower half of discoidal cell (up to base and a straight line through the cell) is dull fuscous-brown; and in the fourth example the whole surface is almost uniformly grey, the rays being faintly shown by a glossier paler clouding, and the intermediate parts by some ochreous tinting,—at extremity of cell an ill-defined blackish spot.

♀ Similar to ♂, except that the markings generally are larger, especially the bar across hind-wing. UNDER SIDE.—More uniform than in ♂, and nearly resembling that of the fourth ♂ example above described, but *with a decided glaucous-green tint* both in hind-wing and in apical region of fore-wing; any trace of paler bars in hind-wing being obsolete, or nearly so.

In the ♂ the fore-wings are more prolonged apically than in the ♀; but the hind-wings are alike in both sexes, the costal prominence beyond middle being very slight, while the projection at the extremity of the first median nervule is very well developed, forming a broad tail or process $\frac{1}{10}$th inch in length. The palpi of the ♀ are longer and more attenuated than those of the ♂.

This *Libythea* is very nearly allied to *L. Labdaca*, Westw., a native of Sierra Leone, Camaroons, Angola, and Congo (see *Genera of Diurnal Lepidoptera*, vol. ii. p. 413 note, pl. lxviii. f. 6). The South-African form may, however, be readily distinguished by the more produced and angulated fore-wings and by the greater prominence of the projection in the hind-wings. The markings of the upper side are all larger, paler, and more fulvous than in *Labdaca*; in the fore-wings the conspicuous disco-cellular bar is a feature wanting or very indistinctly represented in the West-African species, which, moreover, possesses a dull-fulvous or greyish irregular marking (between large discal spot and submedian nervure) absent in *Laius*; and in the hind-wings, the transverse bar is nearly straight instead of arched or concave interiorly, as in *Labdaca*, and the separate spot between the subcostal nervures is peculiar to *Laius*. As regards the under side, the ♂ *Labdaca* (I have not examined the ♀) is very much like the more strongly-marked ♂ s of the South-African insect; but the fulvous-ochreous in fore-wing is limited to a short basi-disco-cellular bar.

The genus *Libythea*, so widely distributed over the earth, yet containing so very few species, was not apparently known to possess any African representative until Westwood (*loc. cit.*) in 1851 described and figured the species from Sierra Leone already mentioned. In 1866 I described (*Trans. Ent. Soc. Lond.*, Ser. iii. vol. v. p. 337) as *L. Cinyras* a scarce *Libythea*, inhabiting Mauritius and Madagascar, and noted at the same time that Mr. Waller, of the Zambesi Mission, had shown me a *Libythea* taken near the Shiré River, which I judged from recollection might be the same species. Since the discovery of the South-African *Libythea*, however, and especially looking to the fact of its occurrence at Quilimane, not far north of the Zambesi Delta, I have little doubt that Mr. Waller's specimen was probably referable to *Laius*, and not to *Cinyras*.

In December 1869, Mr. Walter Morant sent me the first evidence of the occurrence of a *Libythea* in Natal, in the shape of a coloured drawing of a ♀ taken by him on the 9th of that month at Avoca, Victoria County; but I heard nothing more of the species until 1873, when the late Mr. E. C. Buxton met with it near D'Urban, and sent me a photograph and a much injured ♀ specimen. Colonel J. H. Bowker, F.Z.S., in September 1878 landed at Quilimane, and there took six examples of the same species, which he forwarded to the South-African Museum; and he and Mr. P. F. Payn of D'Urban have subsequently taken several specimens of both sexes at Pinetown, Illovo, and other localities in the coast region of Natal. From these latter specimens, mostly in fine condition, my description is drawn up.

Mr. Morant noted his example as taken "on the top of a small tree in a waggon-road through thick bush;" while Colonel Bowker describes the individuals captured by himself in April 1879 as taking short flights, like those of a "Skipper," from one flower to another. The latter further wrote in December 1879 as follows, viz. :—"I think the *Libythea* spoil themselves by their fighting; the eight that I have taken were all about the tops of the same trees, actively flying around, and settling with open wings at the ends of the twigs. Mr. Payn says that his examples were captured under similar circumstances. It is curious how nearly in appearance the *Libythea* resembles *Crenis natalensis* when at rest; both settle head downward with closed wings on the bark of the same species of tree, and it is then hardly possible to distinguish them from each other. This has probably led to the *Libythea* being not often taken by the Natal collectors."

Localities of *Libythea Laius*.

I. South Africa.
   E. Natal.
      a. Coast Districts.—D'Urban (*E. C. Buxton*). Umgeni, Pinetown, and junction of Umlaas and Isipingo Rivers (*J. H. Bowker*). Avoca (*W. Morant*). Illovo (*P. F. Payn*).
II. Other African Regions.
   A. South Tropical.
      b. Eastern Coast.—Quilimane (*J. H. Bowker*).

Family III.—LYCÆNIDÆ.

Lycanida, Leach, " Edinb. Encyc., ix. p. 129 (1815)."
Lycœnidœ, Stephens, "Illustr. Brit. Ent., Haust., i. p. 74 (1827)."
Eumenides and *Lycénides*, Boisd., Sp. Gen. Lép., i. pp. 163-164 (1836).
Erycinidœ (part), Swains., Hist. and Nat. Arrangem. Ins., p. 94 (1840).
Lycœnidœ, Westw., Intr. Mod. Class. Ins., ii. p. 358 (1840); and Gen. Diurn. Lep., ii. p. 468 (1852).

Imago.—*Head* of moderate size or rather small; *eyes* naked or hirsute; *palpi* usually rather long, sometimes very long, scaly, seldom hairy,—the terminal joint slender, distinct, often longer in ♀ than in ♂; *antennæ* variable in length, but usually rather short,—the club commonly somewhat abruptly formed, but sometimes very gradually.

Thorax generally rather slender or moderately stout, sometimes robust. *Wings* large, variable in outline; *fore-wings* usually rather truncate, but occasionally produced,—hind-margin entire, rarely elbowed superiorly; subcostal nervure three- to five-branched, usually four-branched; discoidal cell closed; *hind-wings* rounded, with entire or very slightly denticulated hind-margin, or produced in anal-angular portion, which often bears from one to three longer or shorter tails; inner margins often forming an incomplete groove about the abdomen; discoidal cell closed by very slender nervule. *Legs* rather short, often thick, scaly, rarely hirsute; tibial terminal spurs usually small, sometimes minute, rarely wanting; *fore-legs of ♂* (with rare exceptions) *with tarsus not articulated, but consisting of a single long joint ending in one slightly curved claw;* those of ♀ with the ordinary articulations and terminal hooked claws.

Abdomen usually slender and rather short; rarely thickened or elongate.

LARVA.—More or less onisciform, broadest and thickest about middle, often with dorsal humps or corrugated; sometimes downy or with fascicles of hairs; head and feet very small, inferior, hidden from view above.

PUPA.—Short, thick, usually much rounded; blunt at extremities. Attached by the tail, and (usually) by a girth of silk round the middle; rarely unattached, and lying in the earth or under stones.

This family is a very distinct, compact, and natural one, the character of the unarticulated tarsi of the first pair of legs in the male sex being all but universal, and the principal other points of structure presenting but little variation. This sameness throughout so very numerous an assemblage of species renders the task of classification exceedingly difficult; and no lepidopterist has hitherto found characters adequate to warrant the establishment of divisions or sub-families. To discover natural limitations to the genera is a matter of scarcely less difficulty; and all who have studied the family will admit that, notwithstanding the labours of Westwood, Hewitson, Felder, Moore, and other entomologists, the existing definitions of many of the accepted genera are anything but satisfactory. The work of discriminating species is an arduous one in all large genera, but it becomes specially so in such immense groups of closely-related forms as *Lycæna*, *Thecla*, and *Amblypodia*.

Between fifteen and sixteen hundred species have been recorded, and about forty-seven genera created for their reception,—the three genera just mentioned by themselves including fully half of the entire number of known species.

Among the genera which show more divergence from the common type are specially noticeable the Oriental *Liphyra* and the African *Liptena*, *Pentila*, *D'Urbania*, *Pseuderesia*, *Alæna*, *Mimacræa*, *Deloneura*, *Arrugia*, and *Lachnocnema*, all of which bear some resemblance to butterflies of other families, or even to moths, and the three latter of which present the anomaly of completely articulated and clawed fore-tarsi in the male. This latter character decidedly indicates affinity with the subfamily *Pierinæ* of the *Papilionidæ* ; and a further minor feature of resemblance is found in the reduced or obsolete spurs of the tibiæ.

The great mass of the *Lycænidæ* consists of small butterflies, and it is rare to find any that approach the middle size. In this respect they resemble the *Erycinidæ*, but the range of size is somewhat more extended, the smallest members of the genus *Lycæna* being barely half an inch across the expanded fore-wings, while the giant of the family, *Liphyra Brassolis*, Westw., reaches 3¼ inches. Some of the larger species of *Phytala*, *Epitola*, *Thecla*, and *Amblypodia*, however, reach or slightly exceed an expanse of 2½ inches; but a very large majority of *Lycænidæ* ranges between 1 inch and 1¾ inches. What these butterflies want in stature is more than compensated by their great beauty ; very few of them are of dull colouring (except a good many of the females), and such as are so on one surface of the wings are very often of great splendour on the other. Intense and very generally highly-metallic uniform tints of purple, violet, blue, green, red, or orange prevail on the upper side,—the males, as usual, almost always much exceeding the females in brilliancy ; and the less dazzling but infinitely varied colouring and elaborate patterns of the under side (of close agreement in the sexes of each species), are often rendered still more effective by glittering golden, silvery, or steely spots. No other family of butterflies exhibits such a striking and varied development of the appendages or "tails" borne on the hind-margin of the hind-wings, varying in number from one to three, and in length from a mere dentation to more than an inch, and often curiously widened, fringed, and twisted. In connection with these appendages, it may be noted that all, or nearly all, the *Lycænidæ* (as mentioned above in the general remarks on Rhopalocera, under the heading, " 5. Haunts and Habits "), when settled temporarily, have the custom of moving the hind-wings alternately up and down, so that their upper surfaces rub against each other,—a movement which the "tails," as well as the usual metallic-dotted eye-like spots on the hind-margin, render additionally noticeable.

The Ethiopian Region has not hitherto yielded a very large number of species, only rather under 200 being recorded, but it is rich in genera, 22 of the 47 recognised by most authors being known to occur. Of these 22 genera, no less than 14 (or 13, if, as I think most probable, *Hewitsonia*, Kirby (= *Coryphon*, Boisd.), is not separable from *Epitola*, Westw.) are limited to Africa, viz., *Capys*, *Chrysorychia*, *Pentila*, *Liptena*, *D'Urbania*, *Alæna*, *Phytala*, *Epitola*, *Deloneura*, *Pseuderesia*, *Mimacræa*, *Arrugia*, and *Lachnocnema*. Those peculiar to Tropical Africa are *Phytala*, *Epitola*, *Liptena*, *Pseuderesia*, and *Mimacræa*.

Southern Africa appears to be comparatively richer than the tropical belt of the continent, but this is no doubt partly due to its having been better searched for the smaller butterflies. It has representatives of 17 genera, the species numbering at present 120.[1] Three genera, viz., *Capys*, Hewitson, *Deloneura*, Trim., and *Arrugia*, Wallengr., are peculiar; while *Alæna*, Boisd., and *Lachnocnema*, n.g., do not appear to be known from north of the Equator. *Zeritis* is also a specially South-African genus, 23 of the 28 species known not being found elsewhere. The genus most numerously represented is the cosmopolitan *Lycæna*, of which 47 species are recorded; *Iolaus* has 8, *Aphnæus* 7, *Lycænesthes* 6, and *Hypolycæna* 5 South-African species. Of the remaining genera, there are of *Deudorix* and *D'Urbania* each 4 species, of *Arrugia* 3, *Myrina*, *Chrysorychia*, and *Lachnocnema* 2, while the five others have each but a single representative.

The *Lycænidæ* exhibit no power of sustained flight, although many of them are very active, and some even swift in their motions. They keep very much about particular spots, and many of them (such as *Thecla* and *Lycæna*) are decidedly gregarious. Some of the finest species of *Myrina*, *Iolaus*, &c., remain always about a special bush or tree, returning repeatedly to it when disturbed, and seldom taking wing when unmolested. These and many of the ground-loving species of *Zeritis* can, with caution, be captured by hand. The swiftest and most alert of the South-African species are *Capys Alphæus* and *Deudorix Antalus*, which frequently succeed for some time in evading the collector. *Pentila*, *D'Urbania*, and *Alæna* are exceedingly slow on the wing.

The curious larvæ, shaped like wood-lice for the most part, are extremely sluggish, and look in many cases more like a *coccus* or some vegetable excrescence than caterpillars. Some of them are smooth, many clothed with a short down, some with fascicles of short bristles on regularly disposed tubercles, and a few hairy generally. Several are regularly corrugated dorsally, and others prominently humped in one or two places. Very few of the larvæ of South-African *Lycænidæ* have been discovered; that of *Myrina ficedula*, Trim. (Pl. i. fig. 7), is humped as just mentioned, and coloured protectively in imitation of its food-plant; that of *Iolaus Silas*, Westw. (Pl. i. fig. 8), is very convex dorsally, and slightly forked at the tail; that of *Hypolycæna Lara* (Linn.) (Pl. ii. fig. 1), of almost even width throughout; and that of the aberrant *D'Urbania Amakosa*, Trim. (Pl. ii. fig. 2), unusually slender and hairy.[2]

[1] In the tabular statement given above (in the general remarks on Rhopalocera, under heading, " 7. South-African Butterflies ") I have given the genera as 15 and the species as 116. Since that table was drawn up I have withdrawn *Liptena*, as not possessing a true South-African representative, and added *Chrysophanus* (which I had intended not to keep separate from *Lycæna*), *Alæna* (misplaced in *Acræina*), and *Lachnocnema*, n.g. The additional species are a *Deudorix*, an *Aphnæus*, and two *Lycænæ*.

[2] The caterpillar of *Spalgis Epius* (Westw.) is figured in Moore's *Lepidoptera of Ceylon* as possessing several dorsal erect and lateral horizontally projecting long curved spines (*op. cit.*, pl. 34, fig. 1b.)

The thick blunted *pupæ* are usually attached to the food-plants of their respective larvæ by both the tail and a silken thread at the middle, but the latter support is not always present (ex. *Myrina, Iolaus*, &c.); and some rare cases (as of the European *Thecla Quercus*) are recorded where the chrysalis is simply buried in the ground. I have found the pupa of *Zeritis Thyra* (Linn.) lying unattached beneath a stone.

Genus LYCÆNA.

Lycæna, Fab., Illiger's Mag., vi. p. 285 (1807), Section 3 [part]; Herrich-Schaeffer, Syst. Bearb. Schmett. Europ., i. p. 111 (1843); Westw., Gen. Diurn. Lep., ii. p. 488 (1852).
Polyommatus, Latreille [part], "Hist. Nat. Crust. et Ins., xiv. p. 116 (1805);" and Encyc. Méth., ix. p. 11 (1819).
Lycæna [part], Trim., Rhop. Afr. Aust., ii. p. 233 (1866).

IMAGO.—*Head* small, usually more or less hairy in front; *eyes* hairy or naked (hairy in the majority of South-African species); *palpi* long, compressed, scaly,—second joint densely clothed with long flattened scales, and usually also with bristly hairs, beneath,—terminal joint variable in length, slender, closely scaled, acute, projecting forward obliquely or horizontally; *antennæ* slender, of moderate length or rather short, with each joint ringed with white at its base,—the club abruptly-formed, rather elongate-ovate, not pointed, flattened and often hollowed beneath, slightly curved.

Thorax moderately stout or slender, well clothed with silken down both above and below. *Fore-wings* rather elongate; costa nearly straight beyond basal curve; hind-margin more or less convex, entire; costal nervure short, ending about or a little before middle; subcostal nervure almost always four-branched, but sometimes only three-branched (three instances among South-African species),—the first and second nervules given off before extremity of discoidal cell, and the first sometimes united to costal nervure at about three-fourths of latter's length from base (fourteen instances among South-African species),—third nervule given off about half-way between extremity of cell and apex,—fourth ending at apex; upper radial nervule united to subcostal nervure at extremity of cell, lower to middle meeting point of slender transverse disco-cellular nervules; lower disco-cellular joining third median nervule at a little beyond latter's origin. *Hind-wings* rather elongate, having a very convex hind-margin, but anal-angular portion not prominent (usually much rounded off); costa very slightly arched or almost straight beyond basal prominence; costal nervure ending at apex; subcostal nervure branched considerably before middle; discoidal cell short; disco-cellular nervules very slightly curved outwardly,—radial nervule originating at their middle point of junction; hind-margin often bearing a short almost linear tail at extremity of first median nervule (eighteen South-African instances). *Fore-legs* of ♂ rather long,—femur hairy beneath,—tibia scaly, sometimes armed at extremity with a hook or curved spine superiorly, or with a single inferior straight spine, or with a pair of (or more) straight spines,—tarsus rather curved, spinose beneath, terminating in a single curved claw;—of the ♀ similar, but with tibia more rarely armed, and with fully-developed articulated tarsus; *middle* and *hind legs* rather short and slender, femora hairy beneath, tibiae with short terminal spurs, tarsi very spinose beneath.

LARVA.—Broad and thick, the back very convex, the under side flattened; head and legs very small. Usually of some shade of green or yellow, marked dorsally with longitudinal and sometimes oblique lateral streaks.

PUPA.—Broad, thick, rounded, smooth; anterior extremity somewhat narrowed and depressed, blunt.

(These characters of larva and pupa are derived from the figures given by many authors.)

This genus, of world-wide distribution, is unmanageably numerous in species, but, as in the similar case of *Papilio*, it seems impossible satisfactorily to divide it. In my examination of the forty-seven species known to inhabit South Africa, I have been met with the same failure of distinctive characters in groups and sections that in superficial features seem natural ones, which Westwood (*op. cit.*) commented on thirty-four years ago in his general survey of the species then recognised. Thus the presence of a tail on the hind-wings is found to associate forms otherwise so different as *Bætica*, *Sybaris*, and *Jobates*; the absence of one branch of the subcostal nervure (which only occurs in three species) links to the allied *Cissus* and *Jobates* so very distinct a congener as *Barberæ*, while it separates the latter from such a very close ally as *Metophis*; the junction of the first subcostal nervule with the costal nervure (which is found in the European *Tiresias*, *Fischeri*, and *Alsus*) characterises fourteen species obviously pertaining to five different groups; while naked eyes predominate, no fewer than nineteen species scattered over four groups have hairy ones; and the presence of a hook or straight spine, or both of these, or of several spines at the extremity of the tibia of the fore-legs, in one or both sexes, is equally irregular and misleading as a key to associate allied forms. The only mode of arrangement available seems to be the unsatisfactory one afforded by the colouring and pattern of the under side of the wings, which was adopted by Herrich-Schäffer (*op. cit.*) in tabulating the European species.

All the *Lycænæ* are of small size, the largest not measuring two inches across the expanded fore-wings, while the smallest are the most minute of all butterflies, expanding from half to three-quarters of an inch only. Blue of various tints is the predominant colour in the genus, especially in the males; the females being usually brown or grey shot more or less with blue. The under side varies from pure-white to many shades of grey and brown as its ground, while the darker pattern constantly consists, in both fore and hind wings, of a central spot or lunule, a discal row or chain of spots, and a submarginal and hind-marginal row of spots or lunules. These markings, in those cases where the ground is not white, are edged or ringed with white; and the pattern is traceable in every variety of irregularity and confluence throughout the very numerous species. The hind-wings are further characterised by a sub-basal series of dark round spots, usually ringed with white, and by one or more round black spots centred with metallic silvery-blue or green, and edged inwardly by an orange lunule, near the posterior angle.

It is by no means easy to define the limits of species in this genus, and lepidopterists differ widely as to the limits permissible to simple variation. Between four and five hundred species have been described, and of these probably nearly four hundred will be recognized; while many new forms will certainly be discovered as remote and little-known countries come within the range of systematic collecting. The Palæ-arctic, Oriental, and Australian Regions appear to be approximately about equally rich in *Lycæna*, each possessing between eighty and ninety species, the Oriental being apparently a little richer than the other two. The Nearctic Region comes next, with about seventy species; and then the Ethiopian with fifty-nine. The Neo-Tropical Region is, on the contrary, extremely poor, yielding but fifteen or sixteen kinds; but it is, on the other hand, amazingly rich in the not distantly allied genus *Thecla*, of which fully 450 South-American species have been described. *Lycæna* has an almost universal distribution, ranging in latitude from the far Arctic parallel of 81° 45' (*L. Aquilo*, Boisd.) to Chili (*L. Sibylla*, Kirby), and in longitude literally round the globe. Oceanic islands mostly have one or more representatives of the genus; and even the poverty-stricken (in butterflies) New Zealand possesses two. As far as at present known, the genus is more fully developed in Southern than in Tropical Africa, 47 species being recorded from the former and 32 from the latter; but this is very probably not the real state of the case, as the smaller butterflies are quite unknown from the greater part of the huge tropical area. Of the known South-African *Lycæna*, 27 appear to be peculiar to the sub-region; 19 of the remaining 20 are recorded from South-Tropical Africa; and one (*Messapus*, Godt.) from North (but *not* South) Tropical Africa. Of the 19 just mentioned, 14 extend through both African tropics, and another (*Gaika*, Trim.) inhabits both South-Tropical Africa and Continental India and Ceylon; two (*Telicanus*, Lang, and *Trochilus*, Frey.) range into North Africa, Southern Europe, and the south-western extremity of Asia; *Lysimon*, Hübn., to the latter wide distribution adds India and Java; and *Bætica*, Linn., the most dominant species in the genus, nearly all the Oriental Region and many parts of the Australian, including Victoria and the Sandwich Islands.

The *Lycænæ* are active, flower-frequenting insects, for the most part of short flight not far from the herbage. They are, with few exceptions, lovers of open ground, and many are very gregarious, a large number keeping together about a particular spot and sporting round some favourite cluster of plants. The common and generally distributed species in South Africa are *Bætica, Palemon, Lingeus, Telicanus, Thespis, Cissus, Messapus,* and *Lysimon;* but most of the others are more or less local, and some (such as *Griqua, Macalenga, Bowkeri, Puncticilia, Stellata,* and *Barberæ*) extremely so.

As an aid to determining the South-African species, I have arranged them in the following manner, viz. :—

SECTION A.—*L. Osiris, Asopus, Parsimon, patricia, glauca, Caffrariæ, Asteris, Ortygia, Methymna, puncticilia, hypopolia, Cissus, Jobates, Hippocrates, Niobe, Tantalus, ignota, Letsea, dolorosa, Messapus, Mahallokoana, Lysimon, lurida, stellata, Gaika, Trochilus, Metophis, Barberæ* (28 species).

SECTION B.—*L. Bætica, Sichela, notobia, Tsomo, Noquasa* (5 species).

SECTION C.—*L. Lingeus, Palemon.*

SECTION D.—*L. Telicanus.*

SECTION E.—*L. Jesous, Macalenga, Moriqua, Natalensis* (4 species).

SECTION F.—*L. Hintza, Calice, Melæna, Griqua, Sybaris* (5 species).

SECTION G.—*L. Thespis, Bowkeri.*

These sections are simply characterised by the principal features in the colouring and pattern of the under side of the wings, brief particulars of which will be found below, forming the heading of the several sections.

SECTION A.—Under side variable in tint, from very pale whitish-grey to dark brownish-grey; all the markings white-edged or white-ringed; spots near base of hind-wing constantly black or fuscous and round; spots of discal series often black or fuscous, but sometimes little darker than ground-colour, and usually sufficiently confluent or contiguous to form a chain; a submarginal lunulated dark transverse streak, and a narrow hind-marginal white streak.

L. Osiris, Hopff. [tailed]; *Asopus,* Hopff.; *Parsimon* (Fab.) [tailed]; *patricia,* Trim. [tailed]; *glauca,* Trim. [tailed]; *Caffrariæ,* Trim. [tailed]; *Asteris* (Godt.) [tailed]; *Ortygia,* Trim.; *Methymna,* Trim.; *puncticilia,* Trim.; *hypopolia,* Trim.; *Cissus* (Godt.); *Jobates,* Hopff. [tailed]; *Hippocrates* (Fab.) [tailed]; *Niobe,* Trim.; *Tantalus,* Trim.; *ignota,* Trim.; *Letsea,* Trim.; *dolorosa,* Trim.; *Messapus* (Godt.); *Mahallokoana,* Wallengr.; *Lysimon* (Hübn.); *lucida,* Trim.; *stellata,* Trim.; *Gaika,* Trim.; *Trochilus,* Frey.; *Metophis,* Wallengr.; *Barberæ,* Trim.

119. (1.) **Lycæna Osiris,** Hopffer.

♂ *Lycæna Osiris*, Hopff., "Monatsb. K. Preuss. Akad. Wissensch., 1855, p. 642, n. 21;" and Peters' Reise Mossamb., Ins., p. 409, pl. xxvi. ff. 11, 12 (1862).

Exp. al., (♂) 1 in. 1–3 lin.; (♀) 1 in. $1\frac{1}{2}$–$3\frac{1}{2}$ lin.

♂ *Pale shining pinkish-violaceous with a slight cupreous gloss; a narrow greyish fuscous hind-marginal edging (submacular in hind-wing); cilia shining pale-greyish, near anal angle of hind-wing becoming white. Fore-wing:* a faint, narrow, terminal, disco-cellular greyish-fuscous lunule. *Hind-wing:* a linear, black, white-tipped tail at extremity of first median nervule; above and below the nervule respectively two black hind-marginal spots (of which the upper is the larger), bounded internally by orange-yellow lunules (of which the upper is a good deal broader), and externally by a very fine white line. UNDER SIDE.—*Pale whitish-grey, with a very faint tinge of brownish; in each wing a terminal disco-cellular lunule, and a discal submacular transverse stria, both darker than the ground-colour, and white edged on each side,—two submarginal rows of white lunules (the space between them in fore-wing also darker than the ground-colour), and a hind-marginal white line immediately followed by a brown terminal line. Fore-wing:* markings beyond middle extending from fourth subcostal nervule to submedian nervure; discal row almost regular, becoming more macular and bent slightly inwards inferiorly. *Hind-wing:* a sub-basal row of three small, black, rounded, rather indistinctly white-ringed spots; discal row more irregular than in fore-wing, the seventh spot (between first median nervule and submedian nervure) lunular, farther from base than the rest; the first spot (immediately below costal nervure) nearer base than the rest, black and rounded like the sub-basal ones; inner submarginal row of lunules sagittiform, rather suffused, outer row sub-sagittiform and forming with hind-marginal line a series of imperfect annulets; black hind-marginal spots near anal angle rather smaller than on upper side, incompletely ringed with greenish-silvery, their orange-yellow lunules rather larger.

♀ *Brownish-grey, rather widely suffused from bases with bright pale-bluish; discs dingy-whitish; in hind-wing the rows of white lunules and the white and blackish hind-marginal lines more distinctly marked than on the under side. Fore-wing:* a distinct fuscous terminal lunule; bluish suffusion extends over lower part of discoidal cell along inner marginal area and over disc to beyond middle; in most specimens more or less distinct representation of lower two-thirds of two suffused submarginal rows of white lunules resembling those of hind-wing. *Hind-wing:* bluish suffusion usually fills discoidal cell, but is otherwise more restricted than in fore-wing; upper of the two hind-marginal black spots and its adjacent orange-yellow lunule much larger than in ♂. UNDER SIDE.—As in ♂, but the discal submacular row less regular and more separated into spots in fore-wing, and wanting the eighth (inner-marginal) spot in hind-wing. *Hind-wing:* the lower (anal-angular) of two hind-marginal spots obsolete, or nearly so, but some of the greenish-silvery scales and part of the orange-yellow lunule present.

The under side in both sexes, but especially in the ♀, is sometimes very obscurely marked, the white markings being faint, and the three sub-basal spots and first spot of discal row in the hind-wing scarcely darker than the rest.

As Mr. A. G. Butler has remarked (*Ann. and Mag. Nat. Hist.*, 1875, p. 397), *Osiris*, Hopff., appears to be the African representative of the Indian and Javan *Cnejus*, Fab. (*Suppl. Ent. Syst.*, p. 430, 1798). On comparing specimens from Natal and Delagoa Bay with the Javan ♀ in the Horsfield Collection in the British Museum, I could detect little difference, except that the under-side markings of the *Osiris* ♀ s were all rather wider and less constricted.

Hopffer founded the species on a single ♂ from Querimba. In February and March 1867 I met with a few examples of both sexes on the coast of Natal. Colonel Bowker in June 1880 sent me the paired sexes taken at D'Urban. Both sexes of this pair were remarkably obscure in the under-side markings, but the ♀ more so than the ♂. The specimens that I took frequented long grass in sheltered spots.

Localities of *Lycæna Osiris*.

I. South Africa.
   E. Natal.
      a. Coast Districts.—D'Urban. Verulam. Avoca and Pinetown (*W. Morant*).
   F. Zululand.—St. Lucia Bay (*the late Colonel H. Tower*).
   H. Delagoa Bay.—Hewitson Coll., Brit. Mus.
   K. Transvaal.—Pretoria (*W. Morant*). Lydenburg District (*T. Ayres*).

II. Other African Regions.
   A. South Tropical.
      b. East Coast.—"Querimba."—Hopffer.

120. (2.) Lycæna Asopus, Hopffer.

♂ ♀ *Lycæna Asopus*, Hopff., "Monatsb. K. Preuss. Akad. Wissensch., 1855, p. 642, n. 22;" and Peters' Reise Mossamb., Ins., p. 410, pl. xxvi. ff. 13–15 (1862).
♀ *Lycæna Kama*, Trim., Trans. Ent. Soc. Lond., Ser. iii., i., p. 403 (1862).
♀ *Lycæna Asopus*, Trim., Rhop. Afr. Aust., ii. p. 249, n. 149 (1866).

Exp. al., (♂) 1 in.—1 in. 2½ lin.; (♀) 1 in. 1–3 lin.

♂ *Pale brownish-grey, rather variable in depth of tint, slightly paler discally; on hind-marginal edge a fuscous line; cilia whitish-grey; hind-wing with two submarginal rows of white lunules; no tail. Fore-wing:* a faint, dusky, narrow, terminal disco-cellular lunule; near posterior angle some very indistinct traces of a hind-marginal series of whitish lunules. *Hind-wing*: outer submarginal row of lunules more or less combining with a hind-marginal white line to form imperfect rings; hind-marginal spot between first median nervule and submedian nervure conspicuous, black, bordered inwardly by a broad orange-yellow lunule; in some specimens, a very indistinct, thin, dusky, disco-cellular terminal lunule. UNDER SIDE.—Agreeing very closely with that of *L. Osiris*, Hopff. (*q. v.*); *the ground-colour rather more brownish, rendering the white markings more distinct. Hind-wing: an additional spot in subbasal row*, just below base of first median nervule; inferior hind-marginal spot (next anal angle) either wanting altogether or represented by a few black, greenish-silvery, and (rarely) orange scales.[1]

♀ *Darker, shot from bases over discs with bright pale-blue. Forewing:* disco-cellular lunule darker, broader; whitish lunules near posterior angle usually less indistinct than in ♂ and margined on both sides by darker marks; in one example, beyond these lunules, is a short thin whitish hind-marginal line; blue suffusion occupies lower half of discoidal cell, and extends over disc and along inner margin to beyond middle. *Hind-wing:* submarginal series of white lunules and hind-marginal white line much better marked than in ♂; black hind-marginal spot and adjoining orange lunule larger. UNDER SIDE.—As in ♂, but the markings commonly not so distinct, especially in hind-wing, the sub-basal spots, and the first spot of discal row.

The dull pale-grey ♂ of this species differs very much in appearance from the ♂ *Osiris*; but the ♀ s of the two forms, notwithstanding that *Osiris* has tails and *Asopus* none, are remarkably similar; and in both sexes the under side of the two species are almost identical. *Asopus* is decidedly a smaller insect than *Osiris*; and the minute differences pointed out in the description seem to be constant.

Mr. W. S. M. D'Urban, who first sent me this *Lycæna*, wrote that it was not common near King William's Town in March. I met with it sparingly on the coast of Natal in February 1867; the few specimens I captured were flitting about low shrubs and herbage.

Localities of *Lycæna Asopus*.

I. South Africa.
    B. Cape Colony.
        b. Eastern Districts.—King William's Town (*W. S. M. D'Urban*).
    E. Natal.
        a. Coast Districts.—D'Urban. Victoria County.
        b. Upper Districts.—Estcourt (*J. M. Hutchinson*).
    F. Zululand.—St. Lucia Bay (*the late Colonel H. Tower*).
    H. Delagoa Bay.—Lourenço Marques (*Mrs. Monteiro*).
    K. Transvaal.—Potchefstroom District (*T. Ayres*).

---

[1] Hopffer (*loc. cit.*) mentions a ♀ from Senegal in which this inferior spot is as well developed as the superior one.

II. Other African Regions.
A. South Tropical.
  a. Western Coast.—" Chinchoxo (*Falkenstein*)."—Dewitz.
  b. Eastern Coast.—" Querimba."—Hopffer.
B. North Tropical.
  a. Western Coast.—" Senegal."—Hopffer.

121. (3.) Lycæna Parsimon, (Fabricius).

♂ *Papilio Parsimon*, Fab., Syst. Ent., p. 526, n. 349 (1775).
Papilio Celæus, Cram., Pap. Exot., iv. pl. ccclxxix. ff. K, K (1782).
♂ *Polyommatus Parsimon*, Godt., Enc. Meth., ix. p. 683, n. 209 (1819).
Lycæna Celæus, Trim. [part], Rhop. Afr. Aust., ii. p. 247, n. 148 (1866).
Lycæna Asteris, Wallengr. [♀], K. Sv. Vet.-Akad. Handl., 1857, p. 40, n. 12.

Exp. al., ♂ in. $5\frac{1}{2}$–$6\frac{1}{2}$ lin.; ♀ ♂ in. 6–$9\frac{1}{2}$ lin.

♂ *Dull brownish-grey; a fuscous hind-marginal edging line; cilia brownish-grey at base, but whitish outwardly*,—in hind-wing partly interrupted with narrow fuscous at extremities of nervules. *Fore-wing*: an almost obsolete or very indistinct terminal disco-cellular dusky lunule; in some specimens very faint, paler and darker discal and submarginal marks indicating pattern of under side. *Hind-wing*: much as in fore-wing, the indications of under-side pattern less indistinct, especially towards hind-margin, where an outer submarginal series of paler lunules appears, as well as a thin hind-marginal whitish line, more pronounced, near anal angle; between first and second median nervules, a rather well-marked black spot, bordered inwardly by a pale ochrey-yellow lunule; often indistinct traces of a smaller similar spot below first median nervule; *an extremely short linear black tail, not white-tipped*, at extremity of first median nervule. UNDER SIDE.— Almost exactly as in *Asopus*, but *the ground-colour more tinged with brownish, and the white markings wider and more conspicuous, especially the inner of the two submarginal rows of lunules*. *Fore-wing*: discal row decidedly less regular, bent inward on median nervules, its lowest spot (above submedian nervure) often indistinct, sometimes geminate. *Hind-wing*: the four black white-ringed spots of sub-basal row all conspicuous,—the third touching the second, and just below median nervure at origin of its first nervule; in discal row the last spot (on inner margin), as well as the first spot (close to costa), is conspicuous, rounded, black, white-ringed, and the sixth spot (between first and second median nervules) touching lower extremity of broad terminal disco-cellular lunule; smaller hind-marginal spot, at anal angle itself, much better developed than in *Asopus*, and (except for its much smaller size and the suffused condition of the yellow lunule) like the larger superior one; *between the lowest of the inner submarginal series of lunules and the last inner-marginal spot of discal row a slender longitudinal white streak*.

♀ *Darker brownish-grey; discs paler and broadly shot from bases to beyond middle with violaceous blue; hind-wing with the outer row of submarginal lunules broad, white, and (except towards costa) well-marked. Fore-wing*: terminal disco-cellular dusky marking much broader than in ♂, rather conspicuous; violaceous-blue suffusion just as in *Asopus*, ♀; some very indistinct submarginal traces of paler markings. *Hind-wing*: terminal disco-cellular lunule much thinner than in fore-wing, but distinct; *beyond it, a short, central, dusky, transverse streak, representing discal row;* blue suffusion rather more extended outwardly than in *Asopus* ♀, *mingling with and obscuring the wide whitish suffused lunules of inner submarginal row;* superior hind-marginal black spot considerably larger than in ♂, and its broader adjoining lunule orange-yellow,—inferior one also much better developed. UNDER SIDE.—As in ♂, but discal row rather more regular.

Notwithstanding its very much larger size and its possession of a very short tail on the hind-wings, this butterfly is a close ally of *Asopus*, Hopff., so much so, that it looks like an enlarged copy of the latter. The ♂ in both is remarkable for the entire absence of blue on the upper side, while the ♀ has a considerable blue suffusion from the bases over all the discal area except towards costa. The characters emphasised in the above description are the chief distinguishing features of *Parsimon*.

Fabricius (*op. cit.*) undoubtedly described the ♂ only, making no mention of the conspicuous blue of the ♀, though he notes the variation of the disco-cellular dark spot of the fore-wings and white lunules of the hind-wings, which approximate to the latter sex, in appearance, some ♂ specimens. The supposed type of *Parsimon* (see Butler, *Cat. Fab. D. Lep.*, p. 166) in the Banksian Collection in the British Museum is, however, a ♀ of rather dull colouring, but shot with bluish from the bases; and, after a minute inspection of the specimen, I came to the conclusion that it was not the ♀ of *Parsimon*, but of the closely-allied *L. patricia*, Trim. The latter species has all the upper side of the ♂ of a pale violaceous-blue, which precludes the possibility of that sex being taken for ♂ *Parsimon;* but the ♀s of the two forms are with difficulty separable,—the only constant distinction being in the sub-basal row of spots on the under side of the hind-wing, which in *Patricia* has one spot less than in *Parsimon*, as the third (between median and submedian nervures) is wanting.[1]

I met with a few examples of both sexes in Natal in February and March 1867, both on the coast and inland; they flew actively about long grass on the ridges and sides of hills.

Localities of *Lycæna Parsimon*.

1. South Africa.
    D. Kaffraria Proper.—Bashee River (*J. H. Bowker*).
    E. Natal.
        a. Coast Districts.—D'Urban (*M. J. M'Ken* and *J. H. Bowker*). Avoca (*W. Morant*). Verulam.

---

[1] It is due to the kindness of Mr. Chr. Aurivillius, of the State Museum, Stockholm, who sent me three of the typical specimens of *Asteris*, Wallengr., for inspection, that I have been able to identify the ♀ noted by the latter author to be = *Parsimon*, F., ♀.

 b. Upper Districts.—Udland's Mission Station. Intzutze River. Estcourt (*J. M. Hutchinson*).
 K. Transvaal.—Potchefstroom (*W. Morant*). Potchefstroom District (*T. Ayres*).
 II. Other African Regions.
  A. South Tropical.
   a. Western Coast.—"Kinsembo, Congo (*H. Ansell*)."—A. G. Butler.
  B. North Tropical.
   a. Western Coast.—Sierra Leone.—Coll. Brit. Mus.
   b. Eastern Coast.—"Abyssinia (*Raffray*)."—Oberthür.

122. (4.) Lycæna patricia, *sp. nov.*

Lycæna Asteris, Wallengr. [♂, part], K. Sv. Vet.-Akad. Handl., 1857, p. 40, n. 12.
Lycæna Celæus, Trim. [♂, part], Rhop. Afr. Aust., ii. p. 247, n. 148 (1866).
 ♂ *Cupido Parsimon,* Wallengr., K. Sv. Vet.-Akad. Förhandl., 1875, p. 88, n. 40.

Exp. al., (♂) 1 in. 6–7½ lin. ; (♀) 1 in. 7½–9 lin.

 ♂ Pale rather dull violaceous-blue with a slight pinkish tinge; a narrow hind-marginal fuscous edging, linear in hind-wing; cilia fuscous-grey with white tips,—between extremities of second median nervule and submedian nervure of hind-wing also whitish at origin. *Fore-wing:* a slender, sublunulate, terminal disco-cellular striola; costa very narrowly edged with fuscous, slightly radiating on subcostal nervules. *Hind-wing:* costa bordered with fuscous-grey above first subcostal nervule; between first and second subcostal nervules, a hind-marginal fuscous mark, sometimes succeeded inferiorly by very indistinct traces of a hind-marginal series of similar marks; the spot between first and second median nervules small, rounded, well-defined, black, bordered inwardly by a broad orange-yellow lunule ; *tail* on first median nervule rather short, black, white-tipped. UNDER SIDE.—Almost the same as in *Parsimon*, but *the ground-colour usually darker; the terminal disco-cellular lunule* (*in both wings*) *more or less angulated outwardly; and the discal row narrower, more macular, and not so close to the inner* (*sagittiform*) *submarginal lunules. Hind-wing: only three spots in sub-basal row*,—the spot which in *Parsimon* lies between median and submedian nervures being absent; these spots, as well as those at beginning and end of discal row, always very distinctly marked, but smaller than in *Parsimon*.

 ♀ *Rather broadly bordered costally and hind-marginally with dark brownish-grey, the discal area in both wings being paler and very widely shot from bases with bright lilac-blue inclining to a pink tinge;* markings almost the same as in *Parsimon* ♀. *Fore-wing:* terminal disco-cellular marking very broad, more so than in *Parsimon*, and more or less angulated both inwardly and outwardly. *Hind-wing:* blue suffusion usually extending (less thickly) as far as hind-margin; white lunules of outer submarginal row thinner and less conspicuous; terminal disco-cellular lunule usually indistinct. UNDER SIDE.—As in ♂.

This species is so intimately allied to *L. Parsimon*, Fab., that I should not have separated it except for the extraordinary difference of colouring in the ♂ s; the species named having the upper side of an uniform brownish-grey without a trace of blue, while *L. patricia* has it all light violaceous-blue except along the extreme costal and hind-marginal borders. The longer tail on the hind-wings is, however, characteristic of *Patricia*. A ♂ from Zululand is of a purer, more shining blue upper side than usual, but does not present any other differences. A ♀ from Estcourt, Natal, is darker than usual above, and the blue is less violaceous, occupies a smaller area in the fore-wings, and is almost obsolete in the hind-wings. In a ♂ and ♀ from the Transvaal, the under side is darker than usual, and the submarginal markings are unusually faint; the ♀ is also very obscure on the upper side, with the blue much reduced in both wings, and with an indistinct short discal row of very small fuscous spots. One of the two ♂ s kindly sent to me as typical *Asteris*, Wallengr., by Mr. Chr. Aurivillius from the State Museum at Stockholm, belonged distinctly to this form, only differing in a broader terminal, disco-cellular lunule on the upper side of the fore-wings.[1]

In the Transvaal, Mr. W. Morant took this *Lycæna* early in October 1870; and Wallengren (*loc. cit.*) mentions that Mr. N. Person took, in the same country, during that month, what is described as the ♂ *Parsimon*, but appears to be the ♂ *Patricia*.

Localities of *Lycæna patricia*.

I. South Africa.
   B. Cape Colony.
      b. Eastern Districts.—Uitenhage (*S. D. Bairstow*).
      c. Griqualand West.—Klipdrift, Vaal River (*J. H. Bowker*).
      d. Basutoland.—Maseru (*J. H. Bowker*).
   D. Kaffraria Proper.—Tsomo and Bashee Rivers (*J. H. Bowker*).
   E. Natal.
      b. Upper Districts.—Estcourt (*J. M. Hutchinson*).
   F. Zululand.—Napoleon Valley (*J. H. Bowker*).
   K. Transvaal.—Potchefstroom District (*T. Ayres*). Pretoria (*W. Morant*).

123. (5.) Lycæna glauca, *sp. nov.*

Lycæna Asteris, Wallengr. [♂, part], K. Sv. Vet.-Akad. Handl., 1857, p. 40, n. 12.

Exp. al., (♂) 1 in. 7–9 lin.; (♀) 1 in. $6\frac{1}{2}$–10 lin.

♂ *Glittering silvery-blue, with a slight greenish gloss* (in tint intermediate between upper sides of *L. Corydon*, Scop., and *Daphnis*, W. V.); *a very narrow fuscous hind-marginal edging, linear in hind-*

[1] It is possible that *Lycæna Neyus*, Felder, of which I only know the figures (Nos. 1 and 2) on t. xxxv. vol. ii. of the Lepidoptera of the *Reise der Novara*, may be the ♀ of *L. patricia*. It is, however, smaller than usual, and the blue suffusion is represented as stopping short of the outer disc of the fore-wings. The under side is quite like that of *Patricia*, except that the third (inner-marginal) spot of the sub-basal row on the hind-wings is not given. The locality noted for it is Bogos, in North-Eastern Africa.

wing; cilia as in *L. patricia.* *Fore-wing:* terminal disco-cellular fuscous lunule usually wider and more distinct than in *Patricia;* costa narrowly edged with brownish-fuscous; nervules near apical and hind-marginal border defined with fuscous. *Hind-wing:* costa bordered with very pale fuscous-brownish above first subcostal nervule; a faint hind-marginal mark of the same colour between first and second subcostal nervules; orange lunule inwardly bordering hind-marginal black spot between first and second median nervules broader than in *Patricia,* and slightly suffused; tail at extremity of first median nervule longer and more slender. UNDER SIDE.—*Quite as in Patricia, except for the following slight differences:*—*Fore-wing:* usually an additional small spot at commencement of discal row, between second and third subcostal nervules; terminal disco-cellular marking narrower and much more acutely angulated outwardly. *Hind-wing:* terminal disco-cellular marking narrower, the white edging of its outward angulation centrally prolonged as far as inner edge of discal macular row; the second spot of this row, between the subcostal nervules, narrower and more elongate, and placed more obliquely.

♀ Pattern and markings agreeing with those of ♀ *Parsimon* and ♀ *Patricia,* but the colour of wide suffusion from bases over discs very different, being of a pale silvery-greenish scarcely tinged with blue. *Fore-wing:* terminal disco-cellular spot very conspicuous, broader than in the species named, and relieved on each side by a whitish edging; outer series of submarginal lunules (mesially traversing hind-marginal dark border) less indistinct, and slightly (in one example deeply) tinged with ochre-yellow. *Hind-wing:* angulated terminal disco-cellular striola very slender, more distinct in its upper portion; usually two or three minute fuscous discal spots; both rows of submarginal white lunules well marked, the inner one more so than in the two species named; orange-yellow lunules near anal angle largely developed (more so than in *Parsimon*); in one specimen a third orange lunule between second and third median nervules. UNDER SIDE.—As in ♂, except that the terminal disco-cellular marking in each wing is broader and usually much less angulated.

This remarkably-coloured form was first communicated to me by Mr. Chr. Aurivillius in 1881, as one of the typical examples of Wallengren's *Asteris* in the Stockholm Museum; and I shortly after noted another ♂ and a ♀ in a collection from Kinsembo (Congoland), in the possession of Mr. Doncaster. I have subsequently received a series of both sexes from Mr. F. C. Selous, who took them in the North-West Transvaal in February and March 1882. There is little but the peculiar silvery hue of the upper side to separate this form from *L. patricia:* but this feature is so pronounced in both sexes that I am of opinion it justifies keeping *L. glauca* distinct.

Localities of Lycæna glauca.

I. South Africa.
    K. Transvaal.—Marico and Limpopo Rivers (*F. C. Selous*). Potchefstroom District (*T. Ayres*).

II. Other African Regions.
   A. South Tropical.
      a. Western Coast.—Kinsembo, Congo (*W. Doncaster*). Angola.—Coll. Brit. Mus.
   B. North Tropical.
      a. Western Coast.—Gaboon.—Coll. Brit. Mus.

124. (6.) Lycæna Caffrariæ, *sp. nov.*

♂ ♀ *Lycæna Asteris*, Trim. [part], Rhop. Afr. Aust., ii. p. 247, n. 148 (1866).

Exp. al., (♂) 1 in. 6–7½ lin. ; (♀) 1 in. 6–9 lin.

♂ *Glittering pale violaceous-blue, with a slight pinkish tinge ; narrow blackish hind-marginal edging, linear in hind-wing and at posterior angle of fore-wing ; cilia whiter than in L. glauca or L. patricia, especially in hind-wing.* Fore-wing : terminal disco-cellular mark linear, faintly marked, or nearly obsolete ; costa very thinly edged with greyish-fuscous, radiating on subcostal nervules. *Hind-wing :* costa bordered with dull-greyish above first subcostal nervule ; just before hind-margin from apex, a parallel fuscous streak, becoming widely broken into small, more or less indistinct spots, below second subcostal nervule ; black hind-marginal spot between first and second median nervules, inwardly bounded by an inconspicuous rather pale ochrey-yellow lunule ; often traces of a smaller similar spot below first median nervule ; tail short, edged and tipped with white. UNDER SIDE.—*Brownish-grey ; transverse markings darker, conspicuously edged with white on both sides.* Fore-wing : terminal disco-cellular mark moderately wide, reniform ; in discal row two lowest spots smaller than the rest, and third spot sometimes prolonged inwardly, so as to touch disco-cellular mark ; *outer white edge of discal row fused with inner submarginal row of white lunules ; outer submarginal row constituting an almost straight and continuous denticulated streak.* Hind-wing : a sub-basal row of four small black-ringed spots (of which the third, between median and submedian nervures, is sometimes wanting) ; first and last spots of discal row also black, small, white-ringed, *but the rest enlarged and confluent, and the second and third spots prolonged inwardly, and either confluent with or touching the long, curved, narrow, terminal, disco-cellular mark ;* outer white edge of discal row more or less completely confluent with suffused sagittiform lunules of inner submarginal row ; rings formed by lunules of outer row with hind-marginal white line complete, but thin, and pointed on inner side ; hind-marginal black spot outwardly spangled with bluish-silvery, the lunule inwardly bounding it thin, pointed, inconspicuous, pale-yellow.

♀ *Resembling the ♀ Patricia, but darker ; the blue suffusion more glittering, but less extended outwardly.* Fore-wing : terminal disco-cellular dark spot conspicuous and well developed, but not quite so broad ;

paler traversing streak in hind-marginal dark border extremely indistinct. *Hind-wing:* lunular white marks of inner submarginal row acutely sagittiform, more or less shot with blue, farther than usual from outer row; rings formed by those of outer row with hind-marginal whitish line rather indistinct; black spot and adjoining orange-yellow lunule better marked than in ♂ (in one example a second much fainter orange lunule between second and third median nervules). UNDER SIDE. —As in ♂; hind-marginal white line wider, rather suffused.

The ♂ of this beautiful *Lycæna* is well characterised by the glossy shining surface of its blue upper side, which in tint is almost identical with that of *L. patricia*. As shown by the features of the under side in both sexes emphasised in the above description, *Caffrariæ* diverges from the *Parsimon* group and approximates more towards *Asteris*, Godt.,—a tendency also indicated in the darker upper side of the female.

I have not seen many specimens of this form. Colonel Bowker forwarded a few from Kaffraria Proper in 1863 or 1864, and a single male from King William's Town in 1872. In January and February 1870 I met with several of both sexes in the Albany District of Cape Colony. They frequented the most elevated hills, flying with some swiftness over the bushes and herbage.

Localities of *Lycæna Caffrariæ*.

I. South Africa.
  B. Cape Colony.—Grahamstown, New Year's River, and Zwaartwater's Poort, Albany District. King William's Town (*J. H. Bowker*).
  D. Kaffraria Proper.—Bashee River (*J. H. Bowker*).

125. (7.) **Lycæna Asteris**, (Godart).

PLATE VIII. figs. 3 (♂), 3a (♀).

♂ ♀ *Polyommatus Asteris*, Godt., Enc. Meth., ix. p. 657, n. 137 (1819).
Lycæna Celæus, Trim. [part], Rhop. Afr. Aust., ii. p. 247, n. 148 (1866).

Exp. al., (♂) 1 in. 6–7½ lin.; (♀) 1 in. 7½–9 lin.

♂ *Dark violaceous blue; a blackish hind-marginal border, continuous and rather narrow in fore-wing (rarely traversed close to its outer edge by a slender white line), macular or submacular in hind-wing; cilia white, narrowly fuscous near origin, in fore-wing somewhat irregularly mixed with fuscous, in hind-wing incompletely interrupted with fuscous at extremities of nervules.* Fore-wing: terminal disco-cellular fuscous lunule slender, rarely indistinct; costa very narrowly edged with fuscous. *Hind-wing:* costa bordered with fuscous above first subcostal nervule, and apical area as far as second subcostal nervure; maculæ of hind-marginal border relieved by more or less complete rings of bluish-white; spot between first and second median nervules rather larger than the rest, black, inwardly edged by a faint, usually imperfect, thin, yellow lunule; *tail* on first median nervule extremely short, black. UNDER SIDE. — *Pale greyish-brown, darker in hind-wing; disco-cellular terminal marks very broad, almost cordiform, of a much darker brown, white edged; discal spots of transverse row similarly coloured, more or less confluent, one or two of them prolonged inwardly; inner submarginal row of white lunules (very acuminate in hind-wing) in close contact with outer edge of discal row; space between two submarginal rows also darker in hue;* hind-marginal white line distinct, uniting with the outer row of lunules to form inwardly-pointed white rings. *Fore-wing:* first spot of discal row minute, close to costa, a little nearer base than the second; all the spots below second more or less elongated inwardly, especially the fourth (which almost always touches disco-cellular spot), and the seventh, which is much longer than any other, being extended far towards base by two small spots (one or both of them sometimes completely confluent with it); dark space between two submarginal rows of white lunules broad, pretty even throughout; lowest hind-marginal white ring suffused,'geminate. *Hind-wing:* a small, short, narrow black streak, externally white-edged, at origin of costal nervure; a sub-basal series of four small black white-ringed spots; between the two middle spots of this row and terminal disco-cellular spot some whitish scaling; first spot of discal row rounded and separate from the second, but brown, not black; second spot also rounded; all the rest (except the last, which is separate, small, rounded, black, and inner-marginal) more or less elongate and confluent; the fourth longest, and, with the third, sometimes confluent with disco-cellular spot; hind-marginal black spot between first and second median nervules small, outwardly sealed with bluish-silvery, and imperfectly ringed with dull ochre-yellow; a similar smaller spot close to anal angle. *Cilia* more generally white than on upper side, and regularly though narrowly interrupted with fuscous at extremities of nervules in each wing.

♀ *Dull fuscous, shot from bases over discs with rather bright violaceous-blue. Fore-wing:* blue fills discoidal cell, extends a little beyond it, and from base covers inner-marginal and discal area to rather beyond middle; disco-cellular spot considerably wider than in ♂, but not very broad; rarely from third median nervule downward, indistinct traces of a discal series of elongate spots; occasionally, more or less indistinct indications of the hind-marginal series of whitish rings shown on the under side. *Hind-wing:* rarely the traces of a disco-cellular spot and a discal series of spots about middle; blue occupies much the same area as in fore-wing, but extends farther outward—almost or quite to hind-margin; inner submarginal series of whitish lunules usually hardly distinguishable in blue suffusion; hind-marginal rings more distinct; black spot larger, and its yellow lunule better marked. UNDER SIDE.—As in ♂.

Asteris is well characterised by its dark violaceous-blue colouring on the upper side, and by the broad, partly confluent, dark-brown markings, conspicuously edged with white, on the under side. The actual specimens on which the species was founded are stated by Godart to have been taken by M. Jules Verreaux about Table Mountain, and the description he gives accords very nearly with numerous examples collected by myself in the same locality. The true *Asteris* has not to my knowledge been found away from the Cape peninsula, but it has a very close ally inhabiting various parts of the Colony, —*L. Ortygia*, mihi, - of which I once took an example at Wynberg, near Cape Town, where true *Asteris* is particularly prevalent.

This *Lycæna* appears in the later spring and early summer; I have not noticed it until the later half of October or after the end of December. Both sexes are conspicuous on the wing, and the male has a sustained rather swift and irregular flight. They frequent both hills and low ground in spots well clothed with grass and shrubs; they do not appear to visit flowers very often, but are fond of resting on the stems of grasses,—in which latter position I have sometimes succeeded in catching them with my fingers. They are rather local, but often numerous where they occur.

Localities of *Lycæna Asteris*.

I. South Africa.
 B. Cape Colony.
  a. Western Districts.—Cape Town. Noord Hoch, and Simon's Town.

126. (8.) Lycæna Ortygia, *sp. nov.*

♂ ♀ *Lycæna Asteris*, Trim. [part], Trans. Ent. Soc. Lond., 1870, p. 361.

Exp. al., (♂) 1 in. 4½–6½ lin.; (♀) 1 in. 6½–7¾ lin.

Closely allied to *L. Asteris*, (Godt.)

♂ *Violaceous-blue, considerably lighter than in Asteris;* terminal disco-cellular lunules and hind-marginal markings similar but better defined; *cilia* white, *regularly and broadly interrupted with fuscous at extremities of nervules, throughout. Hind-wing :* no yellow lunule adjoining inner-marginal black spot between first and second median nervules; *no tail.* UNDER SIDE.—*Paler than in Asteris,* the discal and submarginal markings very little darker than the ground-colour; *terminal disco-cellular marks and discal rows much narrower, the latter rather widely separated (except at its lower extremity in fore-wing) from the inner submarginal row of white lunules. Fore-wing :* discal row curving slightly inward as far as third median nervule, but its lowest spot (which has none of the inward prolongation so frequent in *Asteris*) more outwardly placed, so that its external white edge almost meets the lowest white lunule of the inner row; space between two rows of lunules narrower and more broken than in *Asteris;* outer row scarcely forming rings with white hind-marginal line,—no suffusion at posterior angle. *Hind-wing :* basal streak and sub-basal row of four spots as in *Asteris,* but the latter larger and much more conspicuous; *first (costal) spot of discal row,* as well as last, *round and black like the sub-basal ones,* instead of brown; inner submarginal lunules much blunter and thicker, outer ones much blunter but thinner; hind-marginal black spot, ornamented externally with bluish or greenish-silvery and ringed with yellow, larger and much more distinct.

♀ *Not so dark as Asteris, the blue rather paler and more violaceous;* the white cilia regularly interrupted with fuscous, more conspicuous than in ♂. *Fore-wing:* disco-cellular spot narrower than in *Asteris;* blue usually not occupying upper part of discoidal cell,—in one specimen very restricted and indistinct, but crossed discally by a row of elongated fuscous spots. *Hind-wing:* blue less developed externally; discal lunules blunter; in one specimen two indistinct elongate dark discal spots, and in another one spot; hind-marginal rings indistinct and incomplete. UNDER SIDE.—As in ♂, but lunules of inner sub-marginal row (especially in hind-wing) more acute and elongate.

I place with this near ally of *Asteris,* Godt., readily separable by its strongly chequered cilia and non-caudate hind-wings, a ♀ I captured at Mossel Bay in September 1858, which, in the contiguity of the disco-cellular mark and the discal row in the fore-wing, and in their actual confluence in the hind-wing, as well as in the suffusion of the white markings generally, approaches true *Asteris,* but in other respects agrees with *Ortygia.*

I do not know whether this *Lycæna* is really as scarce as it appears to be. Besides the Mossel Bay example just referred to, I have not met with it except in the case of a worn straggler near Cape Town; but two specimens from the neighbourhood of Grahamstown were sent to me by Mrs. Barber and Mr. H. J. Atherstone. Colonel Bowker, however, found it rather prevalent in Basutoland, and forwarded twelve or thirteen examples captured in January 1869. He noted that the butterfly inhabited the tops of hills, and that the ♀s sat quietly in the grass, not moving unless disturbed, while the ♂s coursed actively about. Three ♂s from the Orange Free State were sent to me by Mr. W. Hart at end of 1871; they agreed in all respects with the Basutoland examples.

Localities of *Lycæna Ortygia.*

I. South Africa.
B. Cape Colony.
  a. Western Districts.—Cape Town. Mossel Bay.
  b. Eastern Districts.—Grahamstown (*Mrs. Barber* and *H. J. Atherstone*).
  d. Basutoland.—Maseru and Koro-Koro (*J. H. Bowker*).
C. Orange Free State.—Special locality not noted (*W. Hart*).

127. (9.) Lycæna Methymna, Trimen.

♀ *Lycæna Methymna,* Trim. [part], Trans. Ent. Soc. Lond., 3rd Ser. i. p. 280 (1862).
♂ ♀ *Lycæna Cælæus,* var., Trim., Rhop. Afr. Aust., ii. p. 248 (1866).

Exp. al., (♂) 1 in. 2–5 lin.; (♀) 1 in. $3\frac{1}{2}$–$7\frac{1}{2}$ lin.

♂ *Glossy dark-brown, greyish-brown or warm reddish-ochreous-brown; cilia white, regularly and broadly interrupted with brown. Fore-wing:* disco-cellular marking very faint and slender, or altogether obsolete;

some very indistinct indications of a submarginal row of fuscous spots. *Hind-wing*: a very indistinct (often all but obsolete) small fuscous hind-marginal spot between first and second median nervules; in some of the paler examples faint indications of two or three similar, externally finely whitish-edged spots along lower hind-margin. UNDER SIDE.—*Brownish-grey* (*or greyish-brown in the darker specimens*) *with white-edged darker markings, in arrangement and character like those of Ortygia*, Trimen. *Fore-wing:* terminal disco-cellular mark not very broad, reniform; discal row more or less curved inward to first median nervule, and then outward at its extremity,—commencing with a minute spot on costa; occasionally the third and fourth spots of this row, or one of them, elongated inward so as to touch disco-cellular spot; others of the row sometimes much diminished or wanting altogether; in one specimen the lowest spot but one is prolonged inward; inner submarginal row of white lunules as in *Ortygia*, but occasionally indistinct; outer row forming with hind-marginal white line rings, whose outer edges (opposite the white parts of the cilia) are usually widened and conspicuous, much as in *Asteris*, Godt. *Hind-wing:* short basal streak, usually well developed; three round, black, white-ringed spots in sub-basal row; two similar spots respectively at beginning and end of discal row, but separate from it; second spot of same row small, shaped (and sometimes coloured) like the first; fourth spot almost always (third spot more rarely) confluent with terminal disco-cellular marking; inner row of acutely sagittiform white lunules usually well developed; outer one forming with white hind-marginal line larger, and usually more complete rings than in fore-wing; hind-marginal black spot small, obscure, its bluish-silvery dots and yellow ring very faint or obsolete.

♀ Similar, and equally variable in tint of brown. *Fore-wing:* in a few of the largest and darkest examples, on disc, between second median nervule and submedian nervure, the faint imperfect bluish linear outline of two elongate darker spots like those sometimes found in *Asteris*. *Hind-wing:* hind-marginal fuscous spot often more distinct,—in two of the larger specimens rather conspicuously blue-centred and very faintly yellow-edged inwardly, and in another slightly blue-dusted and faintly bluish-edged inwardly. UNDER SIDE.—As in ♂, but discal row in fore-wing more often imperfect or deficient in parts; outer submarginal row of lunules sometimes very faintly marked, while inner row is occasionally suffused in both wings; and sub-basal row in hind-wing sometimes with four instead of three spots,—the additional one being situated just below the second spot. *Hind-wing:* often some whitish suffused scaling near base.

This species is widely spread in South Africa; it is readily recognised by both sexes being of the same uniform brown on the upper side, and by its being rather smaller in the ♀, and much smaller in the ♂, than the nearly related *Ortygia* and *Asteris*. The darkest (and also the largest) examples I have seen are found in the Cape Peninsula, but some of these have a rufous

tinge;[1] the palest from Grahamstown and Basutoland; while those from Kaffraria Proper and Natal are intermediate in colour. The ochreous-tinged individuals inhabit Malmesbury, N. of the Cape Peninsula, and 1 also took a similar individual in Namaqualand.
I have met with this butterfly only on the summits and slopes of hills. Though not uncommon, it does not exhibit the gregarious tendency of its ally, *L. Asteris*. Though an active insect, it very often settles on low shrubs and herbage, and is easily captured. I have met with it from August to February, but it is prevalent most from October to December.

Localities of *Lycæna Methymna*.

I. South Africa.
  B. Cape Colony.
    a. Western Districts.—Cape Town. Simon's Town. Malmesbury. Bain's Kloof. Spectakel, Namaqualand.
    b Eastern Districts.—Grahamstown. Uitenhage (*S. D. Bairstow*).
    d. Basutoland.—Maseru (*J. H. Bowker*).
  D. Kaffraria Proper.—Tsomo and Bashee Rivers (*J. H. Bowker*).
  E. Natal.—Pinetown (*J. H. Bowker*).

128. (10.) Lycæna puncticilia, Trimen.

PLATE VIII. fig. 4 (♂).

Lycæna puncticilia, Trim., Trans. Ent. Soc. Lond., 1883, p. 350.

Exp. al. 1 in.—1 in. 3½ lin.

Dark-brown, with a slightly brassy surface-gloss; *cilia dark-brown, with rather small but very conspicuous pure-white inter-nervular spots. Hind-wing*: in some ♀ examples, along hind-marginal edge a row of inter-nervular minute white spots, only separated from the white spots of the cilia by the blackish bounding line. UNDER SIDE.—*Dull ashy-brown*; ordinary discal row of darker white-edged spots indistinct, or sometimes obsolete, except for the thin internal white edging; *adjoining the conspicuous white spots of cilia is a row of elongate-ovate white rings*, usually better marked in fore-wing than in hind-wing. *Fore-wing*: dark, white-edged disco-cellular lunule broad, usually quite distinct; spots of discal row confluent into a nearly straight fascia, slightly bent inward on median nervules; beyond discal row, traces of a row of thin sharply sagittate white marks; apical white ring of hind-marginal row thicker and more conspicuous than the rest. *Hind-wing*: first (costal) and last (inner-marginal) spots of discal row distinct, ovate, black, white-ringed; other spots of row confluent, suffused, only their internal white edges marked; a sub-basal transverse row of three circular black spots in thin white rings; *beyond discal row a conspicuous very sharply dentated white transverse line, composed of contiguous sagittiform marks*; the imperfect hind-marginal ovate ring between second and first

---

[1] A dwarf male I captured at Wynberg, near Cape Town, expands barely an inch across the wings.

median nervules is often filled with blackish, centred with a few bluish-white scales.

A single ♀ varies from the rest in possessing on the upper side of the hind-wing a row of four indistinct, small, whitish spots not far from hind-margin, between costa and third median nervule.

A close ally of *L. Methymna*, mihi, but distinguished from it by the following characters, viz. :—(1) The darker upper side, without any reddish tinge; (2) constant absence on upper side of fore-wing of disco-cellular lunule; (3) darker, more ashy under side, with (4) less distinct and less macular discal row; (5) more elongate and distinctly defined white rings of hind-marginal row; and (6) in hind-wing, much more conspicuous and acutely dentated white transverse line beyond discal row.

The only example of this butterfly with which for many years I was acquainted was collected at Genadendal in the Caledon Division, about 1863–1864, and was regarded as a variation of *Methymna*. In the year 1869 (September) I met with another individual flying in company with several of *Methymna* at Malmesbury, about thirty-five miles northward of Cape Town. It was not until September 1879 that, in company with Colonel Bowker, I found the insect common near the village of Malmesbury, and also at Riebeck's Casteel in the Malmesbury District. We captured about thirty specimens, and found them all to agree in the main characters above described. This *Lycæna* flits about, and frequently settles upon, low shrubby plants on rocky *kopjes* and hillsides. On the wing it looks almost black.

Localities of *Lycæna punctieilia*.

I. South Africa.
  B. Cape Colony.
    a. Western Districts.—Malmesbury. Riebeck's Casteel, Malmesbury District. Genadendal, Caledon District (*G. Hettarsch*).

129. (11.) Lycæna hypopolia, *sp. nov.*

Exp. al., (♂) 1 in. 6–7 lin.

♂ *Dull pale-violaceous; a rather narrow brownish-fuscous hind-marginal border emitting short inter-nervular rays; cilia dull-white, uniform. Fore-wing:* a rather narrow but distinct terminal disco-cellular fuscous lunule. *Hind-wing:* costa broadly bordered with brownish fuscous, wider about apex; between first and second median nervules a small and narrow black spot close to hind-margin, inwardly bounded by a suffused pale-yellow lunule, and marked centrally with a few bluish scales; *no* tail. UNDER SIDE.—*Hoary grey; in each wing terminal disco-cellular lunule, discal macular row (except first and last spots, which are small and black), and indistinct row of submarginal lunules, dull ochrey-yellow with whitish edges. Fore-wing:* discal row in two examples ending just above first median nervule, where it is slightly curved inward,—in the third with a faint geminate additional spot below first median nervule; submarginal lunulate row macular, very indistinct. *Hind-wing:* three small, faintly whitish-ringed spots in sub-basal row, of which the first and third are black, the second (in discoidal cell) more or less tinged with ochrey-yellow; in discal row second spot quite separate from first and third,—largest spots are third and fourth,—seventh spot nearer hind-margin than sixth or eighth; submarginal lunulate row more distinct than in fore-wing, with more or less apparent traces of two rows of whitish thin lunules immediately preceding and following it respectively; hind-marginal spot as on upper side, but very much smaller, and with the yellow lunule very thin and indistinct; a second similar but very imperfect spot at anal angle.

Of this fine *Lycæna*, remarkable by the dull-violaceous of its upper side and the hoary-grey of its under side, I have seen only three examples, viz., two taken by Mr. W. Morant on 21st September 1870, in the north-west of Natal, and another captured subsequently in the Transvaal by Mr. T. Ayres. The ♀ is still unknown to me. There is no very near congener that I am aware of; but the species most resembling *Hypopolia* are *L. Niobe* and *L. Tantalus*, Trim., both smaller, and on the under side darker forms.

Localities of *Lycæna hypopolia*.

1. South Africa.
    E. Natal.
        b. Upper Districts.—Blue Bank, near Drakensberg (*W. Morant*).
    K. Transvaal.—Potchefstroom District (*T. Ayres*).

130. (12.) Lycæna Cissus, (Godart).

♂ ♀ *Polyommatus Cissus*, Godt., Enc. Meth., ix. p. 683, n. 210 (1819).
♀ *Agriades Cissus*, Geyer, Forts. Hübner Zutr. Samml. Exot. Schmett., p. 7, n. 206, ff. 811–812 (1837).
♂ ♀ *Lycæna Catharina*, Trim., Trans. Ent. Soc. Lond., 3rd Ser. i. p. 281 (1862).
♂ ♀ *Lycæna Cissus*, Trim., Rhop. Afr. Aust., ii. p. 252, n. 153 (1866).

Exp. al., (♂) 1 in. 2½–5 lin.; (♀) 1 in. 2½–5½ lin.

♂ Dull *violaceous-blue; costa and hind-margin narrowly bordered with greyish-brown in both wings.* *Hind-wing:* on hind-margin, between third median nervule and anal angle, from one to four bright-orange lunules, the largest between second and first median nervules, and with a round black spot touching its outer edge; *no tail. Cilia* whitish-grey. UNDER SIDE.—*Whitish grey;* ocelliform spots blackish or black, white-ringed; *in both wings* a white-edged, blackish streak closing cell; a sinuate row of conspicuous ocelli beyond middle (those in hind-wing smaller and fainter than those in fore-wing); a submarginal row of ill-defined, fuscous, whitish-bordered lunules; and a single small ocellus in discoidal cell. *Hind-wing:* base dusted with blackish and bluish scales; an ocellus near base, just below subcostal nervure and another on inner margin before middle; orange lunules paler, black spot marked outwardly with some bluish-silvery scales.

♀ *Blue paler, less violaceous, than in* ♂. *Fore-wing:* costa, apex, and hind-margin very broadly bordered with greyish-brown; a bluish or whitish mark on hind-margin immediately above posterior angle. *Hind-wing:* costal half of wing greyish-brown, some small outwardly whitish-edged fuscous lunules along hind-margin; orange lunules paler and larger than in ♂, more or less confluent, black spot larger. UNDER SIDE.—Quite like that of ♂; spots larger, more conspicuous.

On the under side, in both sexes, the two rows of whitish submarginal lunules are very ill-defined, and the lower ones of the inner row of the fore-wing much suffused inwardly, while the lowest lunule of the outer row is suffused outwardly and united with a hind-marginal whitish line. The space between the two rows of whitish lunules is enlarged and fuscous below third median nervule in the fore-wing; and the three lower spots of the discal row are considerably larger than the other three and more irregular in form. In the hind-wing the discal spots are often exceedingly small and rather indistinct, and the lunules of the inner submarginal row much prolonged inwardly.

A dwarf ♀ which I captured at Highlands, near Grahamstown, expands scarcely above an inch across the wings.

In the Hope Museum at Oxford in 1867 I noted a ♀ from Sierra Leone in which the discs of the fore-wings were suffused with white. Another ♀ from the Transvaal presents the same suffusion in both fore and hind wings; and some of the same sex which I took in Natal exhibit it in a less degree.

The only near ally of *L. Cissus* known to me is the considerably smaller and tailed *L. Jobates*, Hopff. The former is a handsome species, very prevalent in the Eastern parts of South Africa, and ranging widely to distant points in the African continent. It frequents hillsides, fields, and open ground, and has a low, short flight, keeping mostly about grassy spots. In the Cape Colony and Natal I have met with it commonly from the middle of October to the beginning of April.

Localities of *Lycæna Cissus*.

I. South Africa.
 B. Cape Colony.
  a. Western Districts.—Knysna.
  b. Eastern Districts.—Grahamstown. Kowie River Mouth, Bathurst District (*J. L. Fry*). King William's Town (*W. S. M. D'Urban*).
  d. Basutoland.—Maseru (*J. H. Bowker*).
 D. Kaffraria Proper.—Butterworth and Bashee River (*J. H. Bowker*).
 E. Natal.
  a. Coast Districts.—D'Urban (*M. J. M'Ken*). Verulam. Itongati River. Umvoti. Mapumulo.

b. Upper Districts.—Little and Great Noodsberg. Udland's Mission Station. Greytown. Rorke's Drift (*J. H. Bowker*). Estcourt (*J. M. Hutchinson*).
F. Zululand.—St. Lucia Bay (*the late Colonel H. Tower*).
K. Transvaal.—Lydenburg District (*T. Ayres*).

II. Other African Regions.
A. South Tropical.
 a. Western Coast.—" Chinchoxo (*Falkenstein*)."—Dewitz.
 b1. Eastern Interior.—Bamangwato Country (*H. Barber*).
B. North Tropical.
 a. Western Coast.—Gaboon.—Coll. Brit. Mus. Sierra Leone.—Coll. Hope, Oxon.

131. (13.) Lycæna Jobates, Hopffer.

♂ *Lycæna Jobates*, Hopff., " Monatsb. K. Preuss. Akad. Wissensch., 1855, p. 642, n. 20;" and Peters' Reise Mossamb., Ins., p. 408, pl. xxvi. ff. 9, 10 (1862).
Lycæna Siwani, Trim., Trans. Ent. Soc. Lond., 3rd Ser. i. p. 402 (1862).
Lycæna Jobates, Trim., Rhop. Afr. Aust., ii. p. 245, n. 146 (1866).

Exp. al., (♂) 1 in.—1 in. $2\frac{1}{3}$ lin.; (♀) 1 in. $0\frac{1}{2}$–$2\frac{1}{2}$ lin.

♂ *Pale lilacine-blue, with broad fuscous-grey margins; hind-wing with a short, black, white-tipped tail on first median nervule; cilia greyish, inclining to whitish in parts.* Fore-wing: grey border commences narrowly on costa beyond middle, becomes very broad apically, and gradually narrows along hind-margin to posterior angle. *Hind-wing*: grey border broad costally and apically, bounded inferiorly by subcostal nervure and its second nervule; below the latter it is hind-marginally very narrow, and below radial or third median nervule is represented by small separate spots, of which that between first and second median nervules is larger than the rest, and black; immediately beyond these spots a thin white line, succeeded by a blackish one along hind-marginal edge; immediately before spots a conspicuous, irregular, lunulated, bright-orange band, lying between third (or sometimes second) median nervule and submedian nervure,—rarely reduced to two or three separate lunules adjoining largest spot and spots next anal angle. UNDER SIDE.—*Whitish-grey, the pattern and markings like those of Cissus, Godt., with the exceptions noted below.* Fore-wing: no ocellus in discoidal cell; terminal disco-cellular streak thinner, much more faint; in discal row of spots the two between upper radial and third median nervules not so far beyond the rest, and the lower spots only a little larger than the others; submarginal markings fainter (especially the white ones) and straighter. *Hind-wing*: in discal row all the spots well marked, and the sixth and eighth more displaced outwardly; orange band usually not so much developed, but in some specimens much more so, extending brokenly to near apex; largest hind-marginal black spot, and smaller one below it, outwardly edged with greenish-silvery.

♀ *Blue much paler and duller, inclining to whitish on discs, occupying a smaller space.* Fore-wing: blue fills cell, covers lower disc, and extends along inner margin from base to a little beyond middle; at posterior angle a white mark as in *Cissus* ♀, but more linear in form. *Hind-wing:* blue occupying about the same space as in forewing, but rising higher on disc; hind-marginal blackish spots, white line, and orange bar usually better developed than in ♂, especially the spots. UNDER SIDE.—As in ♂, but spots of discal row usually larger in both wings.

It is remarkable that specimens of both sexes occur in which the ocellated spots of the hind-wing are all filled with orange-fulvous instead of black,— the spots near base being, however, partly blackish. I have before me examples of this kind from Griqualand West, the Transvaal, and Delagoa Bay, and two others from the Free State and Basutoland respectively, which exhibit the peculiarity to a much smaller extent.

The smaller size and tailed hind-wings well distinguish this butterfly from *L. Cissus*, Godt., as well as the distinction indicated in the above description. As regards the ♂, too, the paler tint of the upper side, and its three times as broad apical dark border of the fore-wings, are features readily identifying *Jobates*. The species is not unlike *L. Tiresias* (Rott.), of Central and Southern Europe, on the under side, but does not at all resemble it on the upper side.

I only once met with this butterfly, taking a female flitting about *Acacia* trees in the "Thorn" country, near Greytown, in Natal, on the 12th March 1867. The species has, however, a considerable known range in the east and interior of South Africa, and also occurs at very widely distant spots in the tropical parts of the continent. Its most southern locality known is King William's Town, where Mr. D'Urban took it rarely in February. It is singular that this very delicate-looking *Lycæna* should share with the robuster members of the *Nymphalinæ* a decided partiality for very strong drink; but I am able to record that four examples sent to me by Mr. W. Morant were taken near Hebron, in the Orange Free State, on the 26th October 1870, sucking "at a dead chicken in a bad egg;" and that another, captured by Colonel Bowker at Boshof, in the same State, during September 1872, was busily engaged in "drinking the blood of a freshly killed hartebeast." In Basutoland the latter observer met with *Jobates* among grass near the Caledon River.

Localities of *Lycæna Jobates*.

I. South Africa.
    B. Cape Colony.
        b. Eastern Districts.—King William's Town (*W. S. M. D'Urban*).
        c. Griqualand West.—Klipdrift, Vaal River (*J. H. Bowker*).
        d. Basutoland.—Maseru (*J. H. Bowker*).
    C. Orange Free State.—Hebron (*W. Morant*). Boshof (*J. H. Bowker*).
    E. Natal.
        b. Upper Districts.—Karkloof (*J. H. Bowker*). Greytown. Estcourt (*J. M. Hutchinson*). Ladysmith and Biggarsberg (*J. H. Bowker*). Colenso (*W. Morant*).
    H. Delagoa Bay.—Lourenço Marques (*Mrs. Monteiro*).
    K. Transvaal.—Potchefstroom District (*T. Ayres*).

II. Other African Regions.
    A. South Tropical.
        a. Western Coast.—"Damaraland (*De Vylder*)."—Aurivillius. "Congo: Kinsembo (*H. Ansell*)."—Butler.

b. Eastern Coast.—" Querimba."—Hopffer.
*b*1. Eastern Interior.—Tati River (*F. C. Selous*).
B. North Tropical.
*b*1. Eastern Interior.—" Atbara River " [Soudan].—Butler.

132. (14.) Lycæna Hippocrates, (Fab.)

♂ *Hesperia Hippocrates,* Fab., Ent. Syst., iii. 1, p. 288, n. 105 (1793).
" *Papilio Hippocrates,* Don., Ins. Ind., pl. 45, f. 3 (1800)."

Exp. al., (♂) 9–11 lin.; (♀) 10–11½ lin.

♂ *Dark-brown; apex of fore-wing tipped with white; hind-wing with a rather long, linear, black, white-tipped tail on first median nervule;* cilia brownish mixed with white, becoming pure white tipped with brown at apex of fore-wing and near anal angle of hind-wing. *Fore-wing:* apical white not broad, but conspicuous, below second radial nervule emitting along hind-margin, immediately before a dark bounding line, a very attenuated, interrupted, indistinct white line. *Hind-wing:* costa edged with whitish; between second and first median nervules a small hind-marginal black spot, bounded inwardly by a thin orange-yellow lunule; an extreme hind-marginal black bounding line, preceded by traces of a whitish one, more apparent at anal angle. UNDER SIDE.—*Greyish-white; in each wing the disco-cellular terminal lunule and the two hind-marginal rows of lunules,* and *in the hind-wing* all the spots in the discal row except the first and last, *very thin and faintly-marked, pale-grey,* with edgings scarcely whiter than the ground; hind-marginal black-edging line clearly defined. *Fore-wing:* no spots near base; six spots of discal row very small, thin, black, almost imperceptibly white-edged, the row strongly curved outward superiorly. *Hind-wing:* discal row highly irregular, interrupted on second subcostal and second median nervules, and on submedian nervure,—the first spot (on costa) and the last (on inner margin about middle) black; a sub-basal row of three small but very distinct black spots, viz., one close to costa, another in discoidal cell, and the third (smallest) on inner margin; hind-marginal black spot and orange lunule well marked, the latter a little separated from the former, which sometimes bears a few bluish-silvery scales; at anal angle a short linear black mark just before hind-marginal black line.

♀ *Brown, paler; in each wing a large whitish-grey space, more or less shot with pale blue, from base over inner-marginal and lower discal area;* cilia whiter throughout, except near apex of fore-wing, and in hind-wing with thin nervular fuscous interruptions. *Fore-wing:* pale space covers lower half of cell, and half (or sometimes two-thirds) of the median nervules; a dusky disco-cellular terminal lunule, indistinctly marked. *Hind-wing:* pale space occupies all the wing up to subcostal nervure and its second nervule; disco-cellular lunule and lower par *of* discal row of spots more or less distinctly represented;

along hind-margin the edging white line unites with an inner row of white lunules to enclose a series of dark-brown spots,—the spot between second and first median nervules being, as in ♂, black, with an inward-bounding orange lunule. UNDER SIDE.—As in ♂.

This is a very isolated species, not resembling closely any congener known to me. It is strikingly distinguished by the white apex of the fore-wings in the ♂, forcibly contrasting with the dark-brown ground-colour, and in both sexes by the whiteness of the under side and its very small spots. The pattern and character of the under-side markings perhaps more nearly resemble those of the untailed *L. Messapus*, Godt., and *Mahallokoaena*, Wallengr., than those of any other species. On the upper side the ♀ *Hippocrates* is in pattern and colouring something like the ♀ s of the much larger *Asopus*, Hopffer, and *Jobates*, Hopffer.

The first example known to me as South-African was taken in Zululand by the late Colonel H. Tower in 1866. All the others I have seen—thirteen examples—were captured by Colonel Bowker on the coast of Natal, chiefly in the years 1879 and 1880, the paired sexes being secured on 30th May 1880. Concerning these latter specimens, he wrote that the two were at rest on a creeper in the Park at D'Urban, and that it was the white tips of the fore-wings of the male which attracted his attention. The species was found by the same observer very numerously on the edge of the bush near the mouth of the Umlaas River on the 8th June.

A ♂ specimen from Sherboro Island, near Sierra Leone, in the collection of the British Museum, does not differ from southern examples except in its slightly larger size.

Localities of *Lycæna Hippocrates*.

I. South Africa.
   E. Natal.
      a. Coast Districts.—Mouth of Umlaazi, D'Urban, and Pinetown (*J. H. Bowker*).
   F. Zululand.—St. Lucia Bay (*Colonel H. Tower*).
   H. Delagoa Bay.—Lourenço Marques (*Mrs. Monteiro*).

II. Other African Regions.
   A. South Tropical.
      a. Western Coast.—" Angola."—W. F. Kirby, Cat. Hewits. Coll.
      bb. Eastern Islands.—" Madagascar."—Oberthür.
   B. North Tropical.
      a. Western Coast.—Sherboro Island, near Sierra Leone (*C. S. Salmin*).—Coll. Brit. Mus.
      *b*1. Eastern Interior.—" Abyssinia : Shoa (*Antinori*)."—Oberthür.

133. (15.) **Lycæna Niobe**, Trimen.

♂ ♀ *Lycæna Niobe*, Trim., Trans. Ent. Soc. Lond., 3rd Ser. i. p. 282 (1862); and Rhop. Afr. Aust., ii. p. 253, n. 154, pl. 4, f. 10 [♀].

Exp. al., (♂) 1 in. 0½–4 lin.; (♀) 1 in. 0½–6 lin.

♂ *Dull cupreous-violaceous; base narrowly purplish, and hind-margin rather widely bordered with reddish-brown in both wings; spotless; cilia brownish in fore-wing, but white at apex, in hind-wing brownish, with white tips throughout; no* tail on hind-wing. UNDER SIDE.—*Dusky brownish-grey; ocelliform spots black, with pale-grey rings*, arranged as in *L. Cissus*, but no ocellus in discoidal cell of fore-wing, and the outer row of spots hardly visible on either wing; *space between the two rows marked by a lunulate hoary-greyish band;* close to hind-margin a row of very indistinct, darker, lunular spots. *Hind-wing:* between second and first median nervules, close to hind-margin, a narrow, blackish dot, tipped with ferruginous internally, with greyish-blue externally; row of discal ocelli interrupted, in one specimen nearly obliterated inferiorly.

♀ *Violet brighter and better defined than in ♂, forming a patch on inner-marginal half of both wings*, rising very little above median nervure, and extending a little beyond middle. UNDER SIDE.—As in ♂; spots more conspicuous, especially that on hind-margin between second and first median nervules of hind-wing, in which the ferruginous and blue colouring is distinct.

This is a variable species. Specimens from Grahamstown agree with those above described from Knysna, Cape Colony, except in their larger size, but examples from Kaffraria Proper differ—in the ♂ by a purer (not cupreous) violet upper side and a darker hind-marginal border, and in the ♀ by the entire absence of violet, the upper side being wholly reddish-brown, except for a very slight bluish tinge near the bases. In Natal the ♂ s agree with the Kaffrarian specimens; but of the two ♀ s I have seen, the smaller has only a very small space of basi-inner-marginal rather bright blue, while the other has all the surface, except the broad borders, shot with bright shining-blue.

A constant and conspicuous distinguishing character of *Niobe* is the whitish band across the under side, just beyond the discal row of spots, formed by the inner row of submarginal lunules, in contrast with the dusky-grey ground-colour preceding it, and especially with a dark streak immediately succeeding it.

I founded this species on three examples captured by myself at Knysna in October and March 1858. They were flitting about grass and low plants on the hill-sides, and looked like some small dark *Satyrinæ*. The butterfly was afterwards shown to have a wide range to the eastward and northward, but it seems to be nowhere numerous. I took a single specimen on the coast of Victoria County, Natal, in March 1867; and in February 1870 Mr. H. Barber captured a very fine ♀ near Grahamstown.

Localities of *Lycæna Niobe*.

I. South Africa.
  B. Cape Colony.
    a. Western Districts.—Knysna.
    b. Eastern Districts.—Grahamstown (*M. E.* and *H. Barber*).
    d. Basutoland.—Koro Koro (*J. H. Bowker*).
  D. Kaffraria Proper.—Bashee River (*J. H. Bowker*).

E. Natal.
   a. Coast Districts.—Victoria County. D'Urban (*J. H. Bowker*).
   b. Upper Districts.—Estcourt (*J. M. Hutchinson*). Mooi River (*W. Morant*).
F. Zululand.—St. Lucia Bay (the late *Colonel H. Tower*).
K. Transvaal.—Potchefstroom District (*T. Ayres*).

134. (16.) Lycæna Tantalus, *sp. nov.*

Exp. al., (♂) 1 in. 4–5½ lin.

Dull pale-violaceous ; a narrow brownish-fuscous hind-marginal border,—in hind-wing rather wider, indenting violaceous between nervules ; cilia dingy-whitish, uniform. *Fore-wing*: terminal disco-cellular fuscous mark linear, not very distinct. *Hind-wing*: costa bordered with fuscous-grey as far as subcostal nervure and its first nervule ; no hind-marginal spot, or only the faintest trace of one ; *no tail*. UNDER SIDE.—*Pale or whitish-grey, sometimes with a faint brownish tinge ; all the markings linear or much attenuated ; in each wing* a linear disco-cellular terminal marking, a discal row of dark spots, *a submarginal regular lunulated dark streak*, and a hind-marginal series of very indistinct dark spots; *on hind-marginal edge a series of very small dark spots, adjacent to which the cilia are also dark*. *Fore-wing*: disco-cellular mark and discal row black ; in the latter the third spot is obliquely placed, its lower end inclining outward, and the fifth (between first and second median nervules) is before the rest ; submarginal lunulate streak blackish on both sides, edged faintly and suffusedly with whitish, and regularly interrupted by nervules ; disco-cellular mark and discal row (except first and last spots, which are black) dull-brown ; in the latter the brown spots are sublunulate, the second spot being separate, and the sixth and eighth nearer base than the fifth and seventh spots ; submarginal lunulate streak also brown, its component lunules larger than in fore-wing, and the first and second of them separate both from each other and from the rest.

♀ *Pale shining-blue not violaceous—duller and less developed in hind-wing,—with broad dull fuscous borders*. *Fore-wing*: border of about even width throughout, from base to apex, and from apex to posterior angle ; terminal disco-cellular lunule very much broader and darker than in ♂. *Hind-wing*: blue only thinly covers lower area from base to beyond middle, leaving a very broad costal fuscous border, and a moderately broad hind marginal one ; in the latter a series of blue spots, of which that between first and second median nervules is large and conspicuous, but the rest are very small, indistinct, and lunulate. UNDER SIDE.—As in ♂, but the markings more sharply defined.

The ♂ of this species on the upper side is intermediate in tint, &c., between the ♂ s of *Hypopolia*, mihi, and *Niobe*, mihi, but is rather nearer to the former as well in colour as in size. The ♀ (of which I have seen but one example) resembles the Natalian variation of the ♀ *Niobe*, but differs in possessing a well-developed terminal disco-cellular lunule in the fore-wings, and a blue hind-marginal spot in the hind-wings. On the under side *Tantalus* comes nearer to *Niobe*, but wants the transverse whitish band beyond middle, while the discal row of spots is formed of elongate (not rounded) scarcely whitish edged black marks in the fore-wings, and still thinner dull-brownish marks in the hind-wings, and the submarginal common dark streak is narrowed and more sharply defined.

This *Lycæna* was sent by Colonel Bowker from Kaffraria Proper, but until lately, when other specimens were sent by him from Natal (including a ♀), I had regarded it as a variety of *L. Niobe*. The latter examples were captured by him in August and September.

Localities of *Lycæna Tantalus*.

I. South Africa.
    D. Kaffraria Proper.—Bashee River (*J. H. Bowker*).
    E. Natal.
        a. Coast Districts.—D'Urban and Pinetown (*J. H. Bowker*).

135. (17.) Lycæna ignota, *sp. nov.*

Exp. al., (♂) 1 in. 3–4 lin.; (♀) 1 in. $2\frac{1}{2}$–5 lin.

♂ *Very dull greyish-brown; a narrow dull-fuscous line edging hind-margin; cilia glossy, brownish-grey, becoming whitish outwardly. Fore-wing:* a scarcely perceptible darker, linear, terminal disco-cellular mark. *Hind-wing:* a very indistinct small, narrow, fuscous hind-marginal spot between first and second median nervules, faintly (in one specimen more distinctly) ringed with whitish scales; *no tail.*
UNDER SIDE.—*Dull-grey; markings blackish and pale-brownish, inconspicuously white-edged; submarginal rows of whitish lunules, and hind-marginal whitish line, with which the outer row forms rings, rather faintly marked, but less indistinct in hind-wing. Fore-wing:* terminal disco-cellular mark rather narrow, lunulate, fuscous; slightly curved discal row of six rather small blackish or dark-brownish spots, of which the fifth is rather nearer base than the fourth and sixth. *Hind-wing:* three small rounded black spots in sub-basal row; first and last spots of discal row like them, but the first one rather larger; second spot round, but brown instead of black; other spots of row small, pale-brownish, the sixth rather nearer base than the fifth and seventh; terminal disco-cellular mark sublinear, lunulate, dark- or pale-brown; hind-marginal spot usually distinct, but small, silvery-scaled externally.

♀ *Similar, slightly paler; disco-cellular mark in fore-wing, and hind-marginal spot in hind-wing rather more apparent.* UNDER SIDE. —*More tinged with brown; all the markings less distinct,* the dark ones (except the sub-basal spots and hind-marginal spot of hind-wing, of which the latter is more developed than in ♂) dull-brownish, the whitish ones faint and dull.

On the upper side this obscurely-tinted *Lycæna* is not unlike the male of the South-European *L. Admetus*, Esp., but on the under side both the ground-colour and pattern are very different. On the whole, *L. ignota* comes closest to *L. Letsea*, Trim., but is distinguishable by its darker colouring, and total want of yellowish hind-wing lunules on the upper side, and by the very imperfect development of the whitish markings on the under side; it also exhibits nothing of the tendency so general in *Letsea* to imperfection of the discal row of spots on the under side of the fore-wing.

Mr. T. Ayres discovered this butterfly in the Transvaal, and in 1879 the South-African Museum obtained from him three examples of each sex. The only other specimens that have come under my notice are two males taken by Mr. J. M. Hutchinson, who captured them in the interior of Natal, and kindly presented them to me in the year 1881.

Localities of *Lycæna ignota*.

I. South Africa.
 E. Natal.
  b. Upper Districts.—Estcourt (*J. M. Hutchinson*).
  K. Transvaal.—Potchefstroom and Lydenburg Districts (*T. Ayres*).

136. (18.) Lycæna Letsea, Trimen.

♂, ♀ *Lycæna Letsea*, Trim., Trans. Ent. Soc. Lond., 1870, p. 362, pl. vi., ff. 3, 4.

Exp. al., (♂) 1 in. 3½–4 lin.; (♀) 1 lin. 3½–6 lin.

♂ *Shining brownish-grey; cilia slightly paler, not variegated.* Fore-wing: a terminal disco-cellular streak sometimes faintly visible. Hind-wing: on hind-margin, on each side of the first median nervule, a very faint yellowish lunule, forming a ring with a broken hind-marginal white line, of which the superior is large, and marked externally with a black spot. In two examples these markings are blurred and scarcely traceable; *no* tail. UNDER SIDE.—*Grey; ordinary markings small and neatly defined*, resembling those of *L. Messapus*, Godt. Fore-wing: *lower portion of transverse row of white-ringed black spots beyond middle almost always wanting*, the usual number of spots present being four (in one example there are but three, while in another there are five, with the faint trace of a sixth on one side only). Hind-wing: a faint pale-bluish suffusion over basal portion; yellow lunules more deeply coloured and much better marked than on upper side, the black dot of the superior one more or less dusted with silvery-blue.

♀ *Similar; slightly darker; cilia whiter than in* ♂. Fore-wing: disco-cellular lunule plainer than in ♂, but still indistinct. Hind-wing: yellow lunules broader and brighter, the black dot strongly marked; in one (the largest) example there is a double row of indistinct whitish acute lunular marks along hind-margin, becoming obsolete towards costa, but in the other two the outer portion only of the row is indicated by the very faintest whitish scaling. UNDER SIDE.— All the markings better defined, and with wider white edgings than in the ♂. Fore-wing: discal row composed of six spots in the largest example; of six on one side and five on the other in the smallest; and of five in the third.

In both sexes, when there are more than four spots in the discal row of the fore-wings, the fifth spot is smaller, and (as well as the sixth, when present) placed slightly before the line of the others. The row is but very slightly curved, commencing at a little distance from the costa, immediately above the first discoidal nervule.

Mr. Bowker found this dull-tinted species commonly about the waggon-roads near Rouxville and the Orange River in January 1869, and also in similar situations near Eland's Berg and Klip Spruit in the following month. He noted that it frequented small bushy plants, and, when roused, kept long on the wing.

Localities of Lycæna Letsea.

I. South Africa.
  B. Cape Colony.
    d. Basutoland.—Klip Spruit (J. H. Bowker).
  C. Orange Free State.—Near Orange River, Rouxville, and Eland's Berg (<i>J. H. Bowker</i>).

137. (19.) Lycæna dolorosa, sp. nov.

Exp. al., (♂) 1 in.—1 in. 2 lin.; (♀) 1 in. 2 lin.

♂ *Pale-violaceous; fore-wing with a narrow, even, fuscous hind-marginal border, hind-wing with only a linear black edging and a series of indistinct fuscous spots; cilia greyish, mixed with fuscous in fore-wing, and narrowly interrupted with fuscous at ends of nervules in hind-wing. Hind-wing:* spot between first and second median nervules black, larger than the other hind-marginal spots; a very thin interrupted whitish line between the series of spots and hind-marginal black edging; costa bordered with greyish-brown as far as first subcostal nervule, and at apex as far as second; *no* tail. UNDER SIDE.—*Dull brownish-grey; the markings scarcely darker than the ground-colour, inconspicuously but distinctly edged with whitish on each side; in each wing*, a terminal sub-reniform disco-cellular lunule, a slightly sinuated macular discal row, and two submarginal rows of whitish lunules, indistinct in fore-wing. *Hind-wing:* three small blackish spots in sub-basal row; first and last spots of discal row small, round, and black, and separate from the rest,—sixth spot almost touching lower extremity of disco-cellular lunule; lunules of inner submarginal row broader, more acute than those of outer row, which form more or less incomplete rings with a whitish line immediately before hind-margin; spot between first and second median nervules, on hind-margin, conspicuous, marked externally with brilliant greenish-silvery, and bounded inwardly by a yellow lunule; usually the traces of a similar imperfect spot at anal angle.

♀ *Violaceous paler, duller, restricted to a suffusion from bases, over cellular, inner-marginal, and lower-discal areas; brownish-fuscous border of fore-wing very broad, especially apically; that of hind-wing also very broad costally and apically. Fore-wing*: a linear fuscous terminal discocellular mark. *Hind-wing*: hind-marginal spots better marked, inconspicuously ringed with dingy-whitish; spot between first and second median nervules inwardly bounded by a very dull yellow lunule. UNDER SIDE.—As in ♂, but the whitish edgings of the markings better defined.

Although so much larger and darker an insect, this species shows distinct alliance to *L. Messapus*, Godt., in the character of the under-side markings. Only six males and a single female have come under my notice,—all from the eastern side of South Africa, and four of them from the upper districts of Natal. Of the habits of this butterfly I have no information; it is probably inconspicuous on the wing.

Localities of *Lycæna dolorosa*.

I. South Africa.
    D. Kaffraria Proper.—Bashee River (*J. H. Bowker*).
    E. Natal.
        b. Upper Districts. Estcourt (*J. M. Hutchinson*). Biggarsberg (*J. H. Bowker*). Blue Bank near the Drakensberg (*W. Morant*).
    F. Zululand.—Napoleon Valley (*J. H. Bowker*).
    K. Transvaal.—Potchefstroom District (*T. Ayres*).

138. (20.) Lycæna Messapus, (Godart).

♂ *Polyommatus Messapus*, Godt., Enc. Meth., ix. p. 682, n. 205 (1819).
Polyommatus Sebagadis, Guér., Lefvr. Voy. Abyss., vi. p. 385, pl. 11, ff. 7, 8 (1847).
♂ *Lycæna Acca*, Westw., Gen. D. Lep., pl. lxxvi. f. 1 (1852).
♂ ♀ *Lycæna Messapus*, Trim., Rhop. Afr. Aust., ii. p. 254, n. 155 (1866).

Exp. al., (♂) 8–1 1½ lin.; (♀) 9½ lin.—1 in.

♂ *Blue-violaceous, with a narrow blackish hind-marginal border; cilia greyish, whitish externally. Hind-wing*: a hind-marginal black spot between second and first median nervules (sometimes nearly obsolete), usually inwardly edged by an orange lunule; *no* tail. UNDER SIDE.—*Pale brownish-grey;* in *each* wing,—a streak closing cell (in fore-wing blackish, in hind-wing of the ground-colour) whitish-edged on both sides,—a row of white-ringed spots beyond middle (in fore-wing blackish, in hind-wing of the ground-colour and confluent except the first and rarely the second spot),—and two rows of faint-whitish lunules (separated by darker marks), of which the outer forms annulets with a whitish hind-marginal line; a thin blackish line bordering hind-margin. *Hind-wing:* four white-ringed dark spots before middle,—one (usually the most conspicuous) between costal and subcostal nervures, one in cell and two on inner margin; hind-marginal black spot usually whitish-ringed and often orange-lunuled,—sometimes obsolete; row beyond middle angulated on second subcostal nervule.

♀ *Glistening dark greyish-brown;* very rarely with a few blue scales near bases. *Hind-wing:* besides hind-marginal black spot (which is very rarely indistinct), a row of indistinct pale annulets is usually visible. UNDER SIDE.—As in ♂; markings more distinct.

In some specimens, of both sexes, the markings are very faint beneath, and the ground-colour duller and slightly darker than usual.

I have not found any characters to distinguish *L. Sebagadis* (Guér.) from *Messapus*. A rather large and pale ♂ in the Hewitson Collection in the British Museum, which I examined in 1881, was marked "type," and was thus probably received from M. Guérin as true *Sebagadis*.

In the under-side markings and in the upper side of the ♂, *Messapus* is not unlike a miniature *L. Cissus* (Godt.), but the wholly brown upper side of the ♀ is altogether different from that of the ♀ *Cissus*, and is not unlike that of *L. Alsus* (W. V.), so well-known a native of Europe. In size and appearance generally it most resembles *L. Lysimon* (Hübn.), but the ♂ differs in its very much narrower dark border, and the ♀ in its want of any blue, on the upper side; while both sexes present a darker, much less distinctly spotted under side, wanting in the fore-wing the two sub-basal spots well marked in *Lysimon*.

This *Lycæna* abounds about Cape Town, occurring throughout the year in open ground, especially in grassy spots. Its flight is very weak and close to the ground, and it settles very frequently on low plants. It has a wide distribution in South Africa, but I am not aware of its occurrence in Natal, and have not met with it at all numerously except near Cape Town.

Localities of *Lycæna Messapus*.

I. South Africa.
   B. Cape Colony.
      a. Western Districts.—Cape Town. Vogel Vley, Tulbagh District. Caledon (*J. X. Merriman*). Robertson. Montagu. Knysna. Plettenberg Bay.
      b. Eastern Districts.—Murraysburg (*J. J. Muskett*). Grahamstown. King William's Town (*W. S. M. D'Urban*).
      d. Basutoland.—Maseru (*J. H. Bowker*).
   D. Kaffraria Proper.—Butterworth and Bashee River (*J. H. Bowker*).

II. Other African Regions.
   B. North Tropical.
      b1 Eastern Interior.—"Abyssinia (*Lefebvre*)."—Guérin [*Sebagadis*].

139. (21.) Lycæna Mahallokoæna, (Wallengren).

♂, ♀ *Lycæna Mahallokoæna*, Wallgrn., K. Sv. Vet.-Akad. Handl., 1857; Lep. Rhop. Caffr., p. 41, n. 16.
 „ „ Trim., Rhop. Afr. Aust., ii. p. 257, n. 159 (1866); and Trans. Ent. Soc. Lond., 1870, p. 366, pl. vi. ff. '7, 8 (♂, ♀).

Exp. al., (♂) $8\frac{1}{2}$–$11\frac{1}{2}$ lin.; (♀) 10–11 lin. Closely allied to *L. Messapus*, Godt.

♂ *Blue-violaceous; fore-wing always, hind-wing rarely (and then very slightly), suffused with fulvous-yellow;* narrow fuscous hind-marginal border and greyish white-tipped cilia, as in *Messapus; no tail on hind-wing. Fore-wing:* yellow suffusion extremely variable in extent and development, from a mere costal streak to a broad field occupying all the area except a narrow basal space and broad hind-marginal border of blue,—intermediate examples presenting a broad bar along costa, and strong or moderate suffusion on the median nervure and its branches and on submedian nervure. *Hind-wing:* yellow suffusion never more than a slight tinge about middle of costa and on disc beyond extremity of discoidal cell; hind-marginal black spot between first and second median nervules very distinct, the orange lunule bounding it internally large and conspicuous; a smaller fainter orange lunule between second and third median nervules, and sometimes a still smaller and fainter one immediately below first median nervule. UNDER SIDE.— *Whitish-grey;* markings quite as in *Messapus*, except that in hind-wing the hind-marginal black spot is conspicuous and considerably larger, and there are two well-developed bright-orange lunules instead of a single indistinct or obsolete one.

♀ *Dark-brown, usually rather tinged with greyish; orange-yellow hind-marginal lunules (always two, and usually three) of hind-wing more or less enlarged, so as to form a small conspicuous patch,*[1] *and each externally bounded by a dark spot.* UNDER SIDE.—Quite as in ♂, but markings generally usually rather better defined.

The characters above given easily distinguish this very curious and beautiful form from *Messapus*, but the instability of the fulvous-yellow suffusion on the fore-wings of the ♂ is very noticeable, and seems to indicate that the character, highly peculiar and apparently unique as it is in the genus, is one of comparatively recent acquirement. The accompanying large development of the orange-yellow lunules of the hind-wings (especially marked in the ♀) is to all appearance a feature of much more constancy.

This *Lycæna* was originally discovered by Wahlberg, but no specimens were known to me until Colonel Bowker in 1869 sent several from Basutoland. In the same year a pair taken in the Free State reached me from Mr. W. Morant, who subsequently forwarded examples from the Transvaal, noting the species as plentiful near Potchefstroom on 25th February 1872, and occurring in low, stony ground at Pretoria on the 16th March. Mr T.

---

[1] This character is most largely developed in a specimen taken by Colonel Bowker between the Tugela and Mooi Rivers in Natal. In this, as in a few other examples, there is an incomplete *fourth* lunule.

Ayres almost simultaneously sent me specimens from the Potchefstroom District; and fine examples from Natal and Zululand have more recently been contributed by Mr. J. M. Hutchinson and Colonel Bowker. The species extends into the Tropical Regions, occurring in Damaraland on the west, and centrally in the Bamangwato country. Mr. Morant noted the capture of a ♀ in the Free State at the end of December, and Colonel Bowker that of a ♂ on the Natal coast on 27th April.

Localities of *Lycæna Mahallokoæna*.

I. South Africa.
  B. Cape Colony.
    a. Basutoland.—Maseru (*J. H. Bowker*).
  C. Orange Free State.—Vaal River (*W. Morant*).
  E. Natal.
    a. Coast Districts.—Isipingo, D'Urban, and Pinetown (*J. H. Bowker*).
    b. Upper Districts.—Estcourt (*J. M. Hutchinson*). Between Tuegla and Mooi Rivers, Rorke's Drift, and Biggarsberg (*J. H. Bowker*). Colenso (*W. Morant*).
  F. Zululand.—Isandhlwana and Napoleon Valley (*J. H. Bowker*).
  H. Delagoa Bay.—Lourenço Marques (*Mrs. Monteiro*).
  K. Transvaal.—Potchefstroom (*W. Morant* and *T. Ayres*). Pretoria (*W. Morant*).

II. Other African Regions.
  A. South Tropical.
    a. Western Coast.—" Damaraland (*Du Vylder*)."—Aurivillies.
    *b*1. Eastern Interior.—Bamangwato Country (*H. Barber*). Tauwani River (*F. C. Selous*).

140. (22.) Lycæna Lysimon, (Hübner).

♂ *Papilio Lysimon*, Hübn., Samml. Europ. Schmett., ff. 534-35 (1798?).
    "   "   "Ochs., Schmett. Europ., i. 2, p. 24 (1808)."
♂, ♀ *Polyommatus Lysimon*, Godt., Enc. Meth., ix. p. 701, n. 240 (1819).
♂, ♀ *Lycæna Lysimon*, Herr.-Schff., Schmett. Eur., i. p. 118, t. 5, ff. 28, 29 (1843).
♂, ♀ *Lycæna Knysna*, Trim., Trans. Ent. Soc. Lond., 3rd. Ser., i. p. 282 (1862); and Rhop. Afr. Aust., ii. p. 255, n. 156 (1866).

Exp. al., (♂) 9–10½ lin.; (♀) 10⅓ lin.—1 in. 0½ lin.

♂ *Dull-violet, with a silky gloss; hind-margin of both wings rather widely bordered with blackish; cilia broad, whitish; no* tail on hindwing. *Fore-wing*: costa very narrowly edged with a white line. UNDER SIDE.—*Whitish-grey;* ocelli blackish, whitish-ringed; *in both wings*, a sinuate row of ocelli strongly curved superiorly, beyond middle, a whitish-edged fuscous streak closing discoidal cell, *a distinct ocellus in cell* (sometimes an indistinct ocellus below it), two submarginal rows of pale-fuscous, indistinctly whitish-edged, lunular spots,—and a thin blackish edging line, interiorly faintly whitish-edged. *Hind-wing*: three minute ocelli near base, forming with that in discoidal cell a short row across wing; base blackish-dusted.

♀ *Shining greyish-brown; inner-marginal area of both wings more or less suffused with violet-blue from base. Fore-wing*: a fuscous line closing discoidal cell. UNDER SIDE.—Quite similar to that of ♂, but all the spots more conspicuous, especially the marginal lunular rows. *Fore-wing*: spot below that in discoidal cell always present, often distinct.

The ♂ varies slightly in depth of colour and in the width of hind-marginal dark border;[1] but the ♀ varies very greatly both as regards extent and tint of the blue suffusion, which in some examples is scarcely perceptible, while in others it occupies the larger part of both wings, numerous intermediate grades of development occurring.

It seems probable that this butterfly is the *Otis* of Fabricius (1787), but it is impossible to decide the point from that author's descriptions. Judging from the descriptions and figures (in *Proc. Zool. Soc. Lond.*, 1865, p. 505, pl. 31, f. 7, ♀, and *Lep. Ceylon*, 1881, p. 77, pl. 35, ff. 6, 6a, ♂), I am further of opinion that the *Polyommatus Karsandra* of Moore is identical with *Lysimon*. There were certainly specimens not separable from the latter in Mr. E. L. Layard's Cingalese collection, and I have also seen individuals from various parts of India.

Both sexes of *Lysimon* are readily distinguishable from *Messapus*, Godt., by the more whitish cilia, and the rather paler and more conspicuously spotted under side; and the ♂ by its very much broader dark hind-marginal border on the upper side. On the under side, moreover, *Lysimon* has the discal row of ocellate spots very much more curved superiorly, and possesses a spot in the discoidal cell of the fore-wings wholly wanting in *Messapus*.

This little *Lycæna* is scarce about Cape Town, but commoner further to the eastward. It frequents gardens and waste land, and is fond of settling on the ground in damp grassy depressions or almost dry ditches. It is mainly a species of the late summer and early autumn (February to April), but I have met with it as early as the middle of September, and Mr. D'Urban took it in British Kaffraria in June. The only South-African locality in which I have met the insect at all abundantly is Plettenberg Bay, on the south coast of the Cape Colony. Mr. J. M. Hutchinson in 1882 sent me the sexes captured *in copulâ* at Estcourt, Natal.

The only differences presented by South-African from European examples are an average rather larger size and a more distinct spotting on the under side; and the same is the case as regards Mauritian and Indian specimens, which quite agree with those from South Africa. In Mauritius, *Lysimon* is most abundant; I found it on waste lands all over the island, and it also congregated on grass lawns in gardens.

Localities of *Lycæna Lysimon*.

I. South Africa.
   B. Cape Colony.
      a. Western Districts.—Cape Town. Worcester. Robertson (*J. E. C. Hodges*). Swellendam (*A. C. Harrison*). Knysna. Plettenberg Bay. Clanwilliam (*L. Péringuey*).

---

[1] Two specimens taken in Zululand by Colonel Bowker have the border darker and broader than in any other South-African examples that have come under my notice.

b. Eastern Districts.—Uitenhage. "Grahamstown, King William's Town, and Keishamma River, near Bodiam."—(*W. S. M. D'Urban*). Windvogelberg, Queenstown District (*Dr. Batho*).
　　c. Griqualand West.—Vaal River (*J. H. Bowker*).
　D. Kaffraria Proper.—Bashee and Tsomo Rivers (*J. H. Bowker*).
　E. Natal.
　　a. Coast Districts.—D'Urban (*M. J. M'Ken*).
　　b. Upper Districts.—Maritzburg. Greytown. Estcourt (*J. M. Hutchinson*). Colenso (*W. Morant*). Ladysmith.
　F. Zululand.—St. Lucia Bay (*the late Colonel H. Tower*). Napoleon Valley (*J. H. Bowker*).
　K. Transvaal.—Potchefstroom District (*T. Ayres*).
II. Other African Regions.
　A. South Tropical.
　　a. Western Coast.—" Chinchoxo (*Falkenstein*)."—Dewitz.
　　b. Eastern Coast.—" Querimba."—Hopffer. Lake Nyassa.—Coll. Brit. Mus.
　　bb. Madagascar.—Coll. Brit. Mus. " Bourbon." —Boisduval. Mauritius, Johanna, Comoro Islands (*W. C. Bewsher*).—Coll. Brit. Mus.
　B. North Tropical.
　　a. Western Coast.—" Senegal."—Hopffer.
　C. Extra-tropical North Africa.—" Algeria :—Djebel Aures and Collo."—Oberthür. " Egypt."—Boisduval.
III. Europe.—" Portugal, Spain, South France."—Staudinger, Herrich-Schäffer, &c.
IV. Asia.
　A. Southern Region.—" Coast of Asia Minor."—Staudinger. Ceylon (*E. L. Layard*). Calcutta.—Coll. Hewits. in Brit. Mus. " Bengal."—Godart and Boisduval.
　B. Malayan Archipelago.—" Java."—Westwood.

141. (23.) Lycæna lucida, Trimen.

♂, ♀ *Lycæna lucida*, Trim., Trans. Ent. Soc. Lond, 1883, p. 348.
Allied to *L. Lysimon*, Hübn.

Exp. al., (♂) $8\frac{1}{2}$–11 lin.; (♀) $8\frac{1}{2}$ lin.—1 in. $0\frac{1}{2}$ lin.

♂ *Pale-violaceous, inclining to pink; nervules more or less marked with greyish-brown; fore-wing with a greyish-brown hind-marginal border of variable width*, usually ill-defined inwardly, but outwardly bounded by the ordinary edging black line; *hind-wing with a hind-marginal row of six small fuscous inter-nervular spots*, a little before the ordinary black line edging hind-margin; *cilia* whitish, much obscured with brownish in fore-wing, and varied with it in hind-wing. *Fore-wing:* an indistinct thin brownish lunule at extremity of discoidal cell. *Hind-wing:* tailless. UNDER SIDE.—*Pale-grey, tinged with brownish; spots of bases and discs very distinct, black, with white rings;* beyond ordinary discal row, a row of sagittate white marks, succeeded by two hind-marginal rows of white lunules almost forming inter-nervular rings. *Fore-wing:* a spot in cell, towards extremity; beneath it a similar usually rather larger one; ordinary disco-cellular lunule black with white bordering; discal row of spots strongly incurved on second median nervule; first and second spots of row on costa, respectively before and about middle, minute but very distinct and rather widely apart. *Hind-wing:* a spot at base; a transverse row of four spots before middle; disco-cellular closing lunule narrow, of the ground-colour, white-bordered; discal row strongly elbowed on second subcostal nervule; *a straight white ray runs longitudinally along radial nervule from disco-cellular terminal lunule to row of sagittate marks;* near anal angle, two small blackish spots enclosed by lunules of the two hind-marginal rows.

♀ *Dark-brown, usually more or less marked with violaceous on lower parts of discs and towards bases.* *Hind-wing:* dusky spots of hind-marginal row, as in ♂, more or less apparent in violaceous marked specimens. UNDER SIDE.—Usually a little more brownish than in ♂; the spots even more distinct, and the white ray of hind-wing broader.

VARIETY, ♂ and ♀.—Under side darker than usual; *the hind-wing with basal and discal spots almost obsolete, but with the white ray* very broad and conspicuous. One ♀ example has the violaceous on upper side bluer than usual and largely developed.

Hab.—Pinetown, Natal (♂ *W. Morant,* 1869; ♀ [2] *J. H. Bowker,* 1879).

This species is allied to *L. Lysimon,* but is readily recognised by the much more conspicuous spotting of the under side, with the white ray exhibited by the hind-wing.[1] The male differs also from that of *Lysimon* in the decided pink tinge of the upper side, and the absence of the dusky border of the hind-wing. The female has the upper side much darker than in *Lysimon,* and the violaceous colouring is deeper and not so blue in tint.

This is the butterfly noted in my *Rhopalocera Africæ Australis* (ii. p. 255) as probably a "permanent variety" of the female *L. Knysna,* mihi (= *Lysimon,* Hübn.). At that time (1866) I had not distinguished the male of the form, although the female described was taken *in copulâ* by myself at Plettenberg Bay in February 1859, and so a male (probably worn) must have passed through my hands. From the Tsomo River, in Kafirland Proper, Colonel Bowker forwarded, only a few months later in 1866, a specimen which I could not doubt was the male; and next year, in Natal, I met with several examples of both sexes. It was not, however, till 1870 that I found the butterfly pretty commonly near Grahamstown, and at Highlands (on 30th January) captured the paired sexes.

There is nothing remarkable in the habits of this little species. It is rather sociably disposed, and little groups are found flitting about grassy spots on hill-sides.

---

[1] This white ray, which wholly or in part appears in so many of the European species of *Lycæna,* does not occur in any known South-African representative of the genus except the one under notice. In this one, however, it seems to be always present in both sexes, judging from thirty-two specimens before me, although in one male it is reduced to a line merely.

Localities of *Lycæna lucida.*

I. South Africa.
 B. Cape Colony.
  a. Western Districts.—Plettenberg Bay.
  b. Eastern Districts.—Port Elizabeth. Grahamstown. Zwaartwater Poort.
 D. Kaffraria Proper.—Tsomo River (*J. H. Bowker*).
 E. Natal.
  a. Coast Districts.—D'Urban. Pinetown (*W. Morant* and *J. H. Bowker*). Mapumulo.
  b. Upper Districts.—Estcourt (*J. M. Hutchinson*).
 F. Zululand.—Isandhlwana (*J. H. Bowker*).
 K. Transvaal.—Potchefstroom District (*T. Ayres*).

142. (24.) Lycæna stellata, Trimen.

Lycæna stellata, Trim., Trans. Ent. Soc., Lond., 1883, p. 349.

Exp. al., 7–9 lin.

Greyish-fuscous, with numerous subannular and other white spots arranged in correspondence with those of the under side; hind-wing tailless. Fore-wing: terminal disco-cellular annulet, and discal inferiorly much incurved band of annulets, enclose spots somewhat darker than the ground-colour; near base two similar annulets, the upper one in discoidal cell, the lower immediately below cell; a row of six minute white spots near and parallel to hind-margin; *cilia* broad, fuscous, narrowly but very distinctly interrupted with white between nervules,—the white interruptions close to apex and to posterior angle wider than the rest. *Hind-wing:* terminal disco-cellular annulet and discal band of annulets not so fully developed as in fore-wing, but distinct; an indistinct annulet near base below cell; submarginal row of minute white spots as in fore-wing, but the first spot (nearest costa) considerably larger than the others; cilia broad, white, with only imperfect fuscous interruptions along the inner edge at ends of nervules. UNDER SIDE.—*Pale brownish-grey; the white annulets enclosing fuscous spots.* Fore-wing: the markings very distinct; discal row begins about middle with two small costal annulets; submarginal row of minute white spots black-edged both internally and externally; fuscous of cilia paler than on upper side. *Hind-wing:* an annulet at base; a sub-basal transverse row of four annulets, of which the first (on costa) encloses a darker spot than the rest; terminal disco-cellular annulet rather blurred; first and second annulets of discal band separate from the succeeding ones, and nearer base; submarginal row of minute white spots rather indistinct, but their inner black edges well marked, sub-sagittate; the first and second of these spots (like the corresponding annulets of the discal band) are out of line with and before the others.

This very remarkable *Lycæna* belongs to the group of which *L.*

Lysimon, Hübn., may be regarded as the type, its under side being of similar pattern though more strongly marked. The species, however, with which it best agrees in the under-side markings, is *L. lucida*, Trim.; but, as compared with the latter, it has a somewhat more yellowish tint, and its hind-wing markings are less distinct, not having black centres. The *upper side*, however, is quite unlike that of any species in the *Lysimon* group, and, indeed, that of any other known *Lycæna*, both sexes presenting on a blackish ground the under-side pattern *in finely-depicted white annulets and spots*. In the total absence of blue in both sexes, and in its very small size, *L. stellata* resembles *L. Metophis*, Wallengr., and *L. Barberæ*, Trim.; but its under side is of very different pattern, and quite wants the row of metallic-dotted ocelli so conspicuous in the hind-wing of those two species.

I am indebted for the knowledge of this most interesting little butterfly to Dr. D. R. Kannemeyer, who noticed it for the first time in November 1882, and sent me, in February 1883, two specimens to identify. On receipt of my reply he several times visited the spot (about a mile from the village of Burghersdorp, in the Albert District) which the species frequented, and by the first week in March had captured a considerable number of examples of both sexes. Dr. Kannemeyer describes the insect as being numerous in this special haunt of a few yards in extent. The ground is near a brook, and sedgy; and the little *Lycæna* kept flying about some leguminous and other flowers, close to the ground, in a rapid, jerky manner and in a circular direction. In these respects it evidently much resembles its even minuter congeners, *L. Barberæ*, Trim., and *L. Metophis*, Wallengr.[1]

Locality of *Lycæna stellata*.

I. South Africa.
  B. Cape Colony.
    b. Eastern Districts.—Burghersdorp, Albert District (*D. R. Kannemeyer*).

143. (25.) Lycæna Gaika, Trimen.

♂ *Lycæna Lysimon*, Wallengr., K. Sv. Vet.-Akad. Handl., 1857, Lep. Rhop. Caffr., p. 39.
♂ *Lycæna Gaika*, Trim., Trans. Ent. Soc. Lond., 3d Ser., i. p. 403 (1862).
♂, ♀ *Lycæna Lysimon*, Trim., Rhop. Afr. Aust., ii. p. 256, n. 158, pl. 4, f. 7 [♂] (1866).
Lycæna pygmæa, Snellen, "Tijdschrift voor Ent., xix. p. 153, 6, 7 f. 3 (1876);" Moore [*Zizera pygmæa*], Lep. Ceylon, ii. p. 78, pl. 35, ff. 5, 5 a [♂] (1881).

Exp. al., (♂) 9–10½ lin.; (♀) 9–11 lin.

♂ *Pale-blue; a brownish-grey border (of variable width in fore-wing) on hind-margins; cilia whitish; hind-wing tailless.* UNDER SIDE.—

---

[1] In April 1883 Dr. Kannemeyer wrote that he had found the true home of the *Stellata* in the plateau of the Stormberg range, eight miles east of Burghersdorp. At Botma's Farm it occurred in great abundance, frequenting a minute leguminous plant (probably a *Trifolium*) growing in moist places. This plant is almost certainly the food of the larva, the butterfly keeping about it exclusively. Dr. Kannemeyer also noticed the insect at the farm Kulfontein, six miles further eastward.

Whitish-grey; with minute, whitish-ringed blackish spots; *in both wings* a thin, greyish, whitish-edged mark closing discoidal cell; a transverse row of spots beyond middle (that of fore-wing curved, commencing with *two minute spots on costa before and about middle,* and reaching to submedian nervure; that of hind-wing sharply curved, composed of eight spots, from costa about middle to inner-margin); two dentate, submarginal, lunular, greyish, whitish-edged lines,—the outer one broader, interrupted, macular; and a thin, black, bounding line immediately before cilia. *Hind-wing:* a basal black spot; before middle a transverse row of four spots; no metallic-centred spots near anal angle.

♀ *Dull-brown.* UNDER SIDE.—As in ♂; spots more distinct.

In the ♂ the dark hind-marginal border of the fore-wing is always wider at the apex, but varies considerably in width as well as in tint, in some examples being much darker and with a well-defined inner edge, emitting short nervular rays, while in others it is suffused and without defined inner edge. The border of the hind-wing is constantly narrow and not well-defined inwardly, the costa being also bordered with brownish-grey as far as first subcostal nervule. In the ♀ there is rarely a faint basal and discal suffusion of grey on the upper side. A specimen which I took in Griqualand West, which is larger and paler generally than usual, best exhibits this feature.

This is undoubtedly the same insect that is described and figured by Moore (*op. cit.*) as *Pygmæa,* of Snellen, a native of Java and Ceylon,—specimens that I examined in the British Museum only differing in the less distinctly marked under side. Though apparently belonging to the *Lysimon* group, it is of much more slender structure thoughout, and has remarkably elongate wings. These characters, combined with its whiter under side (which has much more sharply curved discal rows of spots, but is without cellular or subcellular spot near base of fore-wing) readily distinguish *Gaika* from *Lysimon.*

I found this little *Lycæna* in some abundance about D'Urban, in Natal; it flew very feebly, near the ground, among grass and weeds. It seems to be on the wing for the greater part if not the whole of the year; for I took it in June, August, February, and March in Natal, and during September in Griqualand West. It extends to many other parts of Natal, but seems to be scarcer inland. Colonel Bowker sent several examples from Zululand; and northwards the species ranges beyond the Tropic into Damaraland on the Western Coast. Its most southern locality known to me is the coast of Bathurst in the Cape Colony. So small and inconspicuous a butterfly is, however, apt to be overlooked by collectors, and, looking to its wide geographical range, there can be no doubt that it inhabits very many stations as yet unrecorded.[1]

---

[1] In the British Museum there is a ♂ *Lycæna,* ticketed "Pernambuco," which is very closely allied to, if not identical with, *L. Gaika.* The only distinctions I could discover were its hind-marginal border of the fore-wings being broader than usual, and the lunules of the inner submarginal line on the under side being sagittiform instead of nearly straight.

Localities of *Lycæna Gaika*.

I. South Africa.
B. Cape Colony.
  b. Eastern Districts.—Kleinemond River, Bathurst District (*Mrs. Barber*). King William's Town (*W. S. M. D'Urban*).
  c. Griqualand West.—Kimberley. Vaal River (*J. H. Bowker*).
  d. Basutoland.—Maseru (*J. H. Bowker*).
D. Kaffraria Proper.—Tsomo and Bashee Rivers (*J. H. Bowker*).
E. Natal.
  a. Coast Districts.—D'Urban, Umhlanga, Verulam, and Mapumulo. Avoca (*J. H. Bowker*). "Lower Umkomazi."—J. H. Bowker.
  b. Upper Districts.—Greytown. Estcourt (*J. M. Hutchinson*). Kurkloof and Biggarsberg (*J. H. Bowker*). Colenso (*W. Morant*).
F. Zululand.—Napoleon Valley (*J. H. Bowker*). St. Lucia Bay (*the late Colonel H. Tower*).
K. Transvaal.—Potchefstroom District (*T. Ayres*).

II. Other African Regions.
A. South Tropical.
  a. Western Coast.—Damaraland (*Mrs. Latham*).
  b1. Eastern Interior.—"Kilima-njaro (*H. H. Johnston*)."—F. D. Godman.

IV. Asia.
A. Southern Region.—India.—Cutch.—Coll. Brit. Mus. Neilgherry Hills, Madras.—Coll. Hope Oxon. Ceylon (*E. L. Layard*).

144. (26.) Lycæna Trochilus, Freyer.

Lycæna Trochilus, Frey., "Neuere Beitr. Schmett., v. p. 98, t. 440, f. 1 (1844)."
♂, ♀ „ „ Herr.-Schaeff., Schmett. Eur., i. p. 128, t. 48, ff. 224, 225 [♂], t. 49, f. 226 [♀] (1844).
♂ „ „ Wallengr., K. Sv. Vet.-Akad. Handl., 1857; Lep. Rhop. Caffr., p. 41, n. 14.
Polyommatus Trochilus, Kirby, Europ. Butt. p. 99 (1862).
♀ *Lycæna Trochilus*, Trim., Rhop. Afr. Aust., ii. p. 256, n. 157 (1866).
Lycæna parva, R. P. Murray, Trans. Ent. Soc. Lond., 1874, p. 526, pl. x., f. 1 [? ♂.]

Exp. al., (♂) 8–9 lin. ; (♀) 9–10½ lin.

♂ *Brown, with a greyish gloss; a thin, inconspicuous hind-marginal fuscous line; cilia brownish' at origin but outwardly whitish; no tail on hind-wing. Fore-wing*: an indistinct, linear, fuscous, terminal discocellular lunule. *Hind-wing*: a similar, even less distinct, disco-cellular lunule; between third median nervule and submedian nervure, close to hind-margin, three rounded black spots, bounded internally by two rather conspicuous orange-yellow lunules, and one (the lowest) small and indistinct and all but obsolete, and externally by a white linear edging. UNDER SIDE.—*Pale-grey, tinged with brownish; ordinary markings* distinct, with well-defined white edges. *Fore-wing:* no sub-basal spots; terminal disco-cellular lunule scarcely darker than ground-colour; discal row of six black spots, only slightly curved superiorly, the first and fifth spots partly before the rest; two submarginal rows of almost straight white lunulate marks; a white hind-marginal line immediately succeeded by a thin terminal black one. *Hind-wing:* sub-basal row of four round black spots; terminal disco-cellular lunule, and all the spots of irregular discal row except the first and last (eighth)—costal and inner-marginal respectively, which are black,—of the ground colour; lunules of two submarginal rows more acute than in fore-wing, those of outer row forming imperfect rings with hind-marginal white line; three hind-marginal black spots marked outwardly with a semicircle of brilliant greenish-golden, the orange lunules preceding them usually well-developed; usually the nucleus of a fourth spot in the form of greenish-golden scales, between third median nervule and radial nervule.

♀ *Similar, usually darker.* *Hind-wing:* hind-marginal black spots and adjacent orange-yellow lunules larger, the lowest lunule better developed, and rarely a fourth small lunule just above third median nervule; in some European examples a row of small white lunules precedes the orange ones. UNDER SIDE.—As in ♂, but all the markings better developed, especially (in hind-wing) the hind-marginal spots and adjacent orange lunules.

On the upper side this species, especially its ♀, has much the appearance of the ♀ *L. Mahallokoaena*, Wallengr., but the row of from three to five jewelled spots on the under side of the hind-wings constitutes an unmistakeable distinction in *Trochilus*. The Rev. R. P. Murray (*loc. cit.*) has separated a South-African example, under the species name of *L. parva*, distinguishing the new form from *Trochilus* "on account of its much smaller size, and also from its presenting in both wings a series of white markings immediately beyond the discal row of spots." But I find that not only is an expanse of eight lines (which is that given by Mr. Murray for *L. parva*) the *minimum* size in South-African specimens of the ♂, but that European examples of *Trochilus* are often no larger, and sometimes smaller (seven lines); and indeed, on the whole, taking a series of both sexes, the South-African insect appears to be decidedly the larger of the two. The second distinction is not to be found in any South-African specimen that I have seen, the inner of the two ordinary submarginal rows of white lunules succeeding the discal spots, as in typical *Trochilus*, in every case.

When comparing South-African with Northern specimens in 1881, I thought, at first, that I had discovered a difference in the former as regards both the smaller number of jewelled spots (three instead of four or five) and the better developed adjoining orange lunules, but I found that the Northern examples varied much in these very particulars,—two Egyptian ones not differing from South-African individuals in which those characters are best expressed.

An aberration from the White Nile, in the Hewitson Collection of the British Museum, has an orange-yellow bar in the fore-wings near the posterior angle.

The only example that I met with in Natal was flitting about grass in a valley of the Great Noodsberg, on the 18th March 1867. The species has reached me from widely-distant localities, but I do not know of any place where it occurs in any abundance. Specimens sent from Burghersdorp, in the north-east of the Cape Colony, by Dr. Kannemeyer, are darker than usual on the upper side, and have the under-side pattern very strongly marked.

Localities of *Lycæna Trochilus*.

I. South Africa.
  B. Cape Colony.
    a. Western Districts.—Robertson.
    b. Eastern Districts.—Port Elizabeth (*J. L. Fry*). New Year's River, Albany District (*Mrs. Barber*). Burghersdorp (*D. R. Kannemeyer*).
    c. Griqualand West.—Vaal River (*J. H. Bowker*).
    d. Basutoland.—Maseru (*J. H. Bowker*).
  D. Kaffraria Proper.—Bashee River (*J. H. Bowker*).
  E. Natal.
    b. Upper Districts.—Great Noodsberg. Biggarsberg and Rorke's Drift (*J. H. Bowker*).
  F. Zululand.—Napoleon Valley (*J. H. Bowker*).
  H. Delagoa Bay.—Lourenço Marques (*Mrs. Monteiro*).
  K. Transvaal.—Potchefstroom (*T. Ayres*).

II. Other African Regions.
  A. South Tropical.
    a. Western Coast.—Damaraland: "River Kuisip (*Wahlberg*)."—Wallengren.
  B. North Tropical.
    b₁. Eastern Interior.—White Nile.—Coll. Hewitson in Brit. Mus.
  C. Extra-Tropical North Africa.—Egypt.—Coll. Hewitson in Brit. Mus.

III. Asia.
  A. Southern Region.—"Asia Minor and North Persia."—Staudinger.

IV. Europe.—"Turkey."—Doubleday, Herrich-Schäffer, &c. "Balkan Mountains."—Staudinger.

145. (27.) Lycæna Metophis, Wallengren.

Lycæna Metophis, Wallengr., Wien. Ent. Monatschr., 1860, p. 37, n. 17; and K. Sv. Vet.-Akad. Förhandl., 1872, p. 48, n. 21.[1]

Exp. al., (♂) 7–7¾ lin.; (♀) 9–9½ lin.

♂ *Dull-brown, paler and tinged with greyish basally; cilia white, in fore-wing very broadly, in hind-wing extremely narrowly, interrupted with dark-brown at extremities of nervules; hind-wing tailless.* Fore-wing: brown interruptions of cilia at extremities of first and second median nervules exceedingly broad, almost uniting. *Hind-wing*: a hind-marginal series of four small internervular blackish spots, of

---

[1] Through the kind assistance of Prof. Chr. Aurivillius, of the Royal Swedish Museum in Stockholm, I was enabled to obtain a careful coloured drawing of the type of this species, and so more satisfactorily to determine the distinctness of *Metophis* from my more recently named close ally, *L. Barberæ*.

which the two next anal angle are often immediately preceded by two minute whitish spots. UNDER SIDE.—*Pale greyish-brown; markings mostly of the ground-colour, but their white edgings well defined. Fore-wing:* disco-cellular terminal lunule broad, subreniform; discal row of six imperfect annulets between subcostal and submedian nervures, highly irregular,—the second and fourth projecting beyond the rest, and the sixth considerably before them (so as to be in a line with terminal cellular marking); two submarginal rows of thin white lunulate marks, of which the lowest in each row is larger than the rest—so that the two partly unite; a dull whitish hind-marginal line (becoming pure-white and wider at its lower extremity), succeeded by a brown line. *Hind-wing:* a sub-basal series of three to four united rather large white annulets, of which that next costa is oval and much the largest; a broad terminal disco-cellular marking, with both white edges rather suffused; discal series of very imperfect more or less confluent annulets highly irregular,—the second annulet being only partly beyond, and immediately above, terminal disco-cellular marking, while the third is far beyond the second, and the first of a continuous but irregular series of six, terminating on inner margin; between second subcostal and first median nervules, a hind-marginal series of four distinct round black spots, each containing a large crescent of glittering greenish-golden scales, surrounded by a whitish ring, and preceded at a little distance by a more or less suffused whitish lunule; traces of two similarly coloured spots near apex (the second immediately preceded by a rather large and conspicuous whitish spot), and more distinct traces of another at anal angle; cilia dark-brown at its origin.

♀ Similar; on upper side paler. UNDER SIDE.—As in ♂.

This species was discovered by Wahlberg close to Walvisch Bay, in Damaraland, but has since been found to inhabit many localities in Extra-Tropical South Africa. It differs from *L. Trochilus,* Frey, in its smaller size, white cilia interrupted with brown, and want of orange lunules in the hind-wings; while its under side is darker, much more irregularly marked, without black centres to any of the ordinary spots, and with never less than four jewelled spots on the hind-margin of the hind-wings. *Metophis* is, however, much nearer to a *Lycæna* from Ceylon in the British Museum, labelled "*L. Chinga*" (but which has not, I believe, been described), and to the Californian *L. exilis,* Boisd. It differs from both in its broad white cilia interrupted with brown, and from the Californian species in wanting a bluish suffusion at the bases of the wings on the upper side and a shining reddish suffusion over the outer half of the fore-wings on the under side.

I have met with this beautiful little butterfly only at Robertson, in the Cape Colony, where in January 1876 I captured a few specimens flitting about close to the ground on the dry hill-sides. Colonel Bowker has forwarded examples from the north and north-east of the Cape Colony, Mr. E. G. Alston from the north central, and Mr. L. Péringuey from the north-west. The latter found it generally distributed in Namaqualand, though local in its haunts, and collected in that district and the adjoining one of Clanwilliam twenty-four specimens. I am not aware of its occurrence near the east of the Cape

Colony, where, however, the very closely allied *L. Barberæ*, mihi, is known to extend as far as Natal; but its existence at Delagoa Bay probably points to its inhabiting the intervening territory.

Localities of *Lycæna Metophis*.

I. South Africa.
  B. Cape Colony.
    a. Western Districts.—Robertson. Van Wyk's Vley, Carnarvon District (*E. G. Alston*). Varsch River, Clanwilliam District (*L. Péringuey*). Garies, Ookiep, Spectakel, and Klipfontein, Namaqualand District (*L. Péringuey*).
    b. Eastern Districts.—Hope Town (*J. H. Bowker*).
    c. Griqualand West.—Vaal River (J. H. Bowker). Kimberley (H. L. Feltham).
  II. Delagoa Bay.—In Hewitson Coll. in Brit. Mus.

II. Other African Regions.
  A. South Tropical.
    a. Western Coast.—" River Kuisip [Walvisch Bay] (Wahlberg)." Wallengren.

146. (28.) Lycæna Barberæ, Trimen.

Lycæna Barberæ, Trim., Trans. Ent. Soc. Lond., 1868, p. 89, pl. v. f. 7.

Exp. al., (♂) 5½–6¾ lin.; (♀) 7½–9 lin.

Closely allied to *L. Metophis*, Wallengr. ♂ and ♀. *Dark-brown; cilia broad, white, interrupted with brown—in fore-wing broadly, in hind-wing narrowly—at ends of nervules; hind-wing tailless. Hind-wing:* a very indistinct hind-marginal row of small blackish spots. UNDER SIDE.—*Brownish-grey (the hind-wing finely speckled with whitish); in both wings, terminal disco-cellular and discal incomplete white annulets filled with brownish rather darker than ground-colour. Fore-wing:* discal series of annulets arranged much as in *Metophis*, but commencing with a small additional (costal) one,—the lowest annulet rather beyond disco-cellular terminal one; two submarginal rows of white lunules rather farther apart than in *Metophis*, and lunules of inner row more sagittiform; space between two rows rather darker than rest of ground-colour. *Hind-wing:* sub-basal series of four annulets arranged as in *Metophis*, but filled with brownish darker than ground-colour; terminal disco-cellular annulet nearer base, and quite separate from discal row; the latter is much wider, more even, and continuous, forming more of a bar or narrow fascia, interrupted on second subcostal nervule; hind-marginal series of four golden-crescented black spots and other markings as in *Metophis*, except that the conspicuous whitish spot immediately preceding the second hind-marginal marking is wanting.

The sexes do not differ except in size.

Besides the various differences specified above, *Barberæ* has considerably broader and shorter wings than *Metophis;* and I have never found in either sex any trace of the small whitish spots which in *Metophis* often immediately precede the blackish hind-marginal spots on the upper side of the hind-wings. *Barberæ* is also considerably the smaller of the two forms.

In both species, but more especially in *Barberæ*, the discoidal cell is in both fore and hind wings remarkably short, its termination being considerably before the middle.

I had much pleasure in naming this smallest but by no means least beautiful of South-African butterflies after Mrs. F. W. Barber, of Grahamstown, who has rendered important services to entomology as well as to botany, and from whom I first received examples of the insect. In February 1870 I was so fortunate as to make the acquaintance, under Mrs. Barber's guidance, of this minute species. It keeps much to spots of limited extent, usually about stony hillsides, and is usually numerous in such restricted stations. In the bed of the Mill River, on the 10th February, I met with a large number, settling on the large stones, and succeeded in capturing a good series, notwithstanding the difficulty of securing such inconspicuous and fragile little creatures among the stones under a broiling sun. In September 1872 I met with a very few specimens in Griqualand West, and in January 1876 captured four examples at Robertson in the Cape Colony. At Port Nolloth, in August 1873, I took a single ♀, the largest I have seen (exp. nine lines); and Mr. Péringuey has lately brought from Ookiep, in the same district (Namaqualand), a specimen of nearly equal size captured by him in November 1885.

It thus appears that in the Cape Colony and in Griqualand West, *Barberæ* and *Metophis* co-exist in several localities, but, as far as hitherto known, the former is more characteristic of the eastern side of South Africa and the latter of the western. The discovery, however, of *Metophis* so far east as Delagoa Bay, is a fact that may indicate a common range of the two forms throughout the region.

I have seen no smaller butterfly than *Barberæ*—a ♂ that I captured at Robertson expands only five lines, and several others spread less than six lines.

Localities of *Lycæna Barberæ*.

I. South Africa.
 B. Cape Colony.
  a. Western Districts.—Robertson. Ookiep (*L. Péringuey*) and Port Nolloth, Namaqualand District.
  b. Eastern Districts.—Highlands (*M. E. Barber* and *H. J. Atherstone*). Mill River, Mitford Park, and Zwaartwater's Poort, Albany District. Between Somerset East and Murraysburg (*J. H. Bowker*). Murraysburg (*J. J. Muskett*). Burghersdorp (*D. R. Kannemeyer*). Uitenhage (*J. H. Bowker*).
  c. Griqualand West.—Kimberley and Barkly.
 D. Kaffraria Proper.—Tsomo River (*J. H. Bowker*).
 E. Natal.
  b. Upper Districts.—Estcourt (*J. M. Hutchinson*).

SECTION B.—Under side brownish-grey or dull yellowish-grey; markings not or scarcely darker than ground-colour, and more or less completely confluent as fasciæ, so that their well-marked white edges have the appearance of rather irregular striæ on a concolorous ground; a short fascia across middle of discoidal cell of fore-wing.

L. Bætica (Linn.) [tailed], *Sichela*, Wallengr., *notobia*, Trim., *Tsomo*, Trim., *Noguasa*, Trim.

147. (29.) Lycæna Bætica, (Linnæus).

♀ *Papilio Bæticus*, Linn., Syst. Nat., i. 2. p. 789, n. 226 (1767).
♂, ♀ *Polyommatus Bæticus*, Godt., Enc. Meth., ix. p. 653, n. 122 (1819).
Lycæna Bætica, Horsf.. Cat. Lep. E. I. C. Mus., p. 80, n. 14 (1828).
♂ *Lycæna Bætica*, Boisd., Sp. Gen. Lep., i, pl. 7, f. 9 (1836).
♂, ♀ „ „ Trim., Rhop. Afr. Austr., ii. p. 236, n. 138 (1866).
LARVA AND PUPA (South Africa), Trim., *op. cit.* p. 342.

Exp. al., (♂) 1½ lin.—1 in. 4½ lin.; (♀) 1 in. 2–5½ lin.

♂ *Silky violet-blue, with a fuscous hind-marginal edging.* *Hind-wing:* two well-marked black spots, outwardly whitish-edged, near anal angle, of which the larger is above first median nervule. UNDER SIDE.—*Pale-ochreous-grey, with undulated, transverse white striæ;* common to both wings, beyond middle, a fascia composed of a middle broad streak, with a parallel line at equal distances on either side, variable in regularity, angulated in hind-wing below first median,—*a white stripe*, narrow in fore-wing, but wider and *conspicuous in hind-wing*, excepting near inner-margin touching the greater portion of outer line of fascia, —followed by a lunulate stria and a hind-marginal edging. *Fore-wing:* a short triple fascia, similar to the longer one described, across cell, and another like it at extremity of cell. *Hind-wing:* base lightly irrorated with blackish; two to four transverse striæ in basal portion, in places more or less confluent; mixed up with these, a triple streak closing cell; two black spots exteriorly edged with bluish or greenish-silvery, and interiorly bordered by an orange lunule (indistinct in lower spot).

♀ *Dull-brownish, viridly shot with shining blue from base and over disc.* *Hind-wing:* sometimes almost devoid of blue; beyond middle a transverse row of broad, more or less conspicuous white lunules; a row of thinner lunules near margin combine with a hind-marginal line to form bluish-white rings, of which the two next anal angle are complete, enclosing the two black spots. UNDER SIDE.—As in ♂, but marking more conspicuous, especially the white *stripe* in *fore-wing*.

Cilia in both sexes greyish at origin, white on outer edge.

LARVA.—Bright-green; paler on the under surface. A dark-green dorsal line; beneath it, on each side, an indistinct line interrupted on each segment, followed by a row of short, oblique, indistinct streaks of the same dark-green, and a pale-green line just above the legs. Head small, shining, reddish-brown. Two-thirds of an inch in length. Feeds on flowers of *Crotalaria capensis* (a Papilionaceous shrub), in which it lives.

PUPA.—Very pale greyish-ochreous, dusted unequally with blackish; the wing-covers more greenish in tint. A fuscous line down the back; some blackish spots on head and back; two rows of blackish spots on each side of back of abdomen. About half an inch in length; thickest and roundest in abdominal region; head blunt. The pupal state lasts from ten to twelve days in the summer.

Godart (*loc. cit.*) describes the larva in Europe as variegated with red on the back, and as feeding in the pods of *Colutea arborescens*, or of the common green pea. I have not seen any examples so marked at the Cape. Mrs. Wollaston (*Ann. and Mag. Nat. Hist.*, 5th series, vol. iii. p. 224, 1879) mentions the partiality of the "green" larvæ of *Baetica* for the common garden-pea both in St. Helena and Madeira; and I have noticed the butterfly about that plant in Mauritius.[1]

Except in size, this well-known and beautiful *Lycæna* varies but little, the males differing slightly in depth of blue on the upper side, and the females in the development and distinctness of the discal and submarginal white lunules of the hind-wing; while on the under side, in both sexes, the submarginal white stripe and the orange lunule of the superior hind-marginal black spot of the hind-wing present some variation. The specimen of *Damoïtes*, Fab. (*Syst. Ent.*, p. 526, n. 350, 1775), which I examined in the Banksian Collection in the British Museum, is not separable from *Baetica*. Examples that I captured near Algiers, in 1881, are slightly darker than the South-African specimens.

Baetica is generally distributed in Southern Africa, and occurs throughout the year, but is most numerous from October to April. It frequents numerous leguminous plants when in flower, and is fond among others of the "Keurboom" (*Virgilia capensis*). Though able to fly with considerable swiftness, it seldom does so, but flutters about the plants that chiefly attract it, repeatedly settling on the flowers or leaves.

Not only has this butterfly an immensely wide geographical distribution, apparently including nearly the whole of the warmer parts of the Old World, but it is remarkable for having established itself in oceanic islands very remote from any continent. Mrs. Wollaston (*loc. cit.*) observes that it is "the most abundant of the few" [only four species noted] "Diurnal Lepidoptera as yet found in St. Helena," and the Rev. T. Blackburn (as noted above) reared it from the larvæ in the Hawaiian Islands.

Inhabiting all Southern Europe, it extends sparingly into the north of France, and stragglers have been captured on the South Coast of England.

Localities of *Lycæna Baetica*.

I. South Africa.
  B. Cape Colony.
    a. Western Districts.—Cape Town. Genadendal, Caledon District (*G. Hettasch*). Knysna. Plettenberg Bay. Van Wyk's Vley, Carnarvon District (*E. G. Alston*).

---

[1] Mr. A. G. Butler records (*Trans. Ent. Soc. Lond.*, 1882, p. 31) two specimens of the butterfly from the Hawaiian Islands, and gives the Rev. T. Blackburn's note of having bred it from larvæ feeding in pods of what appeared to be a *Melilotus*. These Hawaiian larvæ are described as of an "obscure olive-green," and as having the "dorsal and subdorsal lines and the region included obscurely rosy,"—which latter character accords with Godart's description. The head is described as "testaceous, bearing a V-shaped mark which points backward," and the spiracles as "white."

It thus seems evident that the larva varies considerably more than the imago does.

  b. Eastern Districts.—Grahamstown. King William's Town (*W. D'Urban* and *J. H. Bowker*).
  d. Basutoland.—Maseru (*J. H. Bowker*).
 D. Kaffraria Proper.—Butterworth and Bashee River (*J. H. Bowker*).
 E. Natal.
  a. Coast Districts.—"Lower Umkomazi."—J. H. Bowker. D'Urban (*J. H. Bowker*). Umvoti. Mapumulo.
  b. Upper Districts.—Hermansburg. Greytown. Maritzburg (*Miss Colenso*). Estcourt (*J. M. Hutchinson*).
 F. Zululand.—Napoleon Valley (*J. H. Bowker*). St. Lucia Bay (*Colonel H. Tower*).
 K. Transvaal.—Potchefstroom District (*T. Ayres*). Marico River (*F. C. Selous*).
 L. Bechuanaland.—Motito (*Rev. J. Frédoux*).

 II. Other African Regions.
  A. South Tropical.
   a. Western Coast.—"Damaraland (*De Vylder*)."—Aurivillius. "Kinsembo, Congo (*H. Ansell*)."—A. G. Butler.
   aa. St. Helena.
   b. Eastern Coast.—Zambesi River (*Rev. H. Rowley*).—Coll. Hope Oxon.
   bb. "Madagascar and Bourbon."—Boisduval and Guenée. Mauritius.
   *b*1. Eastern Interior.—Khama's Country (*H. Barber*). Mokloutze River (*F. C. Selous*).
  B. North Tropical.
   a. Western Coast.—Sierra Leone (*W. Cutter*).
   b. Eastern Coast.—"Harkeko and Akeek Island (*J. K. Lord*)."—F. Walker.
   *b*1. Eastern Interior.—"Abyssinia: Shoa (*Antinori*)."—Oberthür.
  C. Extra-Tropical North Africa.
   aa. Western Islands.—Teneriffe and Madeira.—Coll. Brit. Mus.
   b. Mediterranean Coast.—Algiers. "Cairo (*J. K. Lord*)."—F. Walker.

  III. Europe.—Spain.—Coll. Brit. Mus. "France, Switzerland, Austria, Italy, South Russia."—*Auct.* Turkey.—Coll. Brit. Mus.

  IV. Asia.
   A. Southern Region.—"Arabia: Mount Sinai (*J. K. Lord*)."—F. Walker. "West Asia and Persia."—Staudinger. "Kurrachee."—C. Swinhoe. Punjab, North India, and Landoor (Himalaya).—Coll. Brit. Mus. Ceylon.—Coll. Brit. Mus. China: Hongkong.—Coll. Brit. Mus. Japan: "Yamato (*Pryer*)."—O. E. Janson.
   B. Malayan Archipelago.—Java.—Coll. Brit. Mus.

  V. Australia.
   A. Austro-Malayan Archipelago.—"Batchian and Waigiou."—W. F. Kirby, Cat. Hewits. Coll.
   B. Australia Proper.—Victoria: "Melbourne (*Lucas*)."—Butler.
 Sandwich Islands: "Hawaii (*Blackburn*)."—Butler.

148. (30.) Lycæna Sichela, Wallengren.

♂ *Lycæna Sichela*, Wallgrn., K. Sv. Vet.-Akad. Handl, 1857; Lep. Rhop. Caffr., p. 37, n. 4.
Lycæna Emolus, ♂, Trim., Rhop. Afr. Aust., ii. p. 234, n. 136 (1866).

Exp. al., (♂) 9 lin.—1 in. 1½ lin.; (♀) 9 lin.—1 in. 2½ lin.

♂ *Silky dark-violaceous; a fuscous line along hind-marginal edge; cilia dull-greyish, inclining to whitish in hind-wing. Hind-wing:* an indistinct fuscous spot (often obsolete) near hind-margin, between first and second median nervules. UNDER SIDE.—*Pale brownish-grey, with the following very finely on both sides white-edged fasciæ, paler mesially and otherwise very slightly darker than the ground-colour in both wings,* viz., one across middle of discoidal cell, another at its extremity, and the third much longer, discal, irregular, submacular; also two hind-marginal rows of very thin white lunules (those of the inner row much more acute—especially in hind-wing), followed by whitish and very slender black lines along hind-marginal edge. *Fore-wing:* first fascia very short from costa to median nervure, interrupted on costal nervure; second of about the same length; discal fascia strongly curved, from costa to first median nervule. *Hind-wing:* an incomplete basal fascia; the sub-basal fascia from costa to inner-margin interrupted on subcostal and median nervures; discal fascia highly irregular, interrupted on first subcostal and on second and first median nervules; the two lunules of inner row at anal angle and between first and second median nervules respectively each enclose a small but very distinct black spot.

♀ *Greyish-brown with a small rather bright violaceous space in each wing. Fore-wing:* violaceous extends from base along inner-margin to about middle, rises a little above median nervure into discoidal cell, and extends over about basal half only of median nervules. *Hind-wing:* violaceous much more limited than in fore-wing, extending from base only over discoidal cell and for a little way above and below it. UNDER SIDE.—As in ♂; but discal fascia in fore-wing prolonged almost to submedian nervure, not far from posterior angle.

I had not seen any examples of this scarce species when, in Part II. of my *Rhopalocera Africæ Australis*, p. 235, I suggested that it might be the same as my *Lycæna Emolus* (= *Lycænesthes Liodes*, Hewits.). In 1875 two worn ♂s were sent to me in a small collection made by Mr. H. Barber in the NW. Transvaal, but I did not identify them with Wallengren's species until, in 1879, Colonel Bowker sent a series of specimens from the coast of Natal. In order, however, to make sure of what the *Sichela* of Wallengren really was, I obtained in 1881, through the kind assistance of Mr. Chr. Aurivillius, of the Stockholm Museum, a careful drawing of the type specimen (a ♂), and found that I had correctly made out the species.

It requires close examination to detect the actual pattern and arrangement of the under-side markings; as, owing to the fasciæ being paler in their middle and laterally but very little darker than the ground-colour, the whole surface (but especially that of the hind-wing) has the aspect of being simply streaked with fine white irregular transverse lines to beyond the middle. There are no tails on the hind-wings. The markings are very similar on the whole to those of *Lycænesthes Liodes*, and the colour of the upper side is also very like that in the species named. In the possession of a short fascia on the under side of the fore-wing crossing the middle of the discoidal cell, *Sichela* resembles *Batica*, Linn., and *Notobia*, *Tsomo*, and *Noquasa*, mihi. In both sexes of *Sichela*, the first median nervule of the fore-wing is strongly curved downward near its origin.

Most of Colonel Bowker's specimens were taken in January, on a hill close to the Umgeni Railway Station near D'Urban. He noted that the insect settled on bare twigs. In October 1879 Colonel Bowker sent me a worn ♀ taken by him, towards the end of the month, at Port Elizabeth. Both sexes vary greatly in size.

Localities of *Lycæna Sichela*.

I. South Africa.
 B. Cape Colony.
  b. Eastern Districts.—Port Elizabeth (*J. H. Bowker*).
 E. Natal.
  a. Coast Districts.—D'Urban and Pinetown (*J. H. Bowker*).
 H. Delagoa Bay.—Lourenço Marques (*Mrs. Monteiro*).
 K. Transvaal.—Crocodile (Limpopo) River (*H. Barber*).

149. (31.) Lycæna Notobia, Trimen.

PLATE VIII., ff. 6 (♂), 6a. (♀).

♂, ♀ *Lycæna Notobia*, Trim., Trans. Ent. Soc. Lond., 1868, p. 91.

Exp. al., (♂) 1 1 lin.—1 in. 2½ lin. ; (♀) 1 in.—1 in. 2½ lin.

♂ *Shining opalescent-violaceous, rather pale; a very narrow hind-marginal fuscous-brown border; cilia broad, fuscous-brown, with narrow white inter-nervular interruptions well-defined and conspicuous in fore-wing, but rather suffused and ill-defined in hind-wing. Hind-wing:* a small very faint hind-marginal fuscous spot between first and second median nervules. No tail. UNDER SIDE.—*Brownish-grey, with a yellowish tinge; in each wing the terminal disco-cellular striola, and discal submacular fascia, except for their darker margins, are of the ground-colour, but edged rather conspicuously with white. Fore-wing:* across middle of discoidal cell a white-edged broad striola like that at extremity; the latter with its outer white edging prolonged as far as second or first median nervule; macular discal band broad (except just at its beginning on costal edge), regular, curved superiorly, extending to submedian nervure; hind-marginal series of small acute white lunules rather ill-defined, succeeded by the usual whitish and fuscous hind-marginal bounding lines. *Hind-wing:* discal band more macular than in fore-wing, narrower, more irregular, almost interrupted on second subcostal and first median nervules, extending from costa to inner-margin ; a shorter, similar, sub-basal band crossing middle of discoidal cell; also a short imperfect one of three separate spots at base; both before and beyond discal band some whitish scaling; hind-marginal markings as in fore-wing, but lunules larger; a small, black, greenish silvery-dotted spot, faintly ringed with yellow between first and second median nervules; a similar very minute spot at anal angle.

♀ *Greyish-brown, with, in each wing, a very limited space of violaceous*, duller than in ♂, extending from base only over median nervure, lower part of discoidal cell, bases of median nervules, and along inner-margin to about or beyond middle. *Hind-wing:* traces of a hind-marginal row of fuscous spots edged inwardly with bluish-white lunules, of which the two next anal angle are usually tolerably distinct. UNDER SIDE.—As in ♂.

In this species the neuration of the under side is pale and distinct, especially in the hind-wing.

The wings of this very distinct *Lycæna* are rather elongate, but the apical part of the fore-wings is less prominent and the apex itself much blunter than in *Sichela* or *Bætica*. I have not met with any close ally, but think the best position for it is next to *Sichela*, Wallengr.

This butterfly appears to belong to the dry upland districts of the interior. The first specimens I received were taken in 1864–5, near Murraysburg, Cape Colony, by Mr. J. J. Muskett, and examples have since been taken in other parts of the north-eastern districts, in Griqualand West, the Orange Free State, the Transvaal, and Bechuanaland. In 1872, during the month of September, I had the pleasure of capturing a good many specimens in Griqualand West; I found them usually about a tall shrubby lilac-flowered strongly-scented *Salvia*, common in that territory. Their flight was short and feeble, and they were very easily netted.

Localities of *Lycæna Notobia*.

I. South Africa.
  B. Cape Colony.
    b. Eastern Districts.—Murraysburg (*J. J. Muskett*). Between Murraysburg and Somerset East (*J. H. Bowker*). Orange River (*M. E. Barber*).
    c. Griqualand West.—Belmont, Kolberg, Colesberg Kopje, and Waldek's Plant (Vaal River).
  C. Orange Free State.—Boshof (*J. H. Bowker*). Hebron (*W. Morant*).
  K. Transvaal.—Potchefstroom District (*T. Ayres*).

150. (32.) Lycæna Tsomo, Trimen.

PLATE VIII., fig. 7 (♂).

♂ ♀ *Lycæna Tsomo*, Trim., Trans. Ent. Soc. Lond., 1868, p. 91.

Exp. al., (♂) 9–11 lin.; (♀) 9½–11½ lin.

♂ *Pale glossy reddish-brown; in both wings, the base very narrowly tinged with purplish-black, and beyond that a very faint pinkish-violaceous suffusion* along inner-margin, rising into lower part of discoidal cell, and extending over median nervules to beyond middle. *Hind-wing:* between first and second median nervules, a hind-marginal, small, indistinct, blackish spot; no tail. *Cilia* shining greyish. UNDER SIDE.—*Dull yellowish-grey; in both wings, the ordinary transverse markings scarcely darker than the ground-colour, and faintly edged with whitish on each side*, and the hind-marginal ones more or less indistinct; *in hind-wing, the discal macular band immediately succeeded by a suffused whitish one. Fore-wing:* a median and a terminal disco-cellular striola; discal band very regular, even, and continuous from costa to submedian nervure. *Hind-wing:* discal macular band very regular, unbroken, only slightly curved inferiorly; a very indistinct white-edged basal spot, macular sub-basal band, and terminal disco-cellular striola; hind-marginal lunules less indistinct than in fore-wing, more acute; the spot between first and second median nervules black and distinct; whitish submarginal band rather broad.

♀ *Darker, the violaceous very restricted (especially in hind-wing), sometimes obsolete or barely visible.* UNDER SIDE.—As in ♂, but all the markings rather better defined.

This obscure little species on the upper side much resembles the ♀ *L. Lysimon*, Hübn., but is widely different on the under side, which is not unlike that of *L. Bætica*, although on so small a scale. Its only near ally known to me is *L. Noquasa*, Trim., which has an almost identical under side, but has much violaceous-blue on the upper side, especially in the male.

Colonel Bowker discovered this butterfly in January 1865, near the River Tsomo in Kaffraria Proper, and noted it as very numerous during that and the following months in reedy, swampy spots near water. He met with the species again in Basutoland, in March 1869, frequenting similar stations, and found it very numerous on the flowers of a species of mint. Mrs. Barber has sent a specimen from the Stormbergen, and Dr. Kannemeyer several examples from Burghersdorp in the Cape Colony. So dull and small an insect is easily passed over by collectors, and I suspect that its range in the eastern tracts is more general and widely spread than at present appears.

Localities of *Lycæna Tsomo.*

I. South Africa.
   B. Cape Colony.
      b. Eastern Districts.—Stormbergen (*M. F. Barber*). Burghersdorp, Albert District (*D. R. Kannemeyer*).
      d. Basutoland.—Head of Orange River (Drakensberg), Tautjies Berg, and Thaba Bosigo (*J. H. Bowker*).
   D. Kaffraria Proper.—Tsomo River (*J. H. Bowker*).

151. (33.) Lycæna Noquasa, *sp. nov.*

Exp. al., (♂) 10–11 lin.; (♀) 11 lin.

♂ *Bright pale violaceous-blue; fore-wing rather broadly, hind-wing narrowly bordered with fuscous-grey; cilia greyish, paler outwardly.*

Fore-wing: border commencing very narrowly on costa beyond middle, but immediately becoming broad at apex, and thence gradually narrowing along hind-margin almost to a point at posterior angle. *Hind-wing:* costa narrowly greyish; inner-margin greyish mixed with white; border of moderate width at apex, but suddenly narrowed on second subcostal nervule, and thence almost linear to anal angle; hind-marginal spot between first and second median nervules rather large, round, black, conspicuous; no tail. UNDER SIDE.—*Dull yellowish-grey,* slightly browner than in *L. Tsomo,* Trim., but *with the markings almost identical with those in that species,* only their white edges being better marked. *Fore-wing:* discal band more incurved at its lower extremity. *Hind-wing:* discal band straighter inferiorly; white band beyond this not so suffused inwardly, so that its component lunules are distinct; hind-marginal black spot larger, rounder, its outer portion with a few bluish-silvery scales.

♀ *Pale brown with a reddish tinge; a small basi-inner-marginal violaceous-blue space in fore-wing, and a very small basal one in hind-wing.* Fore-wing: blue extends from base over inner-marginal border to a little beyond middle, occupying lower part of discoidal cell, and extending thinly over median nervules near their origin. *Hind-wing:* blue almost limited to discoidal cell, there being only a few scattered blue scales over lower discal area; hind-marginal black spot ringed with bluish-white; above it, traces of two imperfect smaller similar annulets on hind-margin. UNDER SIDE.—As in ♂.

Though so intimately related to *L. Tsomo,* this species differs so very widely from it on the upper side, especially in the ♂, that it is recognisable at the first glance. The dark-bordered blue upper side of ♂, and the brown partly blue one of the ♀, indeed much more resemble those of the respective sexes of *Lysimon,* Hübn., than the subcupreous tints of *Tsomo.*

The first example I received was taken by Mr. W. Morant, on 14th September 1870, in the Upper Districts of Natal,[1] and is the only ♀ I have seen. A worn ♂ reached me in 1878 from Colonel Bowker, who took it near Maritzburg; but it was not until the end of 1884 that I received from him four ♂ s in good condition, taken at the Howick Falls of the Umgeni on the 26th November 1884. He noted the insect as numerous about two small swampy spots some four hundred yards above the Howick Bridge, and as having quite the same habits as *Tsomo.* The specific name given to this *Lycæna* is a slight modification of what Colonel Bowker informs me is the Kafir name for the Umgeni Falls.

Localities of *Lycæna Noquasa.*
I. South Africa.
  E. Natal.
    b. Upper Districts.—Maritzburg and Howick (*J. H. Bowker*).
    "Curry's" (*W. Morant*).

---

[1] Mr. Morant's ticket bore the locality "Curry's," but I have not been able to ascertain where the place intended is situated.

SECTION C.—Under side brownish-grey; discal and submarginal markings considerably darker than ground-colour, white-edged, very irregular, more or less confluent into broken fasciæ; discal fascia of hind-wing so united with a submarginal brownish cloud as to mark off a subquadrate whitish space in anal-angular area.

L. *Lingeus* (Cram.) [tailed], L. *Palemon* (Cram.) [tailed].

152. (34.) Lycæna Lingeus, (Cramer).

♂ *Papilio Lingeus*, Cram., Pap. Exot., iv. pl. ccclxxix. ff. F, G (1782).
♂ *Hesperia Ericus*, Fab., Ent. Syst., iii. 1, p. 281, n. 81 (1793).
♂ ♀ *Polyommatus Lingeus*, Godt., Enc. Meth., ix. p. 656, n. 130 (1819).
♂ ♀ *Lycæna Lingeus*, Trim., Rhop. Afr. Aust., ii. p. 239, n. 140 (1866).

Exp. al., (♂) $11\frac{1}{2}$ lin.—1 in. 1 lin.; (♀) 11 lin.—1 in. $0\frac{1}{2}$ lin.

♂ *Dull violaceous-bluish, very glossy (in some specimens shot with silvery-greenish): a blackish hind-marginal line; cilia white, irregularly varied with blackish, chiefly in fore-wing;* a slender, black, white-tipped tail on first median nervule of hind-wing. UNDER SIDE.—*Pale brownish-grey, with transverse white-edged darker fasciæ; a brown cloud in hind-wing. Fore-wing:* from costa, a single white stria near base, curving inward to origin of median nervure,—a short fascia (of which the outer white reaches below median nervure) crossing cell,—another similar fascia at extremity of cell; beyond middle, a submacular fascia narrowing downward, from costa to first median nervule; along hind-margin a whitish stripe interrupted with brown on discoidal and second and third median nervules, and containing a row of small, dark, lunular spots. *Hind-wing:* base brown; an interrupted macular fascia before, and a very irregular one about, middle; *on disc, a brown cloud,*—on the lower edge of which, between second and first medians, is a conspicuous *subquadrate white spot;* on either side of first median nervule a faintly yellow-ringed, black, bluish-silvery-dusted spot; along hind-margin a very indistinct row of confluent whitish rings.

♀ *Grey-brown, shot with violet-blue from base over disc. Fore-wing:* short *fascia closing cell well-marked on this side,* its white edges suffused; two or three suffused whitish spots in broad hind-marginal brown. *Hind-wing:* a dark streak closing cell; rarely, a row of indistinct bluish-white marks beyond middle; a nearly obsolete hind-marginal row of whitish lunules, that above, and sometimes that below, first median, with an outer adjacent blackish spot. UNDER SIDE.—As in ♂.

A ♀ taken near D'Urban, Natal, by Colonel Bowker, has much whitish suffusion on the disc of both wings, and an unbroken series of submarginal whitish spots (the middle one white) in the fore-wing.

In this species the ♀ is rather smaller than the ♂; and I possess one ♀ which is only 9 lin. across the expanded wings, but this example, which I captured near Cape Town, can only be regarded as abnormally dwarfed, like the still more diminutive ♀ s of L. *Palemon* which are not rarely met with.

Lingeus is a very distinct species, with no near ally except *Palemon*, Cram., from which it is easily distinguished by its curious dull-tinted but exceedingly glossy violaceous-bluish instead of cupreous-violaceous upper side in the ♂, and violaceous-bluish shot instead of plain brown or slightly cupreous-violaceous shot upper side in the ♀. The differences in the under-side markings of the two species are given below in the descriptions of *L. Palemon*.

This is a widely-spread butterfly and not uncommon in woods and bushy spots. It occurs throughout the year, but is frequent only in the summer, and I have not personally taken it before the middle of September or after the end of May. It is fond of hovering about low bushes, and often settles on flowers, where it is very easy of capture. The male is much more numerous on the wing than the ♀. I captured the paired sexes on Table Mountain in February 1865.

Localities of *Lycæna Lingeus*.

1. South Africa.
   B. Cape Colony.
      a. Western Districts.—Cape Town. Piketberg. Triangles Station, Worcester District (*L. Péringuey*). Robertson. Knysna. Plettenberg Bay. Oograbies, Namaqualand. Malmesbury (*L. Péringuey*).
      b. Eastern Districts.—Grahamstown (*M. E. Barber*). King William's Town (*J. H. Bowker*). Keiskamma River (*W. D'Urban*).
   D. Kaffraria Proper.—Bashee River (*J. H. Bowker*).
   E. Natal.
      a. Coast Districts.—D'Urban.
      b. Upper Districts.—Great Noodsberg. Estcourt (*J. M. Hutchinson*).
   F. Zululand.—St. Lucia Bay (*the late Colonel H. Tower*).
   H. Delagoa Bay.—Lourenço Marques (*Mrs. Monteiro*).
   K. Transvaal.—Lydenburg District (*T. Ayres*).
II. Other African Regions.
   A. South Tropical.
      b1. Eastern Interior.—Zambesi River (*F. C. Selous*).
   B. North Tropical.
      a. Western Coast.—Special locality not noted (*E. Bourke*).

153. (35.) Lycæna Palemon, (Cramer).

♂ *Papilio Palemon*, Cram., Pap. Exot., iv. pl. cccxc., ff. E. F. (1782).
♂ ♀ *Lycæna Palemon*, Trim., Rhop. Afr. Aust., ii. p. 240, n. 141 (1866).

Exp. al., (♂) 10 lin.—1 in. 1 lin.; (♀) 10 lin.—1 in. 0½ lin.
Allied to *L. Lingeus* (Cram.)

♂ *Coppery-violaceous, with a brown hind-marginal edging: cilia white, regularly interrupted with brown on nervules. Hind-wing:* a small blackish, pale-edged spot on hind-margin, just above first median nervule; tail rather broad, twisted, blackish, edged and tipped with white. UNDER SIDE.—Very like that of *Lingeus* but *browner*. *Forewing: no* basal stria, outer edge of terminal cellular fascia *not* reaching below median nervure; costal origins of fasciæ more minute and separate than in *Lingeus*; fascia beyond middle more macular, inclining inwards, interrupted on second and first median nervules, and reaching submedian nervure; marginal stripe almost obliterated with brown, except at apex, which is rather conspicuously whitish. *Hind-wing*: fascia about middle darker, much more regular, united at end of cell with discal brown cloud; the white spot enlarged to a *broad whitish inner marginal space*, and more deeply incising the cloud by *a conspicuous acute dash* between third and second medians: hind marginal spots smaller, duller than in *Lingeus*, the lower often obsolete.

♀ *Cupreous brown, more or less shot with violet from bases*; or *dull greyish-brown, without any violet lustre. Fore-wing*: an indistinct terminal disco-cellular spot. *Hind-wing*: black spot whitish-ringed, sometimes dusted with bluish-silvery; indistinct traces of a hind marginal row of whitish rings or lunules. UNDER SIDE.—As in ♂.

The ♂ varies considerably in the depth and intensity of the upper side violaceous, some of the specimens being of remarkable beauty in this respect. A ♂ from the Lydenburg District of the Transvaal has under-side markings very deep rich brown and their white edgings very sharp and distinct. Very small examples of the ♀ are sometimes found; one that I took at Knysna measures little over seven lines across the wings, but the smallest I have seen —captured in Weenen County, Natal, by Mr. Hutchinson—expands 6 lines only.

A very close ally of *Palemon* is found in Australia. In 1881 I saw a specimen of what I believed to be *Palemon*, ticketed "Australia" in the Hewitson Collection; it was placed with Cape specimens, and I noted the locality as in all probability an erroneous one. Finding, however, that Mr. A. G. Butler recorded an example from Melbourne in the *Annals and Magazine of Natural History* for February 1882 (p. 85), I wrote to him on the subject, and he very kindly made a re-examination and comparison of the specimens available, with the result that he was convinced that the Australian species was not identical with *Palemon*, "the chief difference being that on the under-surface of the hind-wings the dark central band is continuous (though in confluent square spots) and not united to the external border by the blackish shade always found in the African species."

This *Lycæna* is common and of wide distribution over South Africa, appearing throughout the year. Its flight is rather weak and always near the ground, and it constantly settles on low flowers. I have often noticed the female fluttering about the leaves of *Pelargonium*, and think it probable that the eggs are laid on them. Though frequenting open ground generally, *Palemon* shows a preference for damp hollows about watercourses; it is often met with in gardens and cultivated spots. I took the paired sexes at Knysna on 20th December 1858.

Localities of *Lycæna Palemon*.

1. South Africa.
    B. Cape Colony.
        a. Western Districts.—Cape Town. Malmesbury. Caledon (*J. X. Merriman*). Knysna. Plettenberg Bay.
        b. Eastern Districts.—Port Elizabeth (*J. L. Fry*). Uitenhage. Grahamstown and Fort Brown, Albany District. King William's Town (*J. H. Bowker*). Murraysburg (*J. J. Muskett*).
        c. Basutoland.—Koro Koro and Maseru (*J. H. Bowker*).

D. Kaffraria Proper.—Bashee River (*J. H. Bowker*).
E. Natal.
  a. Coast Districts. — D'Urban. " Lower Umkomazi." — *J. H. Bowker*.
  b. Upper Districts.—Estcourt (*J. M. Hutchinson*).
F. Zululand.—Napoleon Valley (*J. H. Bowker*).
K. Transvaal.—Lydenburg District (*T. Ayres*).

SECTION D.—Under side brownish-grey; markings darker, confluent, broadly whitish-edged, forming irregular sinuated fasciæ throughout; in fore-wing a longitudinal costal whitish-edged stripe from base, united at its extremity to a similar but transverse stripe on costa before middle; two hind-marginal spots of hind-wing near hind-margin *ringed* with glittering bluish or greenish silvery.
  L. *Telicanus*, Lang [tailed].

154. (36.) Lycæna Telicanus, (Lang).

Papilio Bœticus, Esp., Eur. Schmett., i. Suppl., t. xci. f. 2 ♂ (1784).
Papilio Telicanus, Lang, " Verz. sein. Schmett., ii. p. 47, n. 387-389 (1789)."
♂ ♀ *Papilio Telicanus*, Hübn., Eur. Schmett., i. ff. 371-372, 553-554 (1798-1803).
Papilio Telicanus, Herbst., Nat. Bek. Ins., Schmett., xi. t. 305, ff. 6-9 (1804).
Polyommatus Telicanus, Godt., Enc. Meth., ix. p. 655, n. 128 (1823).
Lycæna Telicanus, Hopff., Peters' Reise n. Mossamb., Ins., p. 406 (1862).
♂ ♀ *Lycæna Telicanus*, Trim., Rhop. Afr. Aust., ii. p. 238, n. 139 (1866).
Var. A.—*Polyommatus Hoffmannseggii*, Zell., " Stett. Ent. Zeit., 1850, p. 312 ♀."
Lycæna Hoffmannseggii, Herr.-Schaeff., Schmett. Eur., i. pl. 133, f. 644 (1853).
Var. B.—*Lycæna pulchra*, R. P. Murray, Trans. Ent. Soc. Lond., 1874, p. 524, pl. x. ff. 7, 8 (♂ ♀).

Exp. al., (♂) 11½ lin.—1 in. 2 lin; (♀) 1 in.—1 in. 3 lin.

♂ *Pale, glossy, bluish-violaceous, with a pinkish tinge;* a blackish line edging hind-margin; cilia whitish, unspotted. *Hind-wing:* on each side of first median nervule an indistinct blackish spot; between spots and hind-marginal edging a white line, often indistinct; tail on first median nervule long, linear, black, white tipped. UNDER SIDE.— *Brownish-grey, with double transverse white striæ enclosing darker-brownish;* common to both wings—a double stria before middle,—another closing cell, and confluent at origin of third median with a third, much arched, submacular similar discal stria quite across wing,—a submarginal, continuous, white, lunular stripe,—a confluent hind-marginal row of white-ringed sublunular spots,—and a linear blackish edging. *Fore-wing:* inner white edge of first stria connected with base by a longitudinal white streak, surmounted by a dark-brown dash on subcostal nervure. *Hind-wing:* base dusky, with two whitish streaks (one along costal edge, the other more transverse); two hind-marginal spots (being the last of the row) conspicuous, black, ringed with brilliant bluish- or greenish-silvery, and in a more or less complete circle of orange or ochreous-yellow.

♀ *Greyish-brown, shot with vivid blue from bases over discs; the white-edged stria closing cell, with the confluent stria beyond it, more or less distinctly marked* (less apparent in hind-wing); submarginal lunular stripe and row of spots very indistinct in fore-wing, well marked in hind-wing: two black spots of hind-wing larger than in ♂. UNDER SIDE.—More conspicuously marked, *especially submarginal lunular white stripe*, which is much broader, particularly in hind-wing, and is sometimes more or less confluent in parts with outer white edging of discal stria.

In a ♀ from D'Urban, Natal, taken by the late Mr. M. J. M'Ken, the dark discal fascia and the row of whitish marks beyond it are both enlarged and more continuous than usual, and this is shown with considerable distinctness even on the upper side. In another, captured at King William's Town by Colonel Bowker, the same dark marking in both wings is completely confluent with the disco-cellular stria (and partly so with the sub-basal one), forming one very broad dark-brownish fascia, more broken in the hind-wing.[1]

LARVA (European).—" Purplish-red, the narrow oblique lines and the dorsal streak darker. On the flowers of *Lythrum salicaria.*"—*W. F. Kirby, from De Villiers and Guenée*, " *Tab. Synopt. Lep. d'Eur.*," 1835.

If Herrich-Schaeffer's figure of the Variety A. actually represents the *Hoffmannseggii* of Zeller, this form must be regarded as differing from the European type-form in its smaller size, and, on the under side, in the sharp definition of the markings, the more macular discal stria of the hind-wing, and in having the second of the hind-marginal row of white-ringed spots no larger than the rest; whereas in *Telicanus* it is much larger, rounded, and sub-ocelliform. Herrich-Schaeffer gives the under side only, and represents no tail on the hind-wing; but doubtless this appendage had been lost in the specimen figured. He does not record the locality; but while Staudinger (*Cat. Lep. Europ. Faunengeb.*, 1871, p. 9) merely says of *Hoffmannseggii*, "Species est Americana" (Hopffer, *loc. cit.*), while expressing great doubt as to the locality of Portugal originally assigned to Zeller's species, states that the butterfly is an African variety of *Telicanus*, and that his specimens from Nubia and the Cape are referable to it.

The South-African specimens (as above described) all more or less approximate to the West-African Variety B. named *Pulchra* by the Rev. R. P. Murray; but in none of the ♀ s that I have seen is there so much white on the upper side as shown in his fig. 8. Mr. Murray notes the "constant smaller size" of the specimens; but the measurements he gives, viz., " ♂ 11'''–1″ 1'''; ♀ 1″ 1'''–1″ 2''',″ of three ♂ and two ♀ examples, are not (except in the case of the smallest ♂) below the average size.

---

[1] An interesting specimen, exhibiting the conjunction of the features of both sexes in a single individual, was presented to me in 1883 by Mr. W. Billinghurst, who took it near Grahamstown. In this example the wings on the right side are of the ♂ coloration on the upper side, while those on the left side are of the ♀ coloration and pattern.

After carefully comparing specimens from different parts of the world, I agree with Hopffer (*loc. cit.*) in considering that *Telicanus*, varying a good deal in size, markings, and colouring wherever it is found, presents nowhere any constant local race which can be satisfactorily defined as a distinct form. There can be no doubt that, as far as is known, the tendency of the butterfly is to be paler and brighter in tint throughout the Ethiopian Region proper than in Southern Europe and extra-tropical North Africa,—this being chiefly shown in the white markings of both sexes on the under side, and the whitish ones of the ♀ on the upper side. Specimens that I captured in Mauritius were, however, as dusky as those usually found on Mediterranean shores, and none of either sex exceeded one inch in expanse of wings. Hopffer notes that specimens from the Cape, Mozambique, and Guinea are largest, and those from Egypt and Arabia the smallest, while European examples hold the middle place.

The nearest ally of *L. Telicanus* is *L. Cassius*, Cram., an abundant species in Central and South America. The under-side dark markings are very similar in *Cassius*, but more attenuated, and the white outer ones largely developed and completely confluent, so as to compose a white ground; the two ocelli of the hind-wing are feebly marked, and there is no tail. On the upper side the ♂ *Cassius* is much like the same sex of *Telicanus*, but the ♀ is broadly suffused with white discally, especially in the hind-wing, where only the costal and hind-marginal borders are dark.

Telicanus is common and widely spread in South Africa, frequenting gardens and open ground, where it flits rather actively about low flowers and grass. I have met with it from the beginning of November to the beginning of June. I observed the paired sexes at Cape Town on 2d April 1882. The ♀ carries the ♂ when flying *in copula*.

Localities of *Lycæna Telicanus*.

I. South Africa.
  B. Cape Colony.
    a. Western Districts.—Cape Town. Robertson. Knysna. Plettenberg Bay.
    b. Eastern Districts.—Port Elizabeth. Grahamstown. East London. King William's Town (*J. H. Bowker*).
    d. Basutoland.—Masern (*J. H. Bowker*).
  D. Kaffraria Proper.—Butterworth (*J. H. Bowker*).
  E. Natal.
    a. Coast Districts.—D'Urban. Verulam. Pinetown (*J. H. Bowker*).
    b. Upper Districts.—Maritzburg. Greytown. Great Noodsberg. Estcourt (*J. M. Hutchinson*).
  F. Zululand.—Napoleon Valley (*J. H. Bowker*). St. Lucia Bay (*the late Colonel H. Tower*).
  H. Delagoa Bay.—Lourenço Marques (*Mrs. Monteiro*).
  K. Transvaal.—Potchefstroom District (*T. Ayres*).

II. Other African Regions.
  A. South Tropical.
    a. Western Coast.—Damaraland (*C. J. Andersson*). "Angola (*Pogge*)."—Dewitz.
    b. Eastern Coast.—"Mozambique."—Hopffer.
    b1. Eastern Central Interior.—Mokloutze, Tati, and Zambesi Rivers (*F. C. Selous*). "Tati, Gwailo River, and near Victoria Falls of Zambesi (*Oates*)."—Westwood.

bb. Eastern Islands.—Madagascar (*F. L. Layard*). "Bourbon."—Boisduval. Mauritius. "Johanna, Comoro Islands (*C. W. Bewsher*)."—Var. B.—A. G. Butler. "Rodriguez (*Gulliver*)."—A. G. Butler.

B. North Tropical.
   a. Western Coast.—"Gaboon River (*Theorin*)."—Aurivillius. "Guinea."—Hopffer.
   b1. Eastern Interior.—"Atbara River" [Soudan].—A. G. Butler.

C. Extra-Tropical North Africa.
   b. Mediterranean Coast.—Algeria: Bougie. "Egypt."—Hopffer.

III. Europe.—Spain and Italy.—Coll. Brit. Mus. "South France, Corsica, Turkey, Central Europe."—*Auct.*

IV. Asia.
   A. Southern Region.—"West Asia."—Staudinger. "Arabia."—Hopffer.

SECTION E.—Under side grey or brownish-grey; all markings white-edged; discal markings not much darker than ground-colour, more or less confluent; hind-marginal series of spots distinct, preceded by a brownish line; fore-wing with a rather ill-defined longitudinal basi-costal blackish or brownish stripe; hind-wing with very distinct sub-basal, round, black, white-ringed spots.

   L. Jesous (Guér.), *Macalenga*, Trim., *Moriqua*, Wallengr., *Natalensis*, Trim.

155. (37.) **Lycæna Jesous,** (Guérin).

♂ *Polyommatus Jesous*, Guér., Lefebv. Voy. Abyss., vi. p. 383, pl. 11, ff. 3, 4 (1847).
♂ ♀ *Lycæna Jesous*, Wallengr., K. Sv. Vet.-Akad. Handl., 1857; Lep. Rhop. Caffr. p. 39.
♂ ♀   ,,   ,,   Trim., Rhop. Afr. Aust., ii. p. 250, n. 150 (1866).
♂ *Lampides Agave*, Walk., Entomologist, 1870, p. 53, n. 48 [as ♀.]

Exp. al., (♂) 11½ lin.—1 in. 1 lin.; (♀) 11½ lin.—1 in.

♂ *Lilac-blue, with a strong pinkish tinge; a very narrow hind-marginal blackish edging;* cilia white, in fore-wing indistinctly spotted with blackish on nervules. *Hind-wing:* two indistinct blackish spots on hind-margin, one on either side of first median nervule; no tail. UNDER SIDE.—*Brownish-white in fore-wing, greyish-white in hind-wing, with transverse, white-edged, brownish fasciæ, and black spots;* in each wing—a short, white-bordered brown streak closing cell,—a rather oblique, inferiorly irregular, submacular fascia beyond middle, from costa to first median nervule,—a brownish lunulate streak, followed by a row of black spots (some indistinct),—and a blackish line on hind-marginal edge. *Fore-wing:* a longitudinal stripe along subcostal to beyond end of cell, black at base, thence suffused and ferruginous-brown; in cell, near extremity, a blackish, white-edged spot, touching median nervure. *Hind-wing:* a basal oblique black stripe; a row of four very conspicuous white-edged black spots before middle, and two similar spots respectively at costal origin and inner-marginal end of macular fascia; two black spots next anal angle outwardly bluish-silvery-dusted, interiorly faintly orange-lunuled; base irrorated with pale-bluish.

♀ *Dull brownish, clouded with whitish on discs, faintly shot with bluish from bases;* in each wing a distinct disco-cellular fuscous lunule, and a marginal row of spots (very faint in fore-wing). UNDER SIDE.— As in ♂; but markings darker and more sharply defined. *Fore-wing:* an additional separate spot at lower extremity of discal fascia, prolonging it almost to submedian nervure.

I have not met with this species in life; though widely distributed in South Africa, it seems to be nowhere numerous. The male is a most lovely butterfly, well distinguished from all its South-African congeners by the exquisite glowing-pink lustre on the upper side of the wings. South-African specimens—as is noticeable in many other butterflies having a wide range through the continent generally—are larger than the North-African typical form. I was enabled to identify *L. Agare* of Walker with Guérin's species by examination of specimens collected by Mr. Lord on the Red Sea coast, in the possession of Mr. R. Meldola. I have not seen any authentic examples of *L. Malathana*, Boisd., a native of Madagascar; but from the brief description in *Faune Ent. de Madag.*, &c., p. 26, it would appear to be very closely allied to *L. Jesous*, but violaceous-blue above, and with only one hind-marginal black spot near anal angle of hind-wing on the upper side, bordered with a fulvous spot.[1]

Localities of *Lycæna Jesous*.

I. South Africa.
  B. Cape Colony.
    b. Eastern Districts.—Grahamstown (*M. E. Barber*). Between King William's Town and Queenstown (*M. E. Barber*). Murraysburg (*J. J. Muskett*).
    c. Griqualand West.—Vaal River (*J. H. Bowker*)
    d. Basutoland.—Maseru (*J. H. Bowker*).
  C. Orange Free State.—Hebron (*W. Morant*).
  D. Kaffraria Proper.—Tsomo and Bashee Rivers (*J. H. Bowker*).
  E. Natal.
    a. Coast Districts.—D'Urban (*J. H. Bowker*).
    b. Upper Districts.—Estcourt (*J. M. Hutchinson*). Blaauwkranz (*C. Hart*). Between Tugela and Mooi Rivers (*J. H. Bowker*).
  K. Transvaal.—Potchefstroom District (*T. Ayres*). Marico and Limpopo Rivers (*F. C. Selous*).
  L. Bechuanaland.—Motito (*the late Rev. J. Frédoux*).

II. Other African Regions.
  A. South Tropical.
    a. Western Coast.—Damaraland (*the late C. J. Andersson*, and *J. A. Bell*). "Angola (*Pogge*)."—Dewitz.

---

[1] In 1871 I noted in the Collection of the British Museum two males of *Jesous* ticketed "Port Natal," which were labelled "*Lampides Malathana*, Boisd."

*b*1. Eastern Interior.—Tauwani and Mokloutse Rivers (*F. C. Selous*).
B. North Tropical.
  *b*1. Eastern Interior.—" Abyssinia (*Lefebvre*)."—Guérin. White Nile.—Coll. Brit. Mus.

156. (38.) Lycæna Macalenga, Trimen.

♂ ♀ *Lycæna Macalenga*, Trim., Trans. Ent. Soc. Lond., 1870, p. 364, pl. vi. ff. 5, 6.

Exp. al., (♂) $10\frac{1}{2}$—$11\frac{1}{2}$ lin.; (♀) $9\frac{1}{2}$—$10\frac{2}{3}$ lin.

♂ *Pale, silky, violaceous-blue, with wide pale sandy-brown pink-tinged borders; bases tinged with deep purplish blue; cilia white throughout. Fore-wing:* blue space separate from basal dark blue, occupying inner margin as far as posterior angle, but leaving a hind-marginal border widening to the apex, and a costal border so greatly widening to the basal dark-blue as almost to touch the inner margin; *inner edge of the discal blue strongly defined by a denticulate raised line of paler blue. Hind-wing:* borders somewhat similar, but that of the hind-margin of even width; the inner edge of discal blue similarly defined about the origin of the subcostal nervules; two indistinct fuscous hind-marginal spots between the second median nervule and anal angle, the superior one the larger, and traces of other dark spots along rest of hind-margin.
UNDER SIDE.—Much resembling that of *L. Jesous. Fore-wing:* the black and ferruginous subcostal stripe ending before the extremity of the cell; *no spot in the cell*; the short streak closing the cell straighter and narrower than in *Jesous*; the submacular transverse white-edged fascia commencing farther from the costa and from the base, not so oblique, curved inwardly rather than outwardly, its terminal separate spot much closer to, and almost immediately below, the closing streak of cell; submarginal lunulate streak more denticulate, obsolete inferiorly; no hind-marginal black dots, but indistinct brownish marks; between the end of the subcostal stripe and the beginning of the submacular transverse fascia, *a longitudinal row of three black white-ringed dots*, of which the middle one is the largest. *Hind-wing:* basal stripe much thinner; third black spot of sub-basal row as large as the first and second, the fourth indistinct (the reverse being the case in *Jesous*); second spot of discal fascia elongate and oblique, instead of round; hind-marginal markings differing as in fore-wing; the two hind-marginal black spots next anal angle without any bluish scaling.

♀ *Pale, glistening sandy-brown, with a faint pinkish gloss*, but without the bluish bases,[1] whitish discs, or terminal cellular spots so well marked in *Jesous. Hind-wing:* two fuscous spots more apparent than in the ♂, or in *Jesous* ♀. UNDER SIDE.—As in ♂, but the ground-colour

---

[1] A ♀ since received from Burghersdorp, Cape Colony, has the bases suffused with dark greyish-blue.

browner throughout. *Fore-wing:* outermost of three subcostal dots wanting; submacular fascia prolonged to submedian nervure (as in ♀ *Jesous*) by an additional separate spot.

The fore-wings are rather markedly elongate in both sexes, being produced apically. It is singular that the under surface should show such decided resemblance to that of *Jesous*, while the upper side differs so widely in both ♂ and ♀. I do not remember to have seen any *Lycana* in which the blue occupies quite the same position as in the ♂ *Macalenga*, or in which it is internally so curiously defined.

The South-Indian *L. Ubaldus* (Cram.), from which the Cingalese *Azanus Crameri*, Moore (*Lep. Ceylon*, p. 80, pl. 36, f. 1), seems barely separable, is a closer ally than even *L. Jesous* or *L. Macalenga*, its under side agreeing almost exactly with that of the latter; but on the upper side the male exhibits no trace of the peculiar restriction and outline of the discal blue so conspicuous in the male *Macalenga*.

This curious species seems to be very scarce. It was originally discovered by Colonel Bowker, who took a male and two females in Basutoland, of which the male and one of the females were noted as captured on flowers near Olifant's Been, on the Cornet Spruit (Makaleng River), in February 1869. In 1871 the same observer sent single examples of the male from north-west of Somerset East and from Griqualand West respectively. The only other specimens I have seen were a male and female captured near Burghersdorp by Dr. Kannemeyer in 1883.

Localities of *Lycana Macalenga*.

I. South Africa.
  B. Cape Colony.
    b. Eastern Districts.—Between Somerset East and Murraysburg (*J. H. Bowker*). Burghersdorp (*D. R. Kannemeyer*).
    c. Griqualand West.—Vaal River (*J. H. Bowker*).
    d. Basutoland.—Maseru and Olifant's Been (*J. H. Bowker*).

157. (39.) Lycæna Moriqua, Wallengren.

Plate VIII. ff. 5 (♂), 5a (♀).

♂ *Lycana Moriqua*, Wallgrn., K. Sv. Vet.-Akad. [Handl., 1857; Lep. Rhop. Caffr., p. 39.
♂ ♀ *Lycana Moriqua*, Trim., Rhop. Afr. Aust., ii. p. 251, n. 151 (1866).
♀ *Lycana benigna*, Möschler, Verh. K. K. Zool.-Bot. Ges. in Wein, 1883, p. 285, t. xvi. f. 1.

Exp. al., (♂) 11 lin.—1 in. 1 lin.; (♀) 11¼ lin.—1 in. 1 lin.

Very nearly allied to *L. Jesous*, Guér.

♂ *Darker, more violaceous; hind-marginal border much broader, ill-defined inwardly. Hind-wing:* spots obsolete, merged in border; no tail. Under side.—Ground-colour *whiter;* disco-cellular streaks *blackish. Fore-wing:* longitudinal stripe *fainter, brownish;* fascia beyond middle *blackish,* more distinctly macular, much less oblique; *usually a small spot below median nervure,* before middle. *Hind-wing:*

no fascia beyond middle, *but an irregular row of eight distinct black spots*, of which the first, fourth, and eighth are larger than the rest; bluish at base almost obsolete; two spots near anal angle faintly blue-dusted, usually without orange lunule.

♀ Whitish discal space *smaller*, *less distinct*, sometimes almost obsolete; blue from base sometimes strongly-marked, sometimes scarcely traceable. UNDER SIDE.—As in ♂; *but in one specimen* the ground-colour is *brownish-grey*, the spots larger than usual, and their white rings and edges conspicuous.

In Natal and at Delagoa Bay a slight *variety* occurs, in which the ♂ has on the upper side a much narrower, almost linear hind-marginal dark-edging, and both sexes have on the under side the discal fascia of the fore-wing more oblique and with its terminal spot more widely separated from it, and the discal series of spots of the hind-wing in places not separate, but contiguous.

On a careful comparison of the description and figure of *L. benigna* given by Mr. Möschler (*loc. cit.*) with ♀ s of *Moriqua*, I cannot find any characters to distinguish it from those of the latter in which the upper-side bluish from base is rather restricted.

The characters above given amply distinguish *Moriqua* from *Jesous;* the most readily observable, in addition to the very different, more violaceous tint of the upper side in the ♂, being, on the under side, the absence of any brownish tinge in the ground-colour, and, in the hind-wing, the presence of a discal series of small black spots instead of a brownish white-edged sub-macular fascia.

I found *Moriqua* common in Natal, both on the coast and inland, and at D'Urban twice captured the paired sexes—on the 1st and 12th February respectively. It is a weak flyer, and delights to settle on the leaves of low shrubs and long grass. Though more prevalent in wooded spots, it is often met with on open ground in sheltered places. Mrs. Barber informs me that this species abounds near the mouth of the Kleinemond River, in the Bathurst District of Cape Colony.

Localities of *Lycæna Moriqua*.

I. South Africa.
   B Cape Colony.
      b. Eastern Districts.—Grahamstown and Fish River Heights (*M. E. Barber*). Kleinemond River (*M. E. Barber*). King William's Town (*W. S. M. D'Urban*).
   D. Kaffraria Proper.—Butterworth, Tsomo River, and Heads of St. John's River (*J. H. Bowker*).
   E. Natal.
      a. Coast Districts.—D'Urban, Verulam.
      b. Upper Districts.—Great Noodsberg. Hermansburg. Greytown.
  II. Delagoa Bay.—Lourenço Marques (*Mrs. Monteiro*).

II. Other African Regions.
   A. South Tropical.
      a. Western Coast.—Kinsembo, Congo (*H. Grose Smith*).

158. (40.) Lycæna Natalensis, sp. nov.

Exp. al., (♂) 1 in.—1 in. 2 lin.; (♀) 1 in. 1–2½ lin.

♂ *Violaceous-blue with a slight pinkish tinge* (very nearly the same tint as in *Moriqua*, Wallengr.), *the discal spots of the under side indistinctly perceptible; a well-defined black line edging hind-margins;* cilia in fore-wing fuscous, except next apex, between third and second median nervules, and between first median nervule and submedian nervure, where white replaces the fuscous,—in hind-wing fuscous basally, but white externally throughout. *Hind-wing:* no tail; usually an indistinct blackish hind-marginal spot at anal angle, and sometimes a second, very indistinct one above it. UNDER SIDE.—*White, with brown stripes and streaks and black spots; in both wings*—a long slender brown terminal disco-cellular lunule; a slightly irregular, submarginal, lunulated, brown, slender streak; and a hind-marginal row of small spots, mostly black, but a few indistinct and brownish. *Fore-wing: a conspicuous longitudinal streak, inferiorly bordering costal nervure*, black from base, but becoming brown and suffused towards extremity of discoidal cell, whence it borders costa and unites with origin of brown discal fascia; *this fascia brown, rather broad, sharply defined, slightly oblique, emitting a prominent dentation outward just above third median nervule*, abruptly interrupted on second median nervule, its widely separated lower part (consisting of a large upper spot and very small inferior one) much nearer base than the rest; a conspicuous rounded black spot in lower part of discoidal cell near extremity; the fourth and sixth spots in hind-marginal row brownish, obsolescent. *Hind-wing:* short, basal, sub-transverse black streak, and sub-basal transverse row of four black spots as in *Moriqua*, but more strongly marked; an extremely irregular and interrupted discal row of very unequal-sized black spots, of which the first and second next costa about middle are large and conspicuous,—the third (a good deal beyond them) minute, —the fourth (still farther beyond) minute or quite obsolete,—the fifth (close to submarginal streak), sixth (considerably nearer base), and eighth (on inner margin about middle) large,—and the seventh very small but usually geminate; spots of hind-marginal row larger than in fore-wing (especially the first, next costa), only the fourth brownish and obsolescent, the fifth and sixth large, spangled with silvery-blue.

♀ *White, with rather broad fuscous-brown costal and hind-marginal borders, and fuscous discal submacular fascia and spots following the pattern of the under side; shot with violaceous-blue from bases. Fore-wing:* terminal disco-cellular lunule much broader, and discal fascia much broader and its lower part less disconnected than on under side of ♂; blue suffusion extending to beyond middle along inner margin and filling discoidal cell. *Hind-wing:* only the lower larger spots of discal row represented; two hind-marginal spots next anal angle blackish, distinct; blue suffusion filling cell and extending thinly below and beyond it over median nervules. UNDER SIDE.—Markings generally larger than in ♂, especially the disco-cellular lunules, the hind-marginal spots, and in fore-wing the discal fascia. *Fore-wing*: a faint brownish suffusion basally.

In a singular *aberration* of the ♀ taken in Natal by Colonel Bowker, the upper side *has no white*, but only a very slightly paler discal cloud on the dusky-brownish general ground-colour; and with the exception of the indistinct terminal disco-cellular lunules, *there are no dark markings*. On the under side the same lunules are conspicuous, but *there are no discal fascia or spots*; the submarginal brown streak is, however, greatly widened and suffused, and there is also a brown suffusion along the hind-margin itself, so that only an irregular lunulate white streak remains between it and the submarginal streak. In the fore-wing the basal brown suffusion is much darker than usual; and in the hind-wing the short basal streak is confluent with the first spot of the sub-basal row, of which row the third and fourth spots are wholly wanting.

This species is nearly allied to *L. Moriqua*, Wallengr., but considerably larger. On the upper side the colouring of the ♂ is very like that of the ♂ *Moriqua*, but the spots of the under side are represented more apparently, and the much more slender hind-marginal edging line is like that of the ♂ *Jesous*, Guér.; while the colouring and pattern of the ♀ are widely different from those of the ♀ *Moriqua* and ♀ *Jesous*,—closely resembling, indeed, the upper side of the ♀ *Hintza*, Trim. On the under side both sexes approximate *Moriqua*, the chief distinctions being the whiter ground generally, the different colour, form, and extent of the discal fascia of the fore-wing, and, in the same wing, the great prolongation costally of the longitudinal stripe; and, in the hind-wing, the totally different respective sizes and positions of the black spots of the discal row. The fore-wings are unusually long, not produced apically, but rather prominent subapically; and the hind-wings are also longer and with a straighter costa than in the allied species. In this respect, and in a much less degree as regards some of the principal under-side markings, *Natalensis* exhibits some indications of alliance to the beautiful West-African *L. Isis* (Drury), in which the ♂ has white discs on the upper side.

I captured a ♂ of this butterfly at Hermansburg, Natal, in March 1867, and regarded it as possibly a variety of *L. Moriqua*. In 1869 a ♀ was sent from Natal, without any note of locality, by the late Mr. M. J. M'Ken; and in 1871 I noted two ♂s and a ♀ ticketed " Port Natal " in the British Museum Collection. The more recent receipt of fine specimens of both sexes from Mr. J. M. Hutchinson, of Estcourt, Natal, has enabled me satisfactorily to describe the insect; but I have no notes as to its special haunts or habits. A few examples from other parts of Natal have also reached me; and in 1881 I examined a ♀ from Delagoa Bay in the Hewitson Collection.

Localities of *Lycæna Natalensis*.

I. South Africa.
  E. Natal.
    a. Coast Districts.—Umgeni Railway Station (*J. H. Bowker*).
    b. Upper Districts.—Maritzburg (*S. A. Windham*). Estcourt and Bushman's River (*J. M. Hutchinson*). Hermansburg.
II. Delagoa Bay.—Coll. Hewitson in Brit. Mus.

SECTION F.—Under side white; markings very conspicuous and sharply defined, black; hind-marginal series (and sometimes also submarginal series) of spots very distinct and regular; discal markings of fore-wing more or less confluent into separate unequal portions; in fore-wing a strongly-marked basi-costal longitudinal stripe and a shorter curved one beneath it; basal area of hind-wing with several elongate spots.

L. Hintza, Trim. [tailed], *Calice*, Hopff. [tailed], *Melæna*, Trim. [tailed], *Griqua*, Trim. [tailed], *Sybaris*, Hopff. [tailed].

159. (41.) Lycæna Hintza, Trimen.

PLATE VIII. ff. 1 (♂), 1a (♀).

♀ *Lycæna Rosimon*, Wallgrn., K. Sv. Vet.-Akad. Handl., 1857; Lep. Rhop. Caffr., p. 38.
♂ ♀ *Lycæna Hintza*, Trim., Trans. Ent. Soc. Lond., 3rd Ser., ii. p. 177 (1864); and Rhop. Afr. Aust., ii. p. 243, n. 144 (1866).

Exp. al., (♂) 11 lin.—1 in. 1 lin.; (♀) 11 lin.—1 in. 1½ lin.

♂ *Shining violet-blue; on discs, the under-side dark spots are indistinctly represented, with (in some cases) a few intermediate whitish ones; a narrow hind-marginal blackish edging. Hind-wing:* the usual hind-marginal blackish spot. UNDER SIDE.—*White, with black striæ and spots:* in each wing, a stria closing cell (in fore-wing from costa and broad, in hind-wing short, thin, and angulated); an irregular, much-interrupted, transverse row of conspicuous spots beyond middle; two submarginal rows of spots, the inner of elongate sublinear, the outer of very small rounded spots; and a black edging line. *Fore-wing:* subcostal stripe continued on costa to end of cell; below it, a broad upward-curving stria from submedian next base; last two spots of transverse row confluent. *Hind-wing:* short subcostal stria divided into two spots; beyond it a curved transverse row of four spots; third spot of transverse row of eight far beyond the rest, almost touching second spot of inner submarginal row; last two spots of outer row large, faintly bluish-silvery-dotted.

♀ *White with fuscous borders; shot with violaceous-bluish from bases; the black markings of under side suffusedly marked. Fore-wing:* costa and hind-margin clouded with fuscous, on the latter obliterating the markings and joining the three outermost spots of transverse row; a whitish spot at posterior angle. *Hind-wing :* costa narrowly fuscous-clouded. UNDER SIDE.—As in ♂; bluish-silvery of two spots of hind-wing brighter.

Cilia in both sexes white, in *fore-wing* regularly interrupted with blackish, in *hind-wing* irregularly and very faintly. Tail of hind-wing rather short, linear, black, white-tipped.

There can be little doubt that the specimen noted by Wallengren as *Rosimon*, Fab., belongs to this species. His description is not detailed, but he particularly notes the cilia as white, spotted *in fore-wings* with black, a character very constant in *Hintza*, while in *Rosimon* the *entire* cilia are as constantly conspicuously black-spotted.[1] The blue ♂ at once distinguishes the species from *Rosimon*, and denotes a relationship to *Theophrastus*. The ♀ of *Rosimon* presents a much broader white discal field, in hind-wing uncrossed by transverse row of spots; the basal blue is more brilliant and with a *greenish* tinge; and with regard to the under side, the spots of the common inner submarginal row are much broader; while in hind-wing the spots of the discal row, instead of being quadrate and nearly central, are *elongate, and all but the two first close to, and usually touching, those of inner submarginal row*,—and the subcostal stria is entire.

This is apparently a scarce species. Mr. W. D'Urban, who discovered it in British Kaffraria, noted it as not common near King William's Town, appearing in November and February; and Colonel Bowker sent a few of both sexes from Kaffraria Proper. I found it scarce during my visit to Natal in 1867, observing it only singly about grassy open spots, in different parts of that Colony, between the end of February and beginning of April.

Localities of *Lycæna Hintza.*

I. South Africa.
   B. Cape Colony.
      b. Eastern Districts.—King William's Town (*W. S. M. D'Urban*).
      c. Griqualand West.—Vaal River (*J. H. Bowker*).
   D. Kaffraria Proper.—Tsomo and Bashee Rivers (*J. H. Bowker*).
   E. Natal.
      a. Coast Districts.—D'Urban. Umhlali. Verulam. Umvoti.
      b. Upper Districts.—Maritzburg. Estcourt (*J. M. Hutchinson*). Tugela and Mooi River Valley (*J. H. Bowker*).
   K. Transvaal.—Marico and Limpopo Rivers (*F. C. Selous*).

160. (42.) Lycæna Calice, Hopffer.

Lycæna Calice, Hopff., "Monatsb. K. Preuss. Akad. Wissensch., 1855, p. 642, n. 18;" and Peters' Reise Mossamb., Ins., p. 405, t. xxvi, ff. 4, 5 [♀] (1862).

Exp. al., (♂) 1½ lin.; (♀) 1 in. 1 lin.

White, with black margins and spots; in each wing, a wide basal black suffusion; a broad costal, apical, and hind-marginal border; a terminal disco-cellular black lunule; and a broken irregular discal row

---

[1] Mr. Chr. Aurivillius, of the Stockholm Museum, to whom I referred the point, informs me that the *Rosimon* of Wallengren is undoubtedly = *Hintza*, Trim.

of spots, in parts confluent with hind-marginal border. *Fore-wing*: basal black sharply defined externally, having almost a straight edge; upper spots of transverse discal row confluent, and, in conjunction with apical border, isolating a conspicuous white spot between upper radial and third median nervules; hind-marginal border, below this, completely confluent with other spots of discal row; at posterior angle a short linear hind-marginal white mark. *Hind-wing*: basal black not so dark or externally so sharply defined as in fore-wing; costal border rather suffused; terminal disco-cellular lunule almost merged in basal black; discal row of spots widely interrupted between second subcostal and third median nervules,—the first and third spots united respectively to costal and hind-marginal borders; an indistinct hind-marginal row of black spots in border, immediately succeeded by a more or less conspicuous interrupted white line usually becoming obsolete superiorly; tail on first median nervule rather long, black, white-tipped. UNDER SIDE.—*White, with narrow black streaks and spots. Fore-wing*: a longitudinal streak from base, bounding costal nervure inferiorly, and ending abruptly on costa before middle; below it a broader, upward-curving streak from submedian nervure close to base, the end of which penetrates discoidal cell near extremity; costa narrowly edged with black; an almost straight transverse streak closing cell, from costa (where it is broader) to origin of third median nervule; an irregular sub-macular discal transverse streak, emitting a more or less marked projection outward between radial nervules, and interrupted on second median nervule; a regular submarginal series of sub-linear or lunulate spots, becoming rather thicker, and approximating very closely to spots of discal row inferiorly; a hind-marginal series of very small rounded spots, succeeded by a black edging line. *Hind-wing*: at base, a very short slightly curved streak, abruptly beginning immediately beyond curve of costal nervure, and ending close to origin of inner margin; a sub-basal curved transverse row of three elongate spots—the lowest on inner-margin; a terminal disco-cellular lunule; a very strongly curved discal row of seven rather small spots, from costal nervure to inner-margin—a wide interruption, from second subcostal to radial nervule, between second and third spots, and two minor interruptions between fourth and fifth, and sixth and seventh spots respectively; of submarginal row of spots, only the first and second (next costa) and the last (on inner margin) distinctly marked, the remainder being faint and linear, and about middle almost obsolete; hind-marginal row of spots very distinct (except the first, which is almost obsolete),—the three lowest usually more or less distinctly including a ring of bluish-silvery scales,—the spot between first and second median nervules largest; a fine hind-marginal bounding black line.

Cilia fuscous at origin throughout; in fore-wing fuscous with well-marked white interruptions at apex, posterior angle, and about middle of hind-margin, and intermediate small whitish ones; in hind-wing white, with dull-fuscous narrow interruptions on nervules.

The sexes do not differ except in size, and in the ♀'s having the black markings narrower and the white field proportionally wider on both upper and under sides.

It was not until July 1879, when Colonel Bowker sent me the paired sexes from Pinetown, Natal, that I identified the true *Calice* of Hopffer, having previously taken for this species the very nearly allied *L. Melæna*, mihi.

The absence of any blue in either sex, and the width and blackness of the dark borders on the upper side, with the very different arrangement of the discal transverse series of spots on the under side, readily distinguish *Calice* from *Hintza*, Trim.

The species appears to be rarer than *Melæna*, and the only examples I have received were taken in Natal. Neither Colonel Bowker nor Mr. Hutchinson noted any particulars of the butterfly's special haunts or habits.

Localities of *Lycæna Calice*.

I. South Africa.
   E. Natal.
   a. Coast Districts.—Pinetown (*J. H. Bowker*).
   b. Upper Districts.—Estcourt (*J. M. Hutchinson*).

II. Other African Regions.
   A. South Tropical.
      a. Western Coast. — "Damaraland (*De Vylder*)." — Aurivillius. "Angola (*J. J. Monteiro*)."—H. Druce.
      b. Eastern Coast.—"Querimba."—Hopffer.
   B. North Tropical.
      a. Western Coast.—"Senegal."—Hopffer.

161. (43.) **Lycæna Melæna,** *sp. nov.*

Lycæna Calice, Trim., Rhop. Afr. Aust., ii. p. 244. n. 145 (1866).

Exp. al., (♂) $10\frac{1}{2}$—$11\frac{1}{2}$ lin.; (♀) 10 lin.—1 in.

White, with black margins and spots. Fore-wing: base and costa broadly and inner margin narrowly suffused with blackish; a large, elongate, black, transverse mark from inner margin near base; a spot closing cell; and several spots of an irregular row beyond middle, joining broad hind-marginal black, and isolating a quadrate white spot not far from apex. *Hind-wing*: a blackish suffusion at base, along costa (leaving part of its edge white), and along hind-margin; beyond middle, a much-curved row of quadrate black spots (usually more or less confluent with hind-marginal blackish), widely interrupted between second subcostal and discoidal nervules; several indistinct black spots at base; in hind-marginal blackish a row of indistinct black spots (of which the two next anal angle are sometimes faintly bluish-silvery-dotted), occasionally whitish-ringed, always followed by a thin white line, indistinct towards costa.. Cilia of *fore-wing* blackish, with a white spot at posterior angle; of *hind-wing* whitish, more or less blackish at origin. UNDER SIDE.—*White, with black striæ and spots;* pattern as in *Rosimon* and *Hintza*, but nearer *Rosimon*. *Fore-wing:* subcostal stripe ending abruptly in a spot on costa before end of cell; striæ from inner margin more regular; stria closing cell straighter; spots of transverse row beyond middle confluent (*the third forming an elongate projection towards hind-margin*),—the sixth and seventh widely disjoined from the rest, and forming a stria in a line with, and almost touching, that closing cell; two submarginal rows of spots,—the inner linear a little below costa, but thence of large quadrate spots (the second touching fifth of row beyond middle),—the outer of rounded spots; a black edging line. *Hind-wing:* at base a spot (*not* a stripe), followed by a transverse row of four spots; other markings very much as in fore-wing, except row beyond middle, which is interrupted as on upper side; four to six spots of outer submarginal row marked with bluish-silvery. Tail of hind-wing long, black, white-tipped.

The ♂ and ♀ of this species are alike, except that in the ♂ the black is more intense, and also rather broader, so that the white discal spaces are more restricted. On the under side of fore-wing it sometimes happens that the projection from transverse row beyond middle joins the spot commencing inner submarginal row, so that a quadrate white spot is isolated on costa. The entire absence of any basal blue readily distinguishes *Melæna* from *Rosimon* and from *Hintza* ♀; but it is by no means so easy to separate it from *Calice*, Hopff., with which I for many years associated it. On the upper side there is scarcely any difference, but *Melæna* constantly presents *in the fore-wing a discal white area almost divided transversely by the approximation of the terminal disco-cellular mark and the lower portion of the irregular discal row, the latter being almost in a straight line beneath the former*. This difference is also very marked on the under side, the two markings in question forming in some specimens an almost continuous stripe, whereas in *Calice* the lower portion of the discal streak is almost continuous of the upper, and so remote from the terminal disco-cellular streak. The principal under-side markings in *Melæna* are thicker than in *Calice*, with the exception of the lower spots of the discal series in the hind-wing; four at least of the hind-marginal black spots of the hind-wing are marked with bluish-silvery; the ground is of a purer white without any slight tinge of yellowish; the submarginal macular series in the fore-wing forms a continuous streak (very attenuated between upper radial and third median nervule), and in the hind-wing an almost continuous one; and the cilia of the hind-wing are without fuscous interruptions.

I found this species not common in Natal in 1867; it frequented wooded spots, and was fond of settling on the upper leaves and twigs of bushes. I observed it on the wing from the beginning of February to the beginning of

April; and Mr. D'Urban recorded its occurrence in British Kaffraria in October, March, and April. Colonel Bowker forwarded a good many examples from Kaffraria Proper.

Localities of *Lycæna Melæna.*

I. South Africa.
   B. Cape Colony.
      b. Eastern Districts.—King William's Town (*W. S. M. D'Urban*).
      c. Griqualand West.—Vaal River (*J. H. Bowker*).
   D. Kaffraria Proper.—Bashee River (*J. H. Bowker*).
   E. Natal.
      a. Coast Districts.—D'Urban. Pinetown (*J. H. Bowker*).
      b. Upper Districts.—Intzutze River. Maritzburg. Estcourt (*J. M. Hutchinson*).
   F. Zululand.—St. Lucia Bay (*the late Colonel H. Tower*).

162. (44.) **Lycæna Griqua,** *sp. nov.*

Exp. al., (♂) 11 lin—1 lin.; (♀) 11½ lin.

Markings and pattern as in *L. Melæna,* Trim., but the basal and marginal black duller, broader, and more suffused, especially in ♀, and the white discal field proportionally diminished, and in ♀ also obscured. *Fore-wing:* isolated subapical whitish spot always reduced in size and obscured, sometimes nearly obsolete, and in one ♂ wanting altogether; the spots much narrower. *Hind-wing:* the spots much narrower and almost obliterated by the broad suffused blackish of the base and margins. UNDER SIDE.—*Very pale creamy-yellowish grey; all the markings arranged as in Melæna, but pale fuscous instead of black, and exceedingly attenuated, especially in hind-wing (where in ♀ they are obsolescent). Fore-wing:* longitudinal basal stripe represented by a very thin short mark at base and (usually) a small spot on costa before middle. *Hind-wing:* only last two spots of hind-marginal row bluish-silvery-dusted, except in one ♂, which has three, and one ♀, which has four so ornamented.

The five ♂s and two ♀s above described are all tailless, but there is the vestige or base of a tail on one hind-wing of a single ♂. As the specimens are in fair condition, it is difficult to believe that all of them could have lost the tails; and yet the form is so very intimately related to *L. Melæna,* Trim., that one would certainly expect it to be caudate.

The only specimens I have seen were collected by Colonel Bowker in 1871 on the Vaal River, Griqualand West. They were accompanied by a single ordinary example of *L. Melæna,* but by no specimens of a character intermediate between the two forms.

Locality of *Lycæna Griqua.*

I. South Africa.
   B. Cape Colony.
      c. Griqualand West.—Vaal River (*J. H. Bowker*).

163. (45.) **Lycæna Sybaris**, Hopffer.

♂ ♀ *Lycæna Sybaris*, Hopff., "Monatsb. K. Preuss. Akad. Wissensch., 1855, p. 642."
♂ *Lycæna Sybaris*, Wallgrn., K. Sv. Vet.-Akad. Handl., 1857; Lep. Rhop. Caffr., p. 37.
♂ ♀ *Lycæna Sybaris*, Hopff., Peters' Reise Mossamb., Ins., p. 408, t. xxvi. ff. 6–8 (1862).
♂ ♀ ,,   ,,   Trim. Rhop., Afr. Aust., ii. p. 242, n. 143 (1866).

Exp. al., (♂) $1\frac{1}{2}$ lin.—1 in. $0\frac{1}{2}$ lin.; (♀) $10\frac{1}{2}$ lin.—1 in.

♂ Shining violaceous-blue; hind-margins rather widely bordered with fuscous, which in hind-wing is submacular and traversed in its outer portion by a fine interrupted white line more pronounced inferiorly; cilia white, irregularly varied with dull fuscous in fore-wing, chiefly about extremities of nervules, not, or very slightly so, varied in hind-wing. *Fore-wing*: a large, broad, conspicuous, terminal disco-cellular, fuscous lunule; sometimes, beyond it, two or three smaller fuscous spots representing some of the irregular discal row of the under side. *Hind-wing*: costa bordered with greyish-fuscous; immediately preceding hind-marginal border a series of more or less indistinct thin lunulate white marks, usually obsolete in upper part of wing; a rather long, twisted, subspatulate, black, white-tipped tail on first median nervule. UNDER SIDE.—*White, faintly tinged with yellowish, with very numerous and well-defined mostly rounded black spots; in each wing a disco-cellular terminal lunule, an irregular discal row, two regular submarginal rows of spots, and a hind-marginal black line*. *Fore-wing*: costa narrowly edged with black from a little before middle to apex; a strongly-marked longitudinal black streak from base inferiorly bounding costal nervure almost to its extremity; below this an oblique upward-curving irregular black streak, from between median and submedian nervures, penetrating discoidal cell rather beyond its middle; inner margin narrowly edged with fuscous-grey; a minute spot immediately above terminal disco-cellular lunule; discal row highly irregular, its six spots (exclusive of a minute one next costa) arranged obliquely in pairs, the spots confluent in the middle and lower pairs; submarginal rows of six spots each, the inner row with larger spots, increasing in size downward. *Hind-wing*: at base a short black streak from costal to submedian nervure; a sub-basal row of four spots; irregular discal row of eight spots, of which the third and fifth are farthest from base; inner submarginal row of eight, outer of seven spots,—the last three or four of outer row spangled with bluish-silvery exteriorly.

♀ *Fuscous, with more or less white on discs, on which the terminal disco-cellular lunules, part of discal row of spots, and (in hind-wing) submarginal rows of spots are more or less conspicuously shown, but more suffusedly than on under side; bases rather thinly and narrowly suffused with violaceous blue, extending along inner margins to a little beyond middle*. UNDER SIDE.—As in ♂.

This *Lycæna* is closely related to *L. Theophrastus* (Fab.), a native of North and West Africa, as well as of India and Ceylon, but readily distinguished by the greater size and blackness of the under-side markings,—the spots being rounder and more distinctly separate, especially those of inner-submarginal row, which in *Theophrastus* are thin, sublunulate, and partly united. On the upper side the ♂ *Sybaris* is bluer, with the hind-marginal border wider and in the hind-wing traversed by a white line, while the ♀ has the basal blue much duller and smaller in extent. These differences also well serve to distinguish *Sybaris* from *L. Balkanica*, Frey. (= *Psittacus*, Herr.-Schaeff.), a native of Turkey and West Asia, in which the under-side markings are thinner and more inclined to be linear than they are in *Theophrastus*, and the upper side of the ♀ has scarcely a trace of whitish on the discs.

I met with only a single ♂ of this very beautiful species, at Klipdrift (now Barkly), on the Vaal River, on 18th September 1872; it was fluttering about low plants, and at length settled on the ground. It appears to be widely distributed, and not uncommon in the interior districts, but is not known to me to have occurred in the Cape Colony proper, or to have been found within forty miles of the coast.

Mr. A. G. Butler, in noting the occurrence of *Sybaris* on the Atbara, Soudan (*Ann. and Mag. Nat. Hist.*, December 1876, p. 482), writes—"There is a ♂ of this species no larger than *L. Barberæ*, Trim."

Localities of *Lycæna Sybaris*.

I. South Africa.
  B. Cape Colony.
    c. Griqualand West.—Klipdrift, Vaal River.
  C. Orange Free State.—Between the Rhenoster and Vaal Rivers (*W. Morant*).
  D. Kaffraria Proper.—Tsomo and Bashee Rivers (*J. H. Bowker*).
  E. Natal.
    b. Upper Districts.—Estcourt (*J. M. Hutchinson*). Rorke's Drift, Ladysmith, and Biggarsberg (*J. H. Bowker*).
  F. Zululand.—Napoleon Valley (*J. H. Bowker*).
  H. Delagoa Bay.—Lourenço Marques (*Mrs. Monteiro*).
  K. Transvaal.—Potchefstroom District (*T. Ayres*). Marico and Limpopo Rivers (*F. C. Selous*).

II. Other African Regions.
  A. South Tropical.
    a. Western Coast.—Damaraland (*J. A. Bell*).
    b. Eastern Coast.—"Querimba."—Hopffer. Zambesi (*Rev. H. Rowley*).
    b1. Eastern Central Interior.—Tauwani River (*F. C. Selous*).
  B. North Tropical.
    b1. Eastern Interior.—"Atbara River" [Soudan].—A. G. Butler.

SECTION G.—Under side white; markings chiefly dull yellow-ochreous with fuscous edges, but partly fuscous in fore-wing; longitudinal basal stripes of fore-wing much less distinct than in Section

F.; spots of discal series in both wings, and spots in basal area of hind-wing, irregularly placed but separate and sub-quadrate.
L. Thespis (Linn.) [tailed], *Boirkeri*, Trim. [tailed].

164. (46.) Lycæna Thespis, (Linnæus).

PLATE VIII. ff. 2 (δ), 2a ($\female$).

♂ *Papilio Thespis*, Linn., Mus. Lud. Ulr., p. 318, n. 136 (1764); and Syst. Nat., i. 2, p. 791, n. 236 (1767).
♀ *Papilio Pitho*, Linn., opp. cit., p. 337, n. 255; p. 795, n. 266.
♂ ♀ *Polyommatus Thespis*, Godt., Enc. Meth., ix. p. 682, n. 207 (1819).
♂ ♀ *Lycaena Thespis*, Trim., Rhop. Afr. Aust., ii. p. 246, n. 147 (1866).

Exp. al., (♂) 1 1—1 1¾ lin.; (♀) 1 in.—1 in. 1 lin.

♂ *Bright shining-blue, with a narrow, blackish, hind-marginal line; cilia broad, white, sharply and regularly intersected with black at ends of nervules.* Fore-wing: a faint line closing cell. *Hind-wing*: two faint-blackish hind-marginal spots, one on either side of first median nervule. UNDER SIDE.—*Whitish, much intersected and chequered with quadrate blackish-grey or ochrey-grey spots.* Fore-wing: costa and hind-margin more or less tinted with ochrey-grey; a transverse blackish-grey spot in cell, another (surmounted by two smaller spots) closing it; a very irregular, interrupted transverse row of spots beyond middle, only separated in parts by white spots from a regular row of spots (parallel to hind-margin) beyond it; a row of minute white lunules edging the latter combined with the ochrey tint to form a pale hind-marginal border, marked with a row of very indistinct dark dots; inner margin dull-grey; from base a narrow blackish streak, inferiorly bounding subcostal nervure, terminating abruptly a little before spot in cell. *Hind-wing*: three irregular transverse rows of ochrey-grey quadrate spots, more or less confluent on inner margin; hind-marginal border as in fore-wing, but more tinted with ochreous superiorly; two lowest hind-marginal spots faintly silvery-centred, the rest of the series less distinct.

♀ *Blackish, shot with blue from bases over discs.* Fore-wing: *inner transverse row of under side distinct, black*, separated from hind-marginal blackish by *conspicuous pure-white spots*; terminal disco-cellular spot large, with a whitish dot on each side; a row of minute indistinct bluish dots near hind-margin. *Hind-wing*: two submarginal rows of bluish lunules, the inner row more or less confused with discal blue, the outer forming three incomplete rings near anal angle; black spots barely distinguishable from hind-marginal blackish edging. UNDER SIDE.—As in ♂; but white ground much brighter and clearer, especially on hind-margins. *Hind-wing*: transverse rows separately traceable to inner-margin; metallic centres of two lowest hind-marginal spots more brilliant.

Has a *very short* tufted tail on first median nervule of hind-wing, not observable in worn specimens.

This very distinct and handsome species is well distinguished by its singularly chequered under side, and, as regards the upper side, in the male by the brightness of the uniform blue field, and in the ♀ by the discal reproduction in part of the chequered pattern of the under side, chiefly in the fore-wings. The cilia in both sexes are unusually broad and conspicuous, pure-white, strongly and evenly interrupted with black. The female is not at all unlike (except for the strong blue suffusion on the upper side) *Pyrgus* in colouring and pattern, and Linnæus (in the *Museum Ludovicæ*, &c.), compares it to the European *P. Malvæ*.

Thespis is tolerably numerous in many parts of the Cape Colony, especially in the South-Western Districts and about Cape Town, but is usually found singly or in pairs. It frequents open ground generally, hovering about low shrubs and plants, and constantly settling; its flight is short and weak, so that it is very easily captured. I have met with it from the beginning of August until the end of April. For so small a butterfly it is unusually conspicuous on the wing. Boisduval (*Voy. de Deleg. l'Afr. Aust.*, p. 588) mentions it as occurring at Port Natal, but I have not seen any examples from that neighbourhood. Colonel Bowker did not send this species from either Kaffraria proper or Basutoland, nor have I received any examples from the northern parts of the Cape Colony or from farther in the interior.

Localities of *Lycæna Thespis*.

I. South Africa.
   B. Cape Colony.
     a. Western Districts.—Cape Town. Stellenbosch. Triangles Station, Worcester District (*L. Péringuey*). Plettenberg Bay.
     b. Eastern Districts.—Port Elizabeth (*W. S. M. D'Urban* and *J. L. Fry*). Grahamstown. Top of Gaika's Kop, Amatola Mountains (*J. H. Bowker*).
   E. Natal.
     a. Coast Districts.—"Bay of Port Natal."—Boisduval.

165. (47.) Lycæna Bowkeri, Trimen.

♂ ♀ *Lycæna Bowkeri*, Trim., Trans. Ent. Soc. Lond., 1883, p. 351.

Exp. al., 1 in. $1\frac{1}{2}$–$2\frac{1}{2}$ lin.

♂ Silky lilacine-blue; each wing with a rather large blackish lunular mark closing discoidal cell, and a moderately-wide blackish macular hind-marginal border; cilia wide, black, conspicuously interrupted with white between nervules. *Hind-wing*: the spots composing hind-marginal border more separated than in fore-wing (especially near anal angle), and immediately preceded by contiguous thin whitish lunules. UNDER SIDE.—Yellowish-white; each wing with a lunular mark closing discoidal cell, an irregular interrupted discal row of spots, and a submarginal row of smaller subquadrate spots,—all pale ochreous-brown, more or less distinctly finely edged internally and externally with blackish; close to hind-margin a row of very distinct sub-lunulate black spots. *Fore-wing*: a longitudinal brown stripe from base (where it is almost black) along subcostal nervure to a little before and above extremity of discoidal cell; three pale ochreous-brown spots near base, viz., a small one on costa almost confluent with terminal disco-cellular lunule, a reniform one in the cell, and an irregularly-shaped one just below the second but outside the cell; irregular discal row of spots abruptly interrupted on second median nervule, between fifth and sixth of its seven spots; costa from before middle to apex edged with blackish. *Hind-wing*: curve of costal edge close to base black; an irregular ochreous-brown basal marking formed of three or four contiguous small spots; three spots near base arranged much as in fore-wing, but the uppermost one much larger; an additional spot on inner margin near base; irregular discal row of spots abruptly interrupted between second and third of its eight spots on lower subcostal nervule, and thence sharply angulated in almost a direct line to about middle of inner margin; the seventh (and very slightly the eighth) black spot of the hind-marginal row dotted with silvery-blue.

♀ Only the basal and inner-marginal region of both wings and the hind-marginal region of hind-wing lilacine-blue, the disc being white; the terminal disco-cellular spot and the transverse irregular discal row as described on under side of ♂, but black and more strongly marked; row of violaceous-whitish lunules internally edging hind-marginal blackish spots more conspicuous than in ♂ in hind-wing, and also indistinctly marked in fore-wing. UNDER SIDE.—As in ♂, but all the spots somewhat sharper and clearer in outline.

This species is a near ally of *L. Thespis* (Linn.), a butterfly hitherto much isolated in its genus by the singular chequered pattern of the under surface of the wings, which is not unlike that of several species of the Hesperide genus *Pyrgus*. In *L. Bowkeri* the under side, owing to its purer white ground, smaller and more neatly-defined markings, and very distinct hind-marginal row of black spots, indicates a departure from that of *Thespis* in the direction of such congeners as *Sybaris*, Hopff., and *Hintza*, Trim., but retains the peculiar pattern of *Thespis* as well as the pale ochreous-brown of the spots generally. On the upper surface the ♂ *Bowkeri* is readily distinguished from the ♂ *Thespis* by its much less vivid, more lilacine blue, and much wider hind-marginal blackish submacular border; while the ♀ may be recognised by the much more developed discal white (especially in the hind-wing and hind-marginal lunulate markings. The tail of the hind-wing is in both sexes longer than in *Thespis*.

Colonel Bowker discovered this interesting butterfly in the earlier part of the year 1881, on the summit of a high table-topped hill overlooking the Inchanga Valley in Natal. He met with four specimens only, two of each sex, and at the time took them to belong to a local variety of *Thespis*. They were flitting about the flowers of a small leguminous shrub growing on rocks at the edge of a high precipice.

Locality of *Lycæna Bowkeri*.

1. South Africa.
   E. Natal.
   b. Upper Districts.—Inchanga (*J. H. Bowker*).

Genus CHRYSOPHANUS.

Chrysophanus, Hüb., Verz. Bek. Schmett., p. 72 (1816); Westw., Gen. Diurn. Lep., ii. p. 497 (1852).
Lycæna, Fab., Illiger's Mag., vi. p. 285 (1807), sect. 3 [part].
Polyommatus, Latr. [part], Encyc. Meth., ix. p. 11 (1819); Herrich-Schäffer, Syst. Bearb. Schmett. Europ., i. p. 130 (1843).

IMAGO.—General characters of *Lycæna*. *Eyes* always naked; *palpi* with bristly hair beneath; *antennæ* rather thicker. *Thorax* stout, as in the robuster species of *Lycæna*. *Fore-wings* more acute apically. *Hind-wings* with anal angle acuter and more prominent (very slightly so in the South-African *C. Orus*), and sometimes with a more or less acute projection at extremity of first median nervule. *Fore-legs* of ♂ longer and stouter than in *Lycæna*,—tibia with scattered spines beneath, besides several at extremity, but without terminal hook,—tarsus very thickly spinose beneath and less so laterally, terminating in a slightly-curved claw; of ♀ similar, but with tarsus fully developed, articulated, with much-curved terminal claws. *Middle* and *hind legs* rather shorter and thicker than in *Lycæna*,—tibiæ with two or three short spines beneath, and with terminal spurs thick,—tarsi very spiny beneath and with a few short spines above; *the long first joint in the ♂ more or less swollen*.

LARVA.—More flattened (less convex dorsally) than in *Lycæna*; more or less finely pubescent.

PUPA.—Shorter, thicker, rounder than in *Lycæna*, especially anteriorly.

(These characters of larva and pupa are gathered from the figures and descriptions of many authors.)

In structure, as well as in pattern of markings, *Chrysophanus* is so intimately related to *Lycæna* that it is very doubtful whether the group is entitled to more than subgeneric rank. The characters of the legs alone seem to present any distinction of importance. At the same time, the beautiful insects referred to *Chrysophanus* have a very distinct facies, and it is perhaps better to keep them apart from the immense genus, to which they are unquestionably most closely allied.

The *Chrysophani* are mostly distinguished by the golden or coppery red of the upper side of the wings, which is specially splendid (and sometimes, as in *C. Orus*, shot with blue-violaceous) in the males, but usually much duller in the black-spotted females, whose hind-wings, too, are sometimes brown or suffused with brown. In some species, both sexes are spotted with black above, and occasionally (*Dorilis*, Hufn.) the male is much darker and with less trace of red than the female. The single South-African species presents but little difference in the sexes, both being metallic orange-red; but the male has a violaceous surface-lustre absent in the female, while the latter has on the hind-wings as well as on the fore-wings a discal series of black spots. On the under side *Chrysophanus* has quite the pattern of the large *Alexis* or *Corydon* group of *Lycaena*; though in the more brilliant species the soft orange tint of the fore-wings gives a very different general aspect, and in some cases (as in *Phlaeas*, Linn.) the markings are in the hind-wings much obscured.

As in *Lycaena*, there has been considerable difficulty in determining the limits of the species, and the synonymy of the European forms is an intricate affair. There would appear, however, to be about forty recognised species, of which rather more than half belong to the Palæarctic Region, ten or eleven inhabiting Europe proper. Nearly all the remainder are recorded from North America; but a few species occur most remarkably at isolated points far remote from each other, viz., four in New Zealand, one in Queensland (Australia), one at the Cape, and one in Chili. In Abyssinia a close ally or variety of *C. Phlaeas* (*C. Pseudophlaeas*, Lucas) occurs, and Mr. Godman has noted (*Proc. Zool. Soc. Lond.*, 1885, p. 539) the capture of a single specimen of a *Chrysophanus* on Kilima-njaro in Tropical Eastern Africa. The outlying species just referred to—with the exception perhaps of *Pseudophlaeas*—are of distinct facies and apparently limited range; but many of the genus roam widely over Europe and Asia, while *Phlaeas* occupies the entire Palæarctic Region and great part of North America, and was one of the five butterflies taken in Grinnell Land (lat. 81° 45′ N.) by Captain Feilden.

Chrysophanus Orus is common and generally distributed in open ground over most of South Africa, and may be found on the wing throughout the year.

166. (1.) Chrysophanus Orus, (Cramer).

? ♀ *Papilio Orus*, Cram., Pap. Exot., iv. t. cccxxxii., ff. E, F (1782).
Papilio Areas, Fab., Mant. Ins., ii. p. 80, n. 728 (1787); and
Hesperia Areas, Ent. Syst., iii. i. p. 311, n. 179 (1793).
Polyommatus Orus, Godt., Enc. Meth., ix. p. 672, n. 172 (1823).
♂ ♀ *Chrysophanus Orus*, Trim., Rhop. Afr. Aust., ii. p. 259, n. 160 (1866).

Exp. al., (♂) 10½ lin.—1 in. 1 lin.; (♀) 11 lin.—1 in. 2½ lin.

♂ *Metallic orange-red, with a blue-violaceous lustre; a narrow blackish hind-marginal edging and some black discal spots; bases slightly dusky. Fore-wing:* a lunule closing cell; usually a small spot *in* cell; beyond middle an irregular row of six or seven spots between costa and submedian nervure ; a black linear costal edging, abruptly widening into a broad apical border, which again grows gradually narrower along hind-margin to anal angle. *Hind-wing:* markings as in fore-wing, but inner edge of hind-marginal border indented on nervules, and spots more minute (those of transverse row being sometimes partly —occasionally wholly—wanting); no spot *in* cell. UNDER SIDE.— *Hind-wing and border of fore-wing brownish-grey. Fore-wing:* spots as above, but larger and with whitish edging; an additional spot in cell, near base ; along inner edge of hind-marginal grey a row of linear blackish lunules, most distinct near inner margin. *Hind-wing:* spots as above, but scarcely darker than ground-colour, only marked by their hoary rings; a transverse row of three minute blackish spots near base, and a little beyond them a row of three larger, duller spots; submarginal row of lunules continued across this wing, tinged with ferruginous. Cilia brownish tipped with white, and with brown interruptions on nervules.

♀ *Similar, but duller, and without violaceous lustre ; basal suffusion and blackish borders wider, darker (especially in hind-wing)* ; *spots larger, all distinct in both wings.* UNDER SIDE.—As in ♂.

A number of specimens taken in Basutoland by Colonel Bowker were paler and larger than usual, and with the blue-violaceous lustre of the males unusually faint. A ♀ captured by the same gentleman in the Biggarsberg, Natal, has all the black spots of the fore-wing enlarged, those of the discal row being all elongated and acuminated inward.

Orus seems to resemble the well-known and widely-ranging *Phlæas* more than any other *Chrysophanus*, the colour and pattern of the fore-wings and the tints of the under side being very similar in the two forms. But *Phlæas* differs conspicuously in its dark-brown upper side of the hind-wings, with only a hind-marginal border of orange-red, and in the absence of any violaceous-blue lustre in the ♂; the fore-wings also are more pointed apically, and the hind-wings anal-angularly, than in *Orus*.

This brightly-coloured little *Chrysophanus* is distributed over the greater part of South Africa, and is particularly prevalent about the vicinity of Cape Town. It flies low, but is very active, frequently sitting on the ground or on low plants, where it suns itself with its wings half expanded. I have taken it in every month of the year except January and February, but it is more numerous from September to November, and again in March and April. It often appears, however, in the winter; and Colonel Bowker found it amongst the very few butterflies that seemed able to bear the severe winter of Basutoland, appearing on sunny days in such fine condition as to induce the belief that they were but just out of the pupa.

Localities of *Chrysophanus Orus.*

I. South Africa.
   B. Cape Colony.
      a. Western Districts.—Cape Town. Stellenbosch. Robertson. Triangles Station, Worcester District (*L. Péringuey*).
      b. Eastern Districts.—Uitenhage (*S. D. Bairstow*). Grahamstown (*H. J. Atherstone*). " King William's Town."—W. D'Urban. Murraysburg (*J. Muskett*). Burghersdorp (*D. R. Kennemeyer*).
      d. Basutoland.—Koro-Koro and Maseru (*J. H. Bowker*).

D. Kaffraria Proper.—Butterworth, and Kei and Bashee Rivers (*J. H. Bowker*).
E. Natal.
   b. Upper Districts.—Maritzburg and Greytown. Estcourt (*J. M. Hutchinson*). Biggarsberg (*J. H. Bowker*).
K. Transvaal.—Potchefstroom District (*T. Ayres*).

GENUS LYCÆNESTHES.

Lycænesthes, Moore, Proc. Zool. Soc. Lond., 1865, p. 773.
Lycæna [part], Trim., Rhop. Afr. Aust., ii. p. 234, A (1866).
Lycænesthes, Hewits., Trans. Ent. Soc. Lond., 1874, p. 343; and Illust. Diurn. Lep., p. 219 (1878).

IMAGO.—*Head* small, rather roughly hirsute in front; *eyes* very finely hairy; *palpi* rather long, slender,—second joint much compressed laterally, longer in ♀ than in ♂, densely clothed with long hair-like scales,—third joint long, slender, acuminate, closely scaled, rising above level of vertex; *antennæ* of moderate length, slender, with a rather abruptly formed club of moderate length, flattened and hollowed superiorly, pointed at extremity.

Thorax decidedly robust, particularly in ♂; its downy clothing dense, especially on breast. *Fore-wings* with apex rather acute; nervures thick, prominent; costal nervure ending about middle of costa; subcostal nervure four-branched,—first and second nervules emitted separately before extremity of discoidal cell,—third very short, emitted about midway between end of cell and apex,—fourth terminating at apex; upper radial nervule united to subcostal nervure at extremity of cell, lower one to disco-cellular nervules about midway between upper radial and third median nervules. *Hind-wings* with costa moderately arched, but very convex at base; slightly produced in anal-angular portion; costal nervure terminating at apex; subcostal nervure branched considerably before middle; on hind-margin three very fine slender tufts or pencils of hair, situated at extremities, respectively, of second and first median nervules and submedian nervure. *Legs* rather long and stout, with the femora very hairy beneath; *fore-legs* of ♂ rather large,—tibia with a pair of very fine terminal spurs,—tarsus closely spinulose beneath, and with a slightly curved terminal claw. *Middle* and *hind legs* with terminal spurs of tibiæ long.

PUPA.—Broad, moderately thick; narrowed to anal extremity, which is recurved and flattened inferiorly; under side much flattened, and slightly hollowed in the middle; head rather blunted, sub-quadrate; sides of thorax slightly sub-angulated; basal portion of abdomen broader than thorax. Attached by tail and silken girth.

(Described from living pupa of *L. Liodes*, Hewits.)

This genus is nearly allied to the robuster forms of *Lycæna*, but differs in its heavier structure generally (especially in the bulk of the thorax), acute fore-wings, slight projection of hind-wings about anal angle, and possession of three tufts of hairs on the hind-margin of the hind-wings. In form and in under-side pattern *Lycænesthes* bears much likeness to *Deudorix*, but wants the anal-angular lobe and linear tail on the hind-wings, and has longer palpi (especially in the ♂). Most of the species exhibit on the upper side various tints of violaceous or purple, much reduced or absent in the duller females, but some of the West-African forms (*Leptines*, *Luzones*, and *Lachares*, of Hewitson) present, instead of any shade of those colours, a large spot of orange-yellow in one or both wings. The under side in these and some other West-African species more resembles that of the *Telicanus* group of *Lycæna*.

Eighteen of the twenty-six known species are African, three Indian, and two Austro-Malayan. The locality "Cayenne," recorded by Butler (*Cat. Fab. Lep.*, p. 188) for *L. Moncus* (Fab.), requires confirmation. Six species have been discovered to inhabit Southern Africa; and of these three (*Amarah*, *Larydas*, and *Sylvanus*) are widely spread over both African Tropics, one (*Otacilia*) extends into the Southern Tropical region, and two (*Liodes* and *Livida*) appear to be peculiar to South Africa. *L. Amarah* is very different from the rest in the singular glittering grey of its upper side; *Larydas*, *Sylvanus*, and *Liodes* all have in the ♂ an upper side of glossy dark-purple, and in the ♀ a discal space of pale violaceous and whitish marked with fuscous spots; *Otacilia* has a rather bright violaceous field in the ♂, replaced by cupreous-brown in the ♀; and *Livida* is in both sexes dull shining ochreous-grey faintly shot with bluish.

All the six species are found on the eastern side of South Africa, but only four of them have been met with in the Cape Colony, and of these but two (*Liodes*, Hewits., and *Otacilia*, Trim.) reach the western districts. *Larydas*, Cram., and *Sylvanus*, Dru., have not occurred to the south and west of Natal, but *Amarah*, Guér., extends to Grahamstown, and *Livida*, Trim., seems hitherto to have been taken only near that place and in some of the adjacent eastern districts.

I have not seen living *Sylvanus* or *Larydas*, but have captured the other species, and found all the four to be very active, alert, little butterflies, resembling altogether in motion and habits the well-known *Thecla* group in Europe. They visit flowers freely, and are also much given to basking in the sunshine on the leaves of bushes or young trees.

167. (1.) Lycænesthes Amarah, (Guérin).

♀ *Polyommatus Amarah*, Guér., Lefebv. Voy. Abyss., vi. p. 384, pl. 11, ff. 5, 6 (1847).
♀ *Lycæna Amarah*, Wallgrn., Lep. Rhop. Caffr. (K. Sv. Vet.-Akad. Handl., 1857), p. 40.

♂ ♀ *Lycaena Amarah*, Trim., Rhop. Afr. Aust., ii. p. 235, n. 137 (1866).
♀ *Lampides Olympusa*, Walk., Entomologist, v. p. 53, n. 49 (1870).

Exp. al., (♂) 1 1/2 lin.—1 in. 2 lin.; (♀) 11 lin.—1 in. 2 lin.

♂ *Pale-grey, with a metallic sub-brassy lustre; a brown hind-marginal bounding line;* cilia greyish-white. *Hind-wing:* traces of two rows of dull-whitish lunules,—those of outer row combining with a whitish inner edging of hind-marginal line to form imperfect rings, and the two last enlarged and *orange* (the upper with an adjacent black spot) on either side of first median nervule; *a slender tail-like tuft of white hairs* at end of first median nervule, *and a similar tuft* at end of submedian nervure. UNDER SIDE.—*Brownish-grey, with white and brownish fasciæ and rows of lunules;* in *each* wing, beyond middle, an irregular, submacular fascia (composed of a broad white central streak, on both sides bordered with brownish and edged with a white line),—a similar short fascia, closing cell, touching edge of long fascia on third median, and two rows of thin white lunules succeeded by a white line. *Fore-wing:* between median and submedian, *a short, wide, abruptly truncate, longitudinal black stripe.* *Hind-wing:* seven conspicuous, black, white-ringed spots, viz., one at base, four in a transverse row near base, and two on costa (respectively immediately before origin of fascia and double row of lunules); orange-lunuled hind-marginal spot bluish-silvery-dusted; a smaller similar spot at anal angle.

♀ *Darker, not so metallic; a coating of bluish-grey shining hair over basal area of hind-wing.* *Fore-wing:* a disco-cellular fuscous line; a faint submacular streak just before hind-marginal line. *Hind-wing:* row of lunules and rings usually conspicuous; orange lunules and spot larger; occasionally a third orange lunule and black spot just above second median. UNDER SIDE.—As in ♂; ground-colour darker.

The peculiar hue of the upper surface at once distinguishes this species from the other South-African *Lycaenesthes*, which present a more or less purple or violaceous colouring in both sexes. The strongly-marked black basal stripe on the under side of the fore-wing is also peculiar to *Amarah*.

Mr. W. D'Urban found this butterfly commonly about King William's Town, frequenting bushy spots, from October to April. Near Grahamstown I saw it but rarely, and it was not common on the coast of Natal from the end of January to the beginning of April 1867. It is a brisk and active insect, and all the specimens that I noticed settled frequently on the leaves of various shrubs. In neither sex does there appear to be any noticeable variation except as regards size. The specimens from the Red Sea Coast, described by Walker (*loc. cit.*) as *L. Olympusa*, which I examined in Mr. R. Meldola's collection, are undoubtedly ordinary *Amarah*.

Localities of *Lycaenesthes Amarah*.

1. South Africa.
    B. Cape Colony.—Grahamstown, New Year's River, and Mitford Park, Albany District. King William's Town (*W. D'Urban* and *J. H. Bowker*). East London.

D. Kaffraria Proper.—Butterworth and Tsomo and Bashee Rivers (*J. H. Bowker*).
E. Natal.
 a. Coast Districts.—D'Urban and Umvoti. "Lower Umkomazi."—J. H. Bowker.
 b. Upper Districts.—Estcourt (*J. M. Hutchinson*). Rorke's Drift (*J. H. Bowker*).
F. Transvaal.—Potchefstroom District (*T. Ayres*). Limpopo and Marico Rivers (*F. C. Selous*).

II. Other African Regions.
 A. South Tropical.
  a. Western Coast.—"Angola."—Kirby (Cat. Hewits. Coll.)
 B. North Tropical.
  b. Eastern Coast.—"Abyssinia (*Lefebvre*)."—Guérin. Hor Tamanib, near Red Sea (*J. K. Lord*).

168. (2.) Lycænesthes Larydas, (Cramer).

? ♀ *Papilio Larydas*, Cram., Pap. Exot., iii. pl. cclxxxii. f. H (1782).
♂ *Polyommatus Larydas*, Godt. [part]. Enc. Meth., ix. p. 619, n. 6 (1819).
♂ *Lycæna Kersteni*, Gerst., "Archiv. f. Naturg., 1871, i. p. 359, n. 27;" and Gliederth.-Fauna Sansib.-Gebiet., p. 373, n. 27, t. xv. f. 5 (1873).
♀ *Lycænesthes Larydas*, Hewits., Ill. Diurn. Lep., p. 222, n. 11, pl. 92, f. 40 (1878).

Exp. al., (♂) 1 in. 1–4 lin.; (♀) 1 in. 1–2 lin.

♂ *Glistening dark-purple; a black hind-marginal line; cilia in fore-wing dark-grey except at posterior angle, where it is white,—in hind-wing paler, becoming white near anal angle;* tufts of hair at ends of second and first median nervules, and submedian nervure thin, rather long, dusky at base, but thence white. UNDER SIDE.—*Pale greyish-brown, with slightly darker, on both sides white-edged, transverse striæ. Fore-wing:* terminal disco-cellular stria broad, beginning on first subcostal nervule; ordinary discal stria abruptly broken on third and first median nervules into three nearly equal portions, of which the uppermost is farthest, and the lowest (in a line with terminal disco-cellular stria) nearest to base; a nearly straight submarginal stria, the white inner edge of which touches outer edge of second division of discal stria on first median nervule; a hind-marginal white line; costa edged with white at base, with ochrey-yellow (indistinctly) elsewhere; *a broad stria from costal to submedian nervure, curved inward inferiorly, crossing middle of discoidal cell;* in cell, at base, a short white longitudinal line curving upward at extremity. *Hind-wing:* terminal disco-cellular stria broad, well-marked; discal stria very irregular, commencing with a costal darker and sub-ocellate spot before the succeeding portion, narrowed in its lower part, and from submedian nervure recurved (and with a central line of white) to inner margin about middle; submarginal stria more irregular than in fore-wing, especially its inner white edge, which inferiorly is recurved (like discal stria) to inner margin; hind-marginal white line as in fore-wing; between it and submarginal stria two black bluish- or greenish-silvery speckled spots,—one between second and first median nervules, circled (except externally) with orange-yellow, the other on and chiefly above submedian nervure internally edged with orange-yellow; costa at base edged with white; a sub-basal transverse row of four dark white-ringed spots (the third of which is brown, and the rest are black) from costal nervure to inner margin.

♀ *Dull-fuscous; bases slightly tinged for some distance with slaty-grey; discs paler, in some instances whitish or white in fore-wing; two submarginal rows of white lunules in hind-wing.* Fore-wing: near posterior angle, between second median nervule and submedian nervure, four whitish marks, of which the two inner are broad and more or less suffused (sometimes merged in discal whitish), the two outer narrow. *Hind-wing:* lunules of inner submarginal row wider than those of outer row, and somewhat suffused; lunules of outer row thin, acute,— that between second and first median nervules immediately succeeded by a sub-trigonate black spot; a very distinct pure-white hind-marginal line immediately followed by a black one. *Cilia* with white parts more developed than in ♂. UNDER SIDE.—Markings arranged as in ♂, but *the white edges of nearly all the striæ—especially the submarginal ones and those near inner-margin of fore-wing—widened and more or less confluent*, so that the greyish-brown ground-colour is considerably reduced.

Cramer's figure represents the under side, and, though rough and enlarged, gives the markings with tolerable fidelity; but the ground colour is much darker than in any examples I have seen,—darker, indeed, than in the ♂, although from the notice in the text (p. 160), that "le dessus des ailes est d'un blanc bleuâtre, les bords en sont d'un [brun] clair," it seems clear that the specimen figured (a West-African one) was a female.

The upper side of the ♂ in this species is the same as that of *L. Sylvanus*, Dru., and not very much darker than in *L. Liodes*, Hewits.; but that of the ♀ is quite peculiar, owing to the absence of any violaceous suffusion, and to the white markings existing near the posterior angle of the fore-wings. On the under side the curious striation in the basal part of the fore-wings distinguishes *Larydus* from all the known South-African species, and shows its alliance to *L. Lysicles* and numerous other West-African species figured by Mr. Hewitson.

The first South-African example of *Larydus* I met with was a worn ♂, taken at D'Urban, Natal, in 1870 by the late Mr. M. J. M'Ken. From 1879 to 1881, however, Colonel Bowker forwarded a good many specimens of both sexes captured in the same locality, including two pairs found *in coitu* on the 17th January 1879 and 25th March 1881 respectively. He noted nothing peculiar in the habits of the butterfly.

Localities of *Lycænesthes Larydus.*

I. South Africa.
  E. Natal.
    a. Coast Districts.—D'Urban (*M. J. M'Ken* and *J. H. Bowker*). Pinetown (*J. H. Bowker*).
II. Delagoa Bay.—Lourenço Marques (*Mrs. Monteiro*).

II. Other African Regions.
A. South Tropical.
   a. Western Coast.—Congo.—Coll. Brit. Mus. " Angola."—Kirby, Cat. Hewits. Coll.
   b. Eastern Coast.—" Mbaramu, Usambara (*O. Kersten*)."—Gerstäcker.
B. North Tropical.—" Old Calabar."—Kirby, Cat. Hewits. Coll.

169. (3.) **Lycænesthes Sylvanus**, (Drury).

♂ *Papilio Sylvanus*, Dru., Ill. Nat. Hist., ii. pl. iii. ff. 2, 3 (1773).
♂ *Polyommatus Larydas*, var. Godt., Enc. Meth., ix. p. 619 (1819).
♀ *Lycæna Emolus*, Gerst., Gliederth.-Fauna Sansib.-Gebiet., p. 373, n. 26, t. xv. f. 4 (1873).
♂ *Lycænesthes Lemnos*, Hewits., Ill. D. Lep., p. 221, n. 8, pl. xc. ff. 13, 14 (1878).
♀ *Lycænesthes Sylvanus*, Hewits., op. cit., p. 222, n. 10, pl. xcii. f. 41 (1878).

Exp. al., (♂) 1 in. 1–3 lin.; (♀) 1 in. 2–3½ lin.

♂ *Glistening dark-purple; a black hind-marginal line; cilia dull-grey, slightly mixed with whitish.* Hind-wing: close to hind-margin, between first median nervule and submedian nervure, an elongate black mark; three caudal tufts, not very slender, dusky at bases, but thence whitish. UNDER SIDE.—*Pale greyish-brown, with scarcely darker, on both sides thinly whitish-edged, transverse striæ.* Fore-wing: no markings before middle; terminal disco-cellular striola short, not produced towards costa, but a minute white-ringed spot just above it; discal stria submacular, slightly irregular, and curved superiorly, but actually interrupted only on first median nervule, beneath which its terminal portion is nearer base than the rest; submarginal stria narrow, faintly marked, obsolescent superiorly; hind-marginal whitish line scarcely visible. Hind-wing: terminal disco-cellular striola well-marked, rather long; discal stria decidedly submacular, scarcely irregular, not interrupted, widened just above submedian nervure, but below it abruptly narrowed, marked mesially with a whitish line, and deflected to inner margin about middle; first (costal) spot of discal fascia much darker than the rest, partly filled with fuscous or fuscous-ferruginous; submarginal stria much better marked than in fore-wing, both its whitish edges sublunulate and rather suffused; a whitish hind-marginal line (obsolescent superiorly) immediately succeeded by a fuscous one; two hind-marginal black greenish-silvery spangled spots,—the upper and larger one between second and first median nervules encircled (except externally) by an ochre-yellow edging,—the lower one, on each side of submedian nervure, edged interiorly with ochre-yellow, which extends a little along inner-marginal edge; a sub-basal transverse row of three round spots in very thin whitish rings, of which the first and second (respectively near costa and in discoidal cell) are ferruginous-red or fuscous-ferruginous, and the third (on inner margin) is fuscous.

♀ *Dull fuscous; a violaceous patch (variable in size) from near base to beyond middle, sometimes extending over lower part of discoidal cell, in fore-wing; hind-wing with two submarginal rows of white lunules, of which the inner is usually somewhat suffused inwardly, and sometimes enlarged into a discal whitish space; all lower half of hind-wing shot with violaceous.* Fore-wing: a terminal disco-cellular blackish striola; in some examples, close to hind-margin, between second median nervule and submedian nervure, a whitish streak tinged with violaceous, followed by a similar line of great tenuity; violaceous patch rarely touches any part of inner margin. *Hind-wing:* terminal disco-cellular blackish striola as in fore-wing, but less distinct; on disc, between second subcostal and origin of second median nervule, a short, blackish macular stria (very conspicuous in specimens with a more or less whitish discal area); upper of two hind-marginal spots, between second and first median nervules, large, black, with a conspicuous orange-yellow lunule bounding it inwardly; a clearly-defined white hind-marginal line, immediately succeeded by a black one, from anal angle as far as second subcostal nervule, where the outer submarginal white lunular row also terminates. UNDER SIDE.—*Much paler than in ♂; markings similar, but their white edges much more developed, those beyond discal stria (which is comparatively darker than in ♂) combining suffusedly, particularly in hind-wing, into a white submarginal band. Hind-wing:* hind-marginal spangled spots larger than in ♂.

Gerstäcker (*op. cit.*), while admitting the difficulty he had experienced, in common with myself and other lepidopterists, in determining what Godart's *Emolus* really is, refers this species to *Emolus*, mihi, which = *L. Liodes*, Hewits., described below. The ♀ figured by him, however, differs from my insects as well as from Godart's description in possessing a sub-basal row of three conspicuous round white-ringed spots in the hind-wing. Godart's *Emolus* (as more fully explained under *L. Liodes*) is in all probability identical with *L. bengalensis*, Moore, the type of the genus *Lycaenesthes*.

I have examined the type of *L. Lemnos*, Hewits., a male from Delagoa Bay, and do not find that it can be separated as a species. The only differences from the ♂ *Sylvanus* that it presents are, on the upper side, a rather paler, more glistening purple, and in the hind-wing a short white (instead of indistinct whitish) line between the black hind-marginal and short preceding lines at anal angle; and, on the under side, rather brighter red in the upper and middle spots of the sub-basal transverse row.

The late Mr. E. C. Buxton was the first to discover this *Lycaenesthes* as South-African, having sent me a pair taken by himself in some part of Natal in 1873. From D'Urban and Pinetown, during the years 1878 to 1884, Colonel Bowker has forwarded nine of each sex, taken at different times of the year. The best locality noted by him was the Park at D'Urban, where he found many specimens on the wing during the last three days of October 1879.

Localities of *Lycænesthes Sylvanus.*

I. South Africa.
  E. Natal.—D'Urban and Pinetown (*J. H. Bowker* and *T. Ayres*).
  II. Delagoa Bay.—Lourenço Marques (*Mrs. Monteiro*).
11. Other African Regions.
  A. South Tropical.
    a. Western Coast.—" Angola."— Kirby, Cat. Hewits. Coll.
    b. Eastern Coast.—" Mombas (*Kersten*)."—Gerstäcker.
  B. North Tropical.
    a. Western Coast.—" Old Calabar."—Kirby, *op. cit.*   " Sierra Leone."—Drury.
    *b*1. Eastern Interior.—" White Nile."—Kirby, *op. cit.*

170. (4.) Lycænesthes Liodes, Hewitson.

♂ ♀ *Lycæna Emolus*, Trim., Rhop. Afr. Aust., ii. p. 234, n. 136, pl. 4, ff. 8, 9 (1866).
♂ *Lycænesthes Liodes*, Hewits., Trans. Ent. Soc. Lond., 1874, p. 349.
♂ *Lycænesthes Sichela*, Hewits., Ill. D. Lep., p. 222, n. 12 (1878).

Exp. al., (♂) 1 in. 0½–2 lin.; (♀) 1 in. 1–2½ lin.

♂ *Glistening dark-violaceous; a slender black, hind-marginal edging line;* cilia whitish. *Hind-wing :* close to hind-margin two to three inconspicuous black spots between second median nervule and submedian nervure; *at extremity of each of these nervures, as well as that of first median, a thin tuft of whitish hairs.* UNDER SIDE.—*Brownish-grey ;* in *each* wing—a double dark streak (enclosing one of ground-colour) closing cell, the whole marking being on both sides white-edged, —a similar fascia, composed of confluent spots, across wing beyond cell, in *hind-wing* bi-angulated near inner margin, in *fore-wing* with the last spot before the rest of the fascia,—two submarginal rows of whitish lunules enclosing a darker space, and a very indistinct interior whitish edging to hind-marginal dark line. *Fore-wing :* rarely a whitish ring in cell. *Hind-wing :* on costa before middle, a white-ringed spot similar to that closing cell; first and last of three black hind-marginal spots always distinct, greenish-silvery-dusted and interiorly orange-lunuled, the second indistinct, silvery-dusted—sometimes obsolete.

♀ *Pale-greyish, shot with violaceous-blue from base ; a blackish disco-cellular spot and transverse macular fascia* in each wing; a blackish hind-marginal border, in *hind-wing* intersected by a row of whitish lunules. *Hind-wing :* hind-marginal white line conspicuous; hind-marginal spots black, the first orange-lunuled. UNDER SIDE.—As in ♂, but all white edgings broader, conspicuous ; outer edging of transverse fascia and inner submarginal row of lunules sometimes suffused and confluent, forming a white band.

Two dwarf ♂ s, from Cape Town and Grahamstown respectively, expand only 10 and 9 lines across the fore-wings.

PUPA.—Above bright yellowish-green; beneath much paler, shining whitish-green; semi-transparent, abdomen more opaque. On back an indistinct median thin fuscous line; on this line, marking junction of thorax and abdomen, a conspicuous, oblong-ovate, salmon-pink, brown-edged spot; on each side of abdomen a row of minute, indistinct, fuscous dots. About 4½ lin. in length.

The remains of a silken girth were attached on each side of the basal segment of the abdomen in the specimen here described, which was sent to me by the late Mr. Kay, on 23d October 1869, with the information that it had been found fastened to the upper side of the leaf of a *Pelargonium* in Cape Town. The imago (a ♂) emerged on the 4th November.

As noted by me (*op. cit.*, p. 235), it was with considerable uncertainty that I referred this butterfly to the *Polyommatus Emolus* of Godart, and that I also suggested that the ♂ might be the same as *Lycæna Sichela*, Wallengren. The late Mr. Hewitson adopted this latter suggestion in his *Illustrations of Diurnal Lepidoptera;* but I have since discovered *Sichela* to be an entirely different insect, not belonging to the group *Lycænesthes*. Godart's *Emolus*, however,—described at p. 656 of *Encyc. Method.*, tom. ix.—is very near to *L. Liodes;* and I think that Mr. F. Moore's type of his genus *Lycænesthes*, viz., *L. Bengalensis*—described in *Proc. Zool. Soc. Lond.*, 1865, p. 773—is almost certainly the same as Godart's species, which is stated to be from Bengal. *Bengalensis* is described as expanding 1¼ in., and so should be a little larger than *Liodes*. Moore points out its alliance to *Dipsas lycænoides* of Felder (1860), and Hewitson (*Ill. D. Lep.*, pp. 214, 219) treats the two as identical. Judging from Felder's figure ("*Reise der Novara*," *Zool., Lepid.*, ii. t. 30, f. 25) of the under side, and his description (p. 218) of the ♂, and Hewitson's figure (*op. cit.*, pl. xcii. f. 39) of the ♀, I consider it very doubtful whether *Lycænoïdes* can be held synonymous with Moore's butterfly. I have examined the specimens of *L. Liodes* in the Hewitson Collection; they are marked as from the Cape, and agree entirely with the Colonial examples above described; and I think it very probable that the locality "Gaboon," assigned to the species in Hewitson's original diagnosis in 1874, and again in 1878, was erroneous.

Liodes belongs to the *Sylvanus* group of the genus; it is considerably smaller than *Sylvanus*, and the ♂ is of a paler tint on the upper side, while the ♀ is much bluer, and has a well-marked discal fuscous fascia in the fore-wings, besides a much more developed one in the hind-wings. On the under side *Liodes* is distinguished by its much less distinct markings in the ♂, and especially by the absence in both sexes of the sub-basal transverse row of round spots in white rings.

This is a common insect in and near Cape Town, frequenting gardens and open places in plantations. It visits many flowers, and is fond of sunning itself on oak-leaves. It is active and wary, and very swift in its short flights, reminding the collector of the species of *Thecla*. I have observed it on the wing throughout the year, except from the beginning of May to the middle of July. It was not uncommon near Grahamstown in January and February 1870. I took it rarely near D'Urban, Natal, in March 1867.

Localities of *Lycænesthes Liodes.*

I. South Africa.
  B. Cape Colony.
    a. Western Districts.—Cape Town. Robertson. Knysna.
    b. Eastern Districts.—Grahamstown. King William's Town (W. D'Urban and J. H. Bowker). Windvogelberg, Queenstown District (Dr. Batho). Murraysburg (J. J. Muskett). Burghersdorp (D. R. Kannemeyer).
  D. Kaffraria Proper.—Bashee River (J. H. Bowker).
  E. Natal.
    a. Coast Districts.—D'Urban.
  K. Transvaal.—Lydenburg District (T. Ayres).

171. (5.) Lycænesthes Otacilia, (Trimen).

PLATE VII., fig. 8 (♂).

♂ *Lycæna Otacilia*, Trim., Trans. Ent. Soc. Lond., 1868, p. 90.

Exp. al., (♂) 1 1 lin.—1 in.; (♀) 1 in.—1 in. 0½ lin.

♂ *Shining pale-violaceous. Fore-wing:* apical area, as far as end of discoidal cell and third median nervule, *brown with a cupreous tinge;* short, rather narrow borders of the same colouring, extending on costa to base and on hind-margin to posterior angle. *Hind-wing:* a cupreous-brown border, wide along costa and at apex, narrow along hind-margin to anal angle; a rather well-defined hind-marginal black spot between second and first median nervules; below the latter nervule the trace of a similar spot. *Cilia* whitish, with faint dusky nervular interruptions on lower part of hind-wing. UNDER SIDE.—*Pale brownish-grey; in each wing, the ordinary terminal disco-cellular spot, and discal and submarginal transverse submacular bands, slightly darker than ground-colour and rather conspicuously white-edged on each side. Fore-wing:* no markings near base; discal band strongly curved inward below cell. *Hind-wing:* a sub-basal transverse row of three round blackish white-ringed spots,—that above cell conspicuous, the other two (respectively in cell and below it) rather faintly marked; discal band so strongly curved as half to encircle terminal disco-cellular spot, and with part of its outer white edging confluent with inner white edging of submarginal lunulated row; hind-marginal spot inwardly edged by an orange-yellow lunule,—as is also a minute spot close to anal angle; an indistinct hind-marginal whitish line forming annulets with the lunulated outer white edging of submarginal row.

♀ *Pale-brownish, with a subcupreous lustre; a very faint basal violaceous suffusion, rather better pronounced in hind-wing; bases narrowly blackish. Hind-wing:* in one example traces of a hind-marginal row of whitish annulets like that on under side; hind-marginal black spot as in ♂. UNDER SIDE.—As in ♂, but rather paler.

In a ♂ from Swellendam, Cape Colony, the discal row is very much narrowed, and its whitish edges, as well as those of the other markings, are everywhere suffused and confluent with the adjacent ones.

The bright-violaceous upper side, with its broad cupreous-brown borders, at once distinguish the ♂ *Otacilia* from the same sex in *Liodes* and *Sylvanus*, and approximate it to *L. livida*, Trim., in which, however, the violaceous is very much duller and more limited in extent. In the ♀ *Otacilia*, on the contrary, the upper side is almost devoid of violaceous. The under side is of a browner less grey tint than in its congeners, and the markings on the whole most resemble those of the ♀ *Liodes*. In size *Otacilia* is the smallest of the South-African species of *Lycænesthes*.

The *Otacilia* of Hewitson (*Illustr. Diurn. Lep.*, pl. 92, ff. 35-37, 1878), which I have examined in that author's collection, is, I think, a distinct species, the ♂ having the upper side violaceous much intenser, and occupying a considerably larger space (especially in the fore-wing), and the ♀ presenting almost as much as in the ♂ *Otacilia*, mihi; both sexes further exhibiting a very conspicuous bright-orange crescent bounding the black hind-marginal spot of the hind-wings. On the under side the markings are darker and more pronounced. Mr. Hewitson's specimens were ticketed as natives of Angola and Sierra Leone.

This little species seems rather widely spread in South Africa, but is not frequent in collections. The first specimen that came under my notice was sent from Swellendam in 1864 by Mr. L. Taats. Mrs. Barber subsequently sent one from Grahamstown, and Colonel Bowker two from Kaffraria Proper, and one, captured on 1st May 1874, from King William's Town. It was not until January 1876 that I met with the species at all numerously. At Robertson, in the Cape Colony, during that month, I observed a good number about the flowers of *Acacia horrida*, and captured examples of both sexes. Like the rest of the genus, they were active and wary, and not very easy to secure among the thorny bushes under the noonday sun of January. I had previously (in March 1867) taken a single female at Greytown, Natal.

Localities of *Lycænesthes Otacilia*.

I. South Africa.
   B. Cape Colony.
      a. Western Districts.—Robertson. Swellendam (*L. Taats*).
      b. Eastern Districts.—Grahamstown (*M. E. Barber*). King William's Town (*J. H. Bowker*).
   D. Kaffraria Proper.—Tsomo River (*J. H. Bowker*).
   E. Natal.
      b. Upper Districts.—Greytown. Estcourt (*J. M. Hutchinson*).

II. Other African Regions.
   A. South Tropical.—Central Interior.—"Victoria Falls, Zambesi River (*F. Oates*)."—Westwood.

172. (6.) Lycænesthes livida, Trimen.

PLATE VII., ff. 7 (♂), 7a (♀).

♂, ♀ *Lycænesthes livida*, Trim., Trans. Ent. Soc. Lond., 1881, p. 443.

Exp. al., (♂) 1 in.—1 in. 2 lin.; (♀) 1 in. 1½ lin.—3 lin.

♂ *Shining greyish-brown, with a cupreous gloss; in both wings a very pale greyish-blue suffusion from base. Fore-wing:* the suffusion vaguely occupies the lower half of discoidal cell, and covers space between median nervure and its first nervule and inner-margin to near posterior angle; an indistinct dark-grey lunular mark at extremity of discoidal cell. *Hind-wing*: the suffusion covers middle field of wing from base, leaving the costa and apical, hind-marginal, and inner-marginal border free; an indistinct dark lunule at extremity of discoidal cell; a little beyond it, a curved macular streak between second subcostal and second median nervules; a thin black line on hind-marginal edge; within it a thin white line, most apparent near anal angle, itself immediately preceded by four to six thin whitish lunules, which join with it to isolate spots of the ground-colour; these spots are darker near anal angle, that between second and first median nervules being black, bounded interiorly by a well-marked orange lunule. *Cilia* in both wings whitish. UNDER SIDE.—*Soft pale-grey; the markings slightly darker, but distinctly edged on both sides with whitish;* in each wing a roughly 8-shaped mark at extremity of discoidal cell, a discal inferiorly-incurved row of more or less confluent similar imperfect rings; a submarginal row of lunules; and a thin hind-marginal whitish edging line. *Fore-wing*: basal area quite spotless as far as extremity of cell. *Hind-wing*: near base, just below costal nervure, a small but distinct round black spot in a whitish ring; the hind-marginal black spot between second and first median nervules, and a smaller similar spot close to anal angle, conspicuously spangled with a few greenish-silvery scales, and interiorly bounded by an orange lunule; between these two spots a few greenish-silvery scales.

♀ *Similar to male, but ground colour paler and duller, while the blue suffusion is considerably brighter in hue.* *Hind-wing*: blue becoming very faint on disc, which bears a transverse row of rather indistinct whitish lunules. UNDER SIDE.—As in male.

This *Lycænesthes* is in several respects intermediate between *L. Liodes*, Hew. (the *Emolus* of my *Rhop. Afr. Aust.*, not the true *Emolus* of Godart), and *L. Otacilia*, mihi. It is at once to be distinguished, however, from both species by the singularly pale and dull hue of the bluish suffusion on its upper surface, which in the male contrasts remarkably with the universal dark purple of *L. Liodes*, and the well-defined bright violaceous of *L. Otacilia*. In size *L. livida* is larger than *L. Liodes*, and very much larger than *L. Otacilia*. The female has, on the upper side of the fore-wing, none of the fuscous spots so strongly marked in the female *L. Liodes*. The under side markings are in both sexes less irregular, and not so dark as in *L. Liodes*, and the ground-colour has none of the yellowish-brown tinge observable on the under side of *L. Otacilia*.

I first noticed this butterfly in Mrs. Barber's collection in February 1870, and made a description of the two female specimens which the collection contained, under the impression that they would probably prove to be the female of *L. Otacilia*, mihi. These examples were taken at Highlands, near Grahamstown, and were kindly presented to me by Mrs. Barber. On the 23d of the same month I captured, at Uitenhage (on Cannon Hill), three males of a *Lycænesthes*, which so closely corresponded with the females mentioned that, upon subsequent comparison, no doubt could be entertained of the identity of species. The males in question were flitting about and settling on the twigs of some bushes at the summit of the hill. A fourth male, taken in Somerset East district, was received from Colonel Bowker in 1871.

I have not seen any further examples of this dull-coloured Lycænid in the collections that I have been able to examine.

Localities of *Lycænesthes livida*.

I. South Africa.
  B. Cape Colony.
    b. Eastern Districts.—Uitenhage. Grahamstown (*M. E. Barber*). Between Somerset East and Murraysburg (*J. H. Bowker*).

Genus DEUDORIX.

Deudorix, Hewitson, Illustr. Diurn. Lep., p. 16 (1862).
Dipsas, Westw. [part], Gen. Diurn. Lep., ii. p. 479 (1852).
Sithon, Trimen, Rhop. Afr. Aust., ii. p. 231 (1866).

IMAGO.—*Head* rather broad; *eyes* clothed with short hair; *palpi* short, slender—second joint long, densely scaly, laterally flattened,—terminal joint acute, slender, directed forward, very short in ♂ but long in ♀; *antennæ* long, slender, with a distinct elongate club, more pronounced in ♂ than in ♀.

Thorax robust (very stout in ♂), densely downy—especially on breast. *Fore-wings* somewhat variable in form, apically usually rather acute (but always less so in ♀ than in ♂); on inner-margin in ♂ almost always a tuft of stiff bristly hairs on under side before middle; subcostal nervure four-branched, and neuration quite agreeing with that of *Aphnæus*. *Hind-wings* more or less produced in anal-angular portion—anal angle itself bearing a very prominent lobe; a rather long linear tail at extremity of first median nervule, and generally a slight acute projection at extremity of second median nervule; costal nervure terminating at apex; subcostal nervure branched just before extremity of cell; neuration generally as in *Aphnæus* and *Hypolycæna*; in ♂ usually a small smooth shining spot near costa before middle, just at base of two branches of subcostal nervure. *Fore-legs* of ♂ much as in *Aphnæus*,—but femur more hairy beneath—tibia only spined at extremity,—tarsus more strongly spined beneath;—of the ♀ generally thicker,—tarsus longer, completely articulate, and with two claws. *Middle* and *hind legs* rather short, moderately thick;—coxæ and femora hairy,—tibiæ smooth, with short terminal spurs,—tarsi thickly spinulose beneath.

LARVA.—Elongate, depressed; set transversely with rows of well-separated fascicles of very short stiff hairs or bristles.

PUPA. — Blunt, thick, rounded; tail considerably incurved. Attached by the tail and a girth round the body.

[These characters of Larva and Pupa are taken from the figures of the early states of *D. Xenophon* (Fab.), and *D. Melampus* (Cram.),— both natives of Java—given in Horsfield and Moore's *Catalogue of the Lepidopterous Insects* in the E. I. Co.'s Museum, vol. i., pl. 1, ff. 2, 2a, and 3, 3a. The larva of the former species is stated to feed on *Schmiedelia racemosa*, and that of the Indian *D. Isocrates* (Fab.), on the interior of the fruit of the common Pomegranate.]

This genus is equivalent to *Sithon*, Hübn. (Verz. bekannt. Schmett., p. 77); but its characters were first defined by Hewitson (*op. cit.*), in 1862, and the latter author's name of *Deudorix* is thus to be preferred. It is well characterised by the robust body, very slender palpi, long slender antennae with long but well-developed clavation, hairy eyes, and very prominent lobate appendage at the anal angle of the hindwings. The forehead—and sometimes also the tip of the abdomen—is commonly adorned with a red or orange patch. The under side is not nearly so elaborately ornamented as in *Aphnaeus*, and is usually of some tint of grey or greyish-brown, with slightly-darker, usually whitish-edged, more or less macular discal bands. The Austro-Malayan *D. Despoena*, Hewits., and allies have, however, a more ornate under side of creamy-yellowish, strikingly barred with black; and the North-Indian *D. Amyntor*, Herbst., has an almost uniform under side of dull-green. In the majority of spines the male is blue on the upper side, but in eight or nine cases intense red or orange-red, and in a few of a bronzy or of an ochrey-yellow tint. The female is almost always of a dull-brownish or greyish on the upper surface, but sometimes exhibits a considerable tinge of the brighter hue of the male, and occasionally (as in *D. Antalus* (Hopff.), and *D. Pheretima*, Hewits.), a different tint from that of her partner.

About forty-two species are on record. The genus ranges from Western Africa to Australia, but finds its principal development in India and the Indo-Malayan Islands, which together possess twenty-two species. The Austro-Malayan Islands have yielded eight, and Australia itself two; while seven are known from Africa. Four of the last-named inhabit Southern Africa, but only one—*D. Diocles*, Hewits.,—seems to be confined to that subregion; the others being found also in the South-Tropical belt, and *D. Antalus* (Hopff.), apparently extending all over the Ethiopian Region. The last-mentioned species is the only form that I have seen in life; both sexes are very active in their frequent short flights, and the male is particularly rapid on the wing. In *Antalus* the ♂ has the upper side of a submetallic bronzy-brown suffused from base with violaceous, while the ♀ is of a paler and more bluish colour; in the other three species, the upper side of the ♂ is more or less occupied with bright-red (not metallic), and that of the ♀ of two of them pale-fuscous with dull-whitish on the discal areas.

173. (1.) **Deudorix Antalus,** (Hopffer).

Dipsas Antalus, Hopff., Monatsb. K. Akad. Wissens. Berlin, 1855, p. 641, n. 15; and (♂ ♀)
Sithon Antalus, Peters' Reise Mossamb.,—Ins., p. 400, pl. xxv. ff. 7-9 [♀] (1862).
♂ ♀ *Lycæna Anta,* Trim., Trans. Ent. Soc. Lond., 3d ser., i. p. 402 (1862).
♂ ♀ *Deudorix Anta,* Hewits., Ill. D. Lep., p. 25, pl. v., ff. 49-51; also *Ialmenus Antalus,* p. 55 (1863 and 1865).
♂ ♀ *Sithon Batikeli,* Trim., Rhop. Afr. Aust., ii. p. 232, n. 135 (1866).

Exp. al., (♂) 1 in. 0½—5 lin.; (♀) 1 in. 3—7 lin.

♂ *Shining æneous-brown, shot with violet from bases; cilia greyish-white. Fore-wing:* inner marginal tuft of hairs black. *Hind-wing:* a rather long, linear, black, white-tipped tail at extremity of third median nervule; two black spots on hind-margin, respectively just above and below origin of tail; lobe of anal angle marked with a greenish silvery-scaled black spot. UNDER SIDE.—*Pale-greyish; in both wings* an incomplete, brownish-grey, whitish-edged ring, closing discoidal cell, a row of similar rings, confluent, forming a rather broad transverse band beyond middle, and a submarginal row of brownish-grey lunular markings, indistinctly white-edged inwardly and outwardly. *Hind-wing:* near base, two or three whitish-ringed fuscous (sometimes dull-ferruginous) spots, forming a short transverse row; hind-marginal spot above tail marked inwardly by a yellowish lunule, that below tail all bluish-silvery; spot on anal lobe inwardly scaled with bluish-silvery.

♀ *Bluer than* ♂, *excepting near hind-margins, which are broadly-brown;* markings similar; a dusky disco-cellular terminal streak in each wing. UNDER SIDE.—Quite similar, the markings more distinct.

From Boisduval's description (*Faune Ent. de Madag., &c.,* p. 24) I was led—as stated in my book above cited—to consider his *Lycæna Batikeli,* as identical with the South-African species which in 1862 I had described as *Lycæna Anta,* but which I subsequently discovered Hopffer had previously received from East Africa and named *Dipsas Antalus.* Boisduval's figure on pl. 3 (*op. cit.*) appeared to me as a rough and highly-coloured representation of *Antalus,* Hopff., of which I had seen several Malagasy specimens. Having lately (1886) seen the figures of *Batikeli* given by Grandidier in the Lepidoptera volume of the *Hist. Physique, Nat. et Polit. de Madag.* (Paris, 1885) on pl. 29, I am, however, satisfied that it is a distinct species from *Antalus.* It is apparently a ♀ that is figured, and the upper side is depicted as considerably darker than in ♀ *Antalus,* especially in the hind-wing, the dull violaceous-blue of the fore-wing being better defined, but that of the hind-wing being reduced to a dull pale longitudinal ray from base between median and submedian nervures. On the under side, the markings generally are redder, less regular, and with their whitish edgings better developed; in the fore-wing there is a linear red hind-marginal edging from apex to second median nervule, and in the hind-wing the three sub-basal white-ringed spots are larger and conspicuously red.

There cannot be any doubt, on comparison, of the identity of Hopffer's East-African *Antalus* and my South-African *Anta.* Both sexes are very variable in size, and this is the case with individuals from the same locality.

This is a near ally of the well-known *D. Isocrates* (Fab.), of India, but is readily distinguished from it by the three small but distinct ocelli near the base of the hind-wings, which are entirely wanting in the Indian species; the ♀, too, is much bluer above, and wants in the fore-wings the ochre-yellow spot immediately beyond the discoidal cell, and in the hind-wings the ochre-yellow lunule edging the upper hind-marginal black spot, which are conspicuous features of the ♀ *Isocrates*.

I met with this species rather sparingly on the Natal coast in February 1867, and again near Grahamstown in January and February 1870. It frequents wooded spots, and often settles on shrubs and low trees; near Grahamstown I found it partial to the blossoms of *Acacia horrida*. Colonel Bowker took it during March and April in Kaffraria, and in July and August in Natal. Both sexes are active and rapid on the wing, but the male especially so. A fine female that I took near D'Urban had just before been pounced upon by a predaceous fly of the *Asilus* group.

Localities of *Deudorix Antalus*.

I. South Africa.
  B. Cape Colony.
    b. Eastern Districts.—Port Elizabeth. Between Zwartkops and Coega Rivers (*J. H. Bowker*). Grahamstown. King William's Town (*Mrs. Drake*). Fort Beaufort: Fish River Randt (*M. E. Barber*).
  D. Kaffraria Proper.—Bashee River (*J. H. Bowker*).
  E. Natal.
    a. Coast Districts.—D'Urban. Verulam.
    b. Upper Districts.—Estcourt (*J. M. Hutchinson*). Maritzburg (*J. Windham*).
  F. Zululand.—St. Lucia Bay (*Colonel H. Tower*).
  G. "Swaziland."—The late E. C. Buxton.
  K. Transvaal.—Potchefstroom (*T. Ayres*). Limpopo River (*F. C. Selous*).

II. Other African Regions.
  A. South Tropical.
    a. Western Coast.—Damaraland (*J. A. Bell*). "Congo: Kinsembo (*H. Ansell*)."—Butler. "Chinchoxo (*Falkenstein*)."—Dewitz.
    b. Eastern Coast.—"Querimba."—Hopffer. "Zanzibar and Tongor (*Raffray*)."—Oberthür.
    *b*1. Eastern Interior.—Tauwani River (*F. C. Selous*).
    bb. Madagascar (*J. Caldwell*). "Johanna, Comoro Islands (*W. C. Bewsher*)."—Butler.
  B. North Tropical.
    a. Western Coast.—Sierra Leone (*Cutter*).—Coll. Trim. Sierra Leone.—Coll. Hope. Mus. Oxon. and Coll. Hewitson. Casamanza, Senegal.—Coll. Boisduval.

174. (2.) Deudorix Diocles, Hewitson.

PLATE VII., fig. 6 (♀).

♂ *Deudorix Diocles*, Hewits., Ill. D. Lep., Suppl., pl. v. ff. 55, 56 (1869); and p. 29, pl. va. f. 57 (1878).

Exp. al., (♂) 1 in. 2—5½ lin.; (♀) 1 in. 5—8 lin.

♂ *Fuscous-brown, with a common, broad, transverse, discal, orange-red band—so much widened in hind-wing as to cover all but a space near* base and an inner-marginal border. *Fore-wing*: band variable in width, commencing abruptly on or a little above median nervure and first median nervule, and widening more or less to inner margin; tuft of hairs on inner margin brown. *Hind-wing*: basal brown not extending nearly to middle, but omitting a thin ray along fold between median and submedian nervures; inner-marginal border dull-greyish; orange-red much paler along costal edge; hind-margin with a linear black edging thicker inferiorly; sometimes a very small blackish spot close to hind-margin between first and second median nervules; a larger black spot scaled with golden-green on anal-angular lobe; tail linear, black, white-tipped; circular badge small, shining-violaceous, just on the branching of subcostal nervure. UNDER SIDE.—*Paler or darker brownish-grey;* in both wings the following rather darker, on both sides white-edged markings, viz., a terminal disco-cellular spot,—a discal transverse submacular, irregular band,—and a submarginal row of lunules. *Fore-wing*: costa narrowly edged with orange from base; inner-marginal area more or less faintly tinged with orange. *Hind-wing*: discal band much more irregular than in fore-wing, angulated sharply between first median nervule and submedian nervure, and thence much narrowed to inner margin; inferior half of hind-margin and terminal third of inner margin with a linear black edging; hind-marginal black spot very well defined, and immediately preceded by a conspicuous orange lunule; spot on anal-angular lobe as above. *Cilia* fuscous mixed with dull-whitish; inferiorly glossed with ochre-yellow. *Abdomen* superiorly tipped with orange-red.

♀ *Dull-fuscous, the discs dusky-whitish; a dull violaceous-bluish gloss, chiefly in basal area; cilia grey.* Fore-wing: inner-marginal area whitish. *Hind-wing*: hind-marginal black edging thicker than in ♂, and bordered anteriorly by a white line, thicker inferiorly; some orange as well as golden-green scales on anal-angular lobe. UNDER SIDE.— Much paler than in ♂; all the markings better developed and defined.

Head in both sexes orange-red in front, edged with white on each side.

This species comes nearest to the Oriental *D. Epijarbas* (Moore), given by Hewitson (*op. cit.*, p. 17) as typical of the genus *Deudorix*, but is smaller, paler beneath, with blunter and less elongate wings; the orange-red band of the ♂ is in the fore-wing transversely instead of longitudinally disposed, and in the hind-wing very much broader costally; while the ♀ is on the upper side very much paler discally and has a bluish suffusion wanting in *Epijarbas*.

The brilliant colouring of the ♂ instantly separates *Diocles* from *Antalus*; but the ♀ s of the two butterflies are very much alike, and the distinguishing characters of *Diocles* ♀ are its larger size, orange-red forehead, whiter more faintly blue-shot discs of upper side, and want of basal ocelli on under side of hind-wings. The last character is common to both sexes, and also marks *D. Isocrates* (Fab.).

It was not until October 1869, when I received a ♂ from Mr. W. Morant, that I was aware of the existence of this butterfly. This example was captured in Natal, on the Umgeni; and in the December following I received from the same gentleman a ♀ from the same locality. These examples were taken respectively in June and July 1868. In 1870 the late Mr. M. J. M'Ken forwarded a ♂, taken in D'Urban Botanic Gardens, on *Poinsettia pulcherrima*, and also a ♀ from the same locality. The late Mr. E. C. Buxton met with the species in Swaziland, and in 1873 sent me a photograph of the ♂. Colonel Bowker has contributed about a dozen examples, taken in March, April, June, and August, at D'Urban, Avoca, and Pinetown, Natal. He and Mr. Morant both describe the habits of *Diocles* as resembling those of *Antalus*.

Localities of *Dendorix Diocles*.

I. South Africa.
 E. Natal.
  a. Coast Districts. D'Urban (*W. Morant, M. J. M'Ken,* and *J. H. Bowker*). Avoca and Pinetown (*J. H. Bowker*).
 G. "Swaziland."—The late E. C. Buxton.

175. (3.) Deudorix Dariaves, Hewitson.

♂ ♀ *Deudorix Dariaves,* Hewits., " Ent. Month. Mag., xiii. p. 205 (1877);" Ill. D. Lep., Suppl., p. 30, pl. Va. ff. 60–62 (1878).

Exp. al., (♂) 1 in. 2—3 lin.; (♀) 1 in. 6½ lin.

♂ *Fuscous-brown; fore-wing without marking; hind-wing very broadly orange-red exteriorly. Fore-wing:* inner-margin rather convex near base; sexual tuft of hair long, grey, mixed with fuscous-brown. *Hind-wing:* orange-red occupying most of the field as in *D. Diocles,* Hewits., but extending subcostally (below sexual violaceous badge) nearer to base; fuscous-brown of basal area reaching to extremity of discoidal cell, and emitting a broad streak all along fold between first median nervule and submedian nervure; hind-marginal black linear edging, tail, anal-angular lobe, and inner-marginal greyish border, as in *Diocles.* UNDER SIDE.—*Fore-wing:* dull pale-grey; white-edged markings as in *Diocles,* but discal macular band more curved near costa. *Hind-wing:* a sub-basal series of four large, conspicuous, white-ringed spots, of which three are ferruginous-red and the fourth (close to inner-margin) dull brown; terminal disco-cellular mark and discal band arranged as in *Diocles,* but whitish mesially, and with their white edges (as well as inner edge of submarginal lunular streak) enlarged; costal spot of discal band interiorly tinged with ferruginous; hind-marginal black spots strongly marked, especially that on anal-angular lobe, the lunule interiorly edging upper spot pale-yellow; fuscous space between spots spangled with bluish-silvery scales.

♀ *Dull-greyish brown, with a paler discal space in both wings; hind-wing with a submarginal macular whitish band. Fore-wing:* paler space suffusedly covering basal half of median nervules. *Hind-wing:* paler space less apparent, but extending towards base; whitish band lying between second subcostal and first median nervules; hind-marginal spots, black and white edging, and tail quite as in *Diocles.*

(This description of the ♀ is made from Hewitson's figure; the under side is not figured, but is stated by Hewitson to be like that of the ♂, but paler.)

This close ally of *D. Diocles*, Hewits., may at once be recognised by the red sub-basal ocelli on the under side of the hind-wings, and, as regards the ♂, by the uniform dark-brown of the fore-wing. In the latter sex of *Dariaves* the forehead and tip of abdomen are red, but the former is of a duller tint. The female, to judge from Hewitson's figure and description, is darker than that of *Diocles*, and without any violaceous-blue gloss, while possessing a whitish band in the hind-wings not represented in *Diocles*. On the under side the hind-wings, except near base, are much whiter than the fore-wings (owing to the enlargement of the white edges of the principal markings),—a character which also distinguishes the species from *Diocles*.

D. Dariaves was discovered at Delagoa Bay by the late Mr. J. J. Monteiro, and I have received a specimen of the ♂ taken by Mrs. Monteiro in the same locality.

Localities of *Deudorix Dariaves*.
I. South Africa.
  II. Delagoa Bay.—Lourenço Marques (*Mrs. Monteiro*).
II. Other African Regions.
  A. South Tropical.
    b. East Coast.—" Zanzibar."—Cat. Hewitson Coll.

176. (4.) Deudorix Licinia, (Mabille).

♂ *Thecla Licinia*, Mab. in Grandid. Hist. Phys., &c., de Madag., pl. 30A, ff. 5, 5a (1885).

Exp. al., 1 in. 3 lin.

♂ *Orange-red; fore-wing with a rather narrow fuscous-brown border.* Fore-wing: base of inner margin suffused with fuscous-brown; border commencing rather widely at base, narrower about middle, wide at apex, and thence gradually narrowing along hind-margin to posterior angle; inner-marginal sexual tuft reddish-brown. *Hind-wing*: a narrow blackish-brown suffusion at base; sexual badge small, shining leaden-grey; inner margin with a dull-greyish border, set with whitish hairs; orange-red extends to hind-margin itself, except close to anal angle, where there is a fine linear black edging; tail black, white-tipped; a small hind-marginal black spot immediately below tail; a larger black spot, sprinkled with greenish-silvery scales, on anal-angular lobe. UNDER SIDE.—*Brownish-grey* (except inner-marginal area of fore-wing, which is whitish tinged superiorly with ochre-yellow); *in both wings, terminal disco-cellular mark and irregular discal band outlined with dark-red*, and with faint white outer edges,—submarginal lunulate streak dark-grey and white, indistinct. *Hind-wing*: a sub-basal row of three round dark-red white-ringed spots; a linear black edging, immediately preceded by a white line, along inferior half of hind-margin; upper hind-marginal black spot bounded interiorly by an orange-yellow lunule; anal-angular one inferiorly edged with orange-yellow; between the two spots some fuscous and greenish-silvery scales. *Cilia* on upper side fuscous in fore-wing, red in hind-wing; on under side reddish generally, but mixed with white near anal angle of hind-wing.

I have not seen the ♀ of this *Deudorix*. The ♂, both in the elongated wings and in the pattern and colouring of the upper side, closely resembles *D. Melampus* (Cram.), except that the red is paler, inclining to orange; but the under side markings are very dissimilar, those of *Melampus* being very greatly narrowed, indistinct, and not at all red, but of a tint scarcely separable from that of the ground-colour; the Indian species also wants the sub-basal ocelli in the hind-wing.

Mr. Henley Grose Smith, who kindly sent me two specimens and drawings of this butterfly, writes that it closely resembles a Madagascar species named *Licinia* by M. Mabille, but that this gentleman and M. Grandidier, having examined a specimen forwarded to the latter by Mr. Smith, had pronounced it to be distinct from that species.

On comparison subsequently, however, of several South-African specimens with the figures given in the work of Grandidier above quoted, I find the former to agree too closely with the latter to admit of their separation as species. The South-African examples have a rather narrower dark border to the fore-wing, especially at base and apically, and a paler, almost obsolete, brownish inner-marginal cloud in the hind-wing; while the under side is somewhat darker in ground-colour and has the red markings rather brighter.

Compared with the allied *D. Livia* (Klug), from Upper Egypt and Arabia, the South-African specimens of *Licinia* differ in the other direction, being larger, of a deeper (less orange) red, and with the border of the fore-wing broader (especially on costa and hind-margin) and better defined inwardly; but on the under side the markings are very much redder—those of *Livia* having scarcely a tinge of that colour.

In 1879, I received from Mrs. Barber a ♂ taken in Matabeleland by Mr. H. Barber, which quite agrees with the examples above described.

Localities of *Deudorix Licinia*.

I. South Africa.
  II. Delagoa Bay (*Mrs. Monteiro*).
II. Other African Regions.
  A. South Tropical.
    *b*1. Eastern Interior.—Matabeleland (*H. Barber*).
    bb. Eastern Islands.—"Madagascar."—Grandidier.

Genus CAPYS.

Capys, Hewits., Illustr. Diurn. Lep., p. 59 (1865).
Zeritis, Trim. [part], Rhop. Afr. Aust., ii. p. 270 (1866).

IMAGO.—Closely allied to *Deudorix*. *Head* rather broader; palpi in ♂ shorter, the terminal joint being minute,—in ♀ longer, the terminal joint being very long and slender, and porrected far in front of the head; antennæ with a longer club.

Thorax considerably longer and stouter in both sexes, but especially in ♂. *Fore-wings* in ♂ more produced in apical region, in ♀ more convex on hind-marginal border; neuration as in *Deudorix*; no tuft of hairs on inner margin. *Hind-wings* more rounded, especially in ♀, not produced in anal-angular portion, but at anal angle itself a marked sublobate projection, more prominent in ♂; hind-margin regularly dentated; neuration as in *Deudorix*. *Legs* as in *Deudorix*, but thicker, and the tibiæ of middle and hind legs considerably shorter.

Abdomen larger and thicker than in *Deudorix*, especially in ♀.

Hewitson rightly removed the noble Lycænide on which he founded this genus from its questionable association with *Zeritis*; but he admits it to be "very nearly allied to the genus *Deudorix*," and it is perhaps hardly separable from the latter. Besides the distinctions above given, *Capys* wants the rather long linear tail on each hind-wing, so characteristic of *Deudorix*.

The only species known is the *Alphæus* of Cramer, a butterfly which on the upper side is black, with a broad metallic-red band across the wings, and on the under side chiefly pale-grey crossed by a bar of darker grey and ferruginous. With the exception of *Zeritis Thero* (Linn.), its expanse across the fore-wings is the largest among the South-African *Lycænidæ*, and in bulk of body it exceeds them all. Though widely distributed throughout Southern Africa, it is very local in its haunts, and seems more prevalent in the vicinity of Cape Town than elsewhere. It occurs in the Transvaal, but has not hitherto been recorded from any tropical locality. The butterfly is fond of rocky elevated spots; and several males usually sport about in company, taking frequent short flights of extreme rapidity; while the female, though well able to fly, is rarely seen on the wing.

177. (1.) Capys Alphæus, (Cramer).

PLATE VII., f. 5 (♀).

♂ *Papilio Alphæus*, Cram., Pap. Exot., ii. t. clxxxii. ff. E, F. (1779).
♂ *Polyommatus Alphæus*, Godt., Enc. Meth., ix. p. 663, n. 155 (1819).
♂ *Zeritis? Alphæus*, Westw., Gen. D. Lep., ii. p. 500, pl. lxxvii. f. 3 (1852).
♂ ♀ *Zeritis Alphæus*, Trim., Rhop. Afr. Aust., ii. p. 270, n. 168 (1866).

Exp. al., (♂) 1 in. 4—9½ lin.; (♀) 1 in. 6—11 lin.

♂ *Glossy-black, with a broad, discal, sub-metallic red band*, from fourth subcostal nervule, or from upper radial nervule of fore-wing to submedian nervure of hind-wing, near anal angle; a mixed golden and purplish gloss over basal region; *cilia* white, with black spots at ends of nervules. *Fore-wing* : band exteriorly indented with black on nervules, and narrowed on inner margin. *Hind-wing :* a narrow costal blackish border; inner-marginal border hairy, dull grey; anal angle bluntly produced, marked with a red spot; on subcostal nervure, at origin of nervules, *a small, subovate, glistening space*. UNDER SIDE.—*Hind-wing and border of fore-wing* (except inner margin) *hoary grey*, clouded with darker. *Fore-wing :* bright orange, paling into dull-yellowish on inner margin ; at end of cell two short, ferruginous, blackish-edged, transverse marks, between which is a greyish space ; between them and apex two longer similar, more widely apart, crenelated streaks from costa, converging as far as orange where the inner one ends, but the outer is dimly prolonged along external edge of orange. *Hind-wing :* a broad, central, irregular, dark-grey, ferruginous- and black-edged, transverse stripe, on its inner edge deeply pierced upwardly by a streak of ground-colour; a bright ferruginous hind-marginal edging and parallel submarginal streak, both obsolete near apex; on anal angle a black spot.

♀ Wings rounder, especially hind-wing; fore-wing not apically prominent, but with hind-margin rather convex. *Red less metallic and paler, but occupying a larger field*, reaching nearer to base; dark borders narrower except on costa, not so black; cilia broader. UNDER SIDE.— As in ♂. *Fore-wing*: streaks from costa not convergent.

The central band on under side of hind-wing is sometimes quite divided by the grey intersection on discoidal nervule.

In two ♀s from Port Elizabeth, taken by Mr. S. D. Bairstow, the red field is more restricted than usual, especially in the hind-wing, where the costal border is broadly black as far as second subcostal nervule. A ♀ from the Lydenburg District of the Transvaal exhibits quite the opposite tendency, having the red in both wings much enlarged and paler than usual. The three ♂s accompanying this ♀ show a slight or moderate enlargement of the red, but a ♂ from Natal has it as much developed as in the Transvaal ♀, and is also remarkable for acuter wings. The five examples just mentioned all possess a feature not noticed in any specimens from the Cape Colony, viz., an orange-red base to the cilia of the lower half of the hind-wing.

This splendid Lycænide frequents hill ridges and rocky "kopjes" on mountain sides, seldom occurring in low-lying situations. Both sexes are rapid on the wing, but the male extremely so; female specimens are, however, rarely met with, and no doubt are habitually inactive, while the males keep flying about a particular spot of limited extent, darting away in pursuit of each other or of different butterflies, and quickly returning to some favourite perch. Near Cape Town I have found it settling most frequently on the leaves of young *Proteaceæ* and of *Watsonia*; I have only twice noticed it on flowers, and never saw it settle on the ground. Mr. T. D. Butler, the Museum taxidermist, brought me a female which he found on the Devil's Mountain, Cape Town, sitting on damp ground in a slight hollow. The long hill lying between Wynberg and Protea is the best locality for *Alphæus* near Cape Town, and on one occasion I found it rather numerous near the highest block-house on the Devil's Mountain. In this neighbourhood it is apparently on the wing all the year round, though October and March seem to be the months most favourable for it, and I have not captured it during November, May, or June.

Localities of *Capys Alphæus*.

I. South Africa.
B. Cape Colony.
 a. Western Districts.—Cape Town. Montagu. Knysna.
 b. Eastern Districts.—Port Elizabeth (*S. D. Bairstow*). Grahamstown (*M. E. Barber*).
E. Natal.—Special locality not noted (*M. J. M'Ken*).
K. Transvaal.—Potchefstroom and Lydenburg District (*T. Ayres*).

GENUS HYPOLYCÆNA.

Hypolycæna, Felder, Wien. Ent. Monatschr., vi. p. 293 (1862); Hewitson, Ill. Diurn. Lep., p. 48 (1865).
Myrina [part], Westw., Gen. Diurn. Lep., ii. p. 475 (1852).
Amblypodia [part], Trim., Rhop. Afr. Aust., ii. p. 226 (1866).

IMAGO.—Structure more slender than in *Iolaus*. *Head* small; *eyes* smooth; *palpi* long, ascendant, divergent,—the second joint much flattened laterally, with a dense clothing of long stiff scales laterally and inferiorly,—the terminal joint long, slender, smooth, sharply pointed; *antennæ* quite slender, white-ringed, not gradually incrassated, but with a distinct elongated club.

Fore-wings rather variable in shape,—in the typical (*Erylus*) group more elongate and pointed apically,—in the *Faunus* group more truncate and with a more convex costa: subcostal nervure with only three nervules, of which the first and second are emitted at some distance apart towards the end of discoidal cell, and the third from the end itself of the cell, and ending at the apex; first radial nervule originating from same point as third subcostal, second from junction of curved middle and lower disco-cellular nervules. *Hind-wings* more or less produced in the anal-angular portion, bearing a lobe at anal angle itself, and a more or less developed tail at extremity of submedian nervure; almost always a second sublinear tail on first median nervule, and sometimes a third on second median; costa more or less convex; costal nervure terminating at apex (except in *H. Cæculus*, Hopffer, where it ends about middle of costa); discoidal cell short, truncate; radial nervule originating at meeting-point of disco-cellular nervules, the lower of which joins median nervure at origin of third median nervule. *Fore-legs* of ♂ rather long and slender,—the femur hairy beneath,—the tibia scaly, with a few fine hairs,—the tarsus very indistinctly articulated, finely spiny beneath, and terminating in a single curved claw;—of ♀ somewhat stouter and shorter, with tarsus longer and thicker, more spiny beneath, distinctly articulate, and terminating in two claws. *Middle* and *hind legs* rather short,—the tibia considerably shorter than the femur, and its terminal spurs long and stout,—the tarsus, owing to the length of the first joint (which is swollen in the hind-legs of the ♂), considerably longer than the tibia.

LARVA.—Very broad and thick, slightly narrower and thinner posteriorly; head very small.

PUPA.—Robust, rounded, rather tapering posteriorly; head and back of thorax but slightly prominent.

(The characters of larva and pupa are from drawings by Mrs. Barber of those of *H. Lara* (Linn.))

There is considerable diversity among the butterflies of this genus, as shown by the characters above given, but their slender structure and only three-branched subcostal nervure of the fore-wings are features which readily distinguish them from their allies the *Myrinæ* and *Iolai*. The upper surface of the males, though less metallic than in the genera just named, is usually of some deep rich purplish or violaceous-blue, while that of the females is dull grey or brown with more or less discal white. The under surface resembles that of the genus *Iolaus*, being white or greyish with neatly-defined discal transverse stripes, sometimes more or less broken up into separate spots. In the typical group (*H. Erylus*, Godt., *Philippus*, Fab., and allies) there are two linear black tails of only moderate length on each hind-wing; but in the West-African group represented by *H. Faunus* (Drury), *H. Antifaunus* (Doubl.), and *H. Lebona* (Hewits.), the corresponding two tails, and especially that on the submedian nervure, are greatly elongated, broad, and white, while there is a similar but shorter additional tail on the second median nervule.

About thirty species are recorded. North India has yielded seven, and the Indo-Malayan Islands nine; while only three are known from the Austro-Malayan Islands. Africa has as many as thirteen, but of these five only have been discovered in Southern Africa. The most widely distributed of the five are *H. Philippus* and *H. Lara* (Linn.), inhabiting both North and South Tropical Africa; *H. Cæculus* (Hopff.) is really South Tropical, only just entering the South-African Sub-Region at Delagoa Bay. *H. Hirundo*, Wallengr., and *H. Buxtoni*, Hewits., extend over a large part of Eastern South Africa, but do not appear to be recorded from any place within the Tropics; both range into the eastern districts of the Cape Colony, but while *Hirundo* is not uncommon there, only one capture of *Buxtoni* so far to the south and west is known to me. *H. Lara* is the only species generally distributed throughout South Africa; it is common about Cape Town.

The last-named species is very unlike nearly all its congeners, the upper side colouring being in both sexes of a glistening pale-ferruginous, shot basally with a pearly gloss. At the posterior angle of each wing there are two or more conspicuous black spots in white rings. These ocelli recur less distinctly in *H. Rabe* (Boisd.), from Madagascar, and in *H. Hirundo*; and these two species—but especially the latter, with its single long white-fringed tail at the posterior angle of the hind-wing—serve to connect *Lara* with the rest of the genus.

178. (1.) **Hypolycæna Cæculus**, (Hopffer).

Iolaus Cæculus, Hopff., Monatsb. d. K. Akad Wissensch. Berl., 1855, p. 642, n. 17; and Peters' Reise Mossamb.-Ins., p. 402, pl. xxv. ff. 12–14 (1862).
Hypolycæna Cæculus, Hewits., Ill. D. Lep., p. 52, n. 14 (1865).

Exp. al., 1 in. 2—4 lin.

♂ *Bright submetallic blue; fore-wing with very broad apical hind-marginal, hind-wing with moderately broad costal-apical, black border.* Fore-wing: costal edge before middle pale-reddish; blue only thinly covering costal border before middle, and leaving inner-marginal lobe close to base uniform grey, but extending from base to beyond middle,—its outer edge rather deeply indented with black on median nervules; hind-marginal black border (in two out of three specimens) extending broadly and evenly to posterior angle (in the third narrowing to a point). Hind-wing: on costa at base, a large subovate glistening grey patch, containing a transverse dull-fuscous mark, partly overlapped by lobe of fore-wing; above and beyond this patch the costal black border runs pretty evenly from base to apex; close to hind-margin, two rather large black spots between second median nervule and submedian nervure; a smaller black spot, marked outwardly with some greenish-silvery scales, on anal-angular lobe; a thin but very distinct black linear edging all along hind-margin; bases of cilia *pure-white*, forming a very distinct line immediately beyond the black linear edging of hind-margin, and conspicuously margining and tipping the tails (which are mesially rufous) on first median nervule and submedian nervure. UNDER SIDE.—*Very pale-grey, with a faint yellowish tinge; the following rufous-ochreous, very thinly fuscous-edged, narrow transverse striæ common to both wings, viz.*:—one near base, not extending below median nervure in fore-wing, and angulated and interrupted near inner margin in hind-wing; a short stria marking extremity of discoidal cell; two (not parallel) beyond middle, becoming fuscous near inner margin of fore-wing, and biangulated towards that of hind-wing; and a hind-marginal edging stria; all these striæ more or less faintly margined with whitish, except the hind-marginal one, which in hind-wing is internally bounded by a well-defined white line. *Fore-wing* : inner-marginal area before middle smooth, silvery. *Hind-wing* : basal lobe very prominent, and a sub-vesicular swelling near base; some fuscous irroration near hind-margin; the middle hind-marginal black spot obsolete, but the upper one and that on anal-angular lobe well-marked, conspicuously edged with greenish-silvery, and inwardly bounded with golden-yellow scaling; base of cilia conspicuously white, as on upper side.

♀ *Much paler and duller; the blue in both wings becoming obscurely whitish in disc.* Fore-wing : a dusky striola marking extremity of discoidal cell. *Hind-wing* : a white line inwardly bounding linear black hind-marginal edging; black spots near anal angle large; above them, just before white line, two or three smaller more obscure similar spots. UNDER SIDE.—As in ♂.

In one ♀ from Delagoa Bay, the blue of the upper side is scarcely visible, the whole surface except for some very obscure bluish-grey scaling being pale fuscous-brownish. The under side is quite as usual.

The last-named specimen together with a normal representative of each sex were kindly lent to me by Mr. H. Grose Smith, who received them from Delagoa Bay. From the same locality Mrs. Monteiro, in 1878, was so good as to send me a pair; and a fine ♂, now in the South-African Museum, was also one of her captures in the year 1883.

Hewitson (*op. cit.*) notices that his example of this butterfly from the Zambesi had the under side darker than usual and of a rufous-grey, but he does not mention the sex of this specimen.

H. Cæculus ♂, in its deep-blue black-bordered upper-side colouring and glistening badge of the hind-wing, has quite the appearance of an *Iolaus*, but the general structure and under-side pattern in both sexes justify Hewitson's location of the species in the genus *Hypolycæna*. On the under side, the well-marked stria before the middle is a good distinguishing character, neither *H. Philippus* nor *H. Buxtoni* presenting it.

Localities of *Hypolycaena Cæculus.*

I. South Africa
II. Delagoa Bay.—Lourenço Marques (*Mrs. Monteiro*).
II. Other African Regions.
   A. South Tropical.
      a. Western Coast.—"Angola (*Pogge*)."—Dewitz.
      b. Eastern Coast.—"Zambesi."—Hewitson. "Querimba."—
         Hopffer. "Tchouaka (*Raffray*)."—Oberthür.
      *b*1. Interior. "Tete."—Hopffer. "Lake Nyassa."—Kirby, Cat.
         Hewits. Coll.

179. (2.) **Hypolycæna Philippus,** (Fabricius).

♀ *Hesperia Philippus,* Fab., Ent. Syst., iii. 1, p. 283, n. 87 (1793).
Iolaus Orejus, Hopff., Monatsb. K. Akad. Wiss. Berl., 1855, p. 641;
   and Peters' Reise Mossamb., Ins., p. 401, pl. xxv. ff. 10, 11 [♀]
   (1862).
♂ ♀ *Thecla Orejus,* Wallgrn., K. Sv. Vet.-Akad. Handl., 1857; Lep.
   Rhop. Caffr., p. 35.
♂ *Hypolycæna Philippus,* Hewits., Ill. Diurn. Lep., p. 50, pl. 22, ff. 15, 16
   (1865)
♂ ♀ *Amblypodia Erylus,* Trim., Rhop. Afr. Aust., ii. p. 228, n. 132 (1866).

Exp. al., 1 in. 1–4 lin.

♂ *Dull-brown, with a more or less intense changing pink-violet lustre;* a brown line along hind-marginal edge. *Fore-wing:* covering bases of median nervures, an ill-defined dusky patch. *Hind-wing:* hind-marginal line edged with white on both sides (except near apex); touching it internally, between second median nervule and anal angle, three black spots, of which the first is inwardly edged by an orange, the second by a whitish lunule, and the third, on lobe of angle, small, broadly orange, with an inward white lunule; tails black, white-edged and tipped. *Cilia* greyish, paler in hind-wing. UNDER SIDE.—*Whitish-grey, with glistening white-edged orange-ochreous transverse striæ;* common to both wings, a well-marked stria beyond middle, in hind-wing interrupted, and acutely angulated beyond first median nervule,—a submarginal, sublunulate thin stria, also angulated in hind-wing,—and a hind-marginal edging line; in each wing a double striola closing cell. *Hind-wing:* a conspicuous spot, coloured like striæ, near base, between costal and subcostal nervures; orange lunules of first and third spots near anal angle more conspicuous, the third marked with bluish or greenish-silvery scales, a patch of which also marks the interval between these two spots.

♀ *Brownish-grey, without* violet lustre; a faint bluish tinge near bases. *Fore-wing:* costa with a narrow ochreous edge; sometimes an indistinct paler fascia on disc, widening downwards from its origin on third median nervule. *Hind-wing: two submarginal rows of white lunular markings* (of which the inner is broader) between second subcostal and submedian nervure; the outer row ends with the three black spots, which are more conspicuous (especially as regards the orange lunules edging two of them) than in ♂; hind-marginal line and its white edges very distinct. UNDER SIDE.—*Whiter;* the markings brighter, clearer, and more conspicuous.

This butterfly is allied to *H. Erylus* (Godt.), but considerably smaller. Its under side is much paler, and the markings of a brighter tint and more clearly defined; while the upper side of the ♂ is of a much less intense, more pinkish-violaceous than purple-blue lustre, and has the patch in fore-wing very much smaller and less conspicuous.

An unusually small ♂, which I took near D'Urban, Natal, has an expanse of wings of only $10\frac{1}{2}$ lines.

II. Philippus has a wide African distribution, but does not appear to penetrate the Cape Colony far beyond its eastern border. I found it a common insect on the coast of Natal, where I captured the paired sexes on 21st February 1867. I observed it on the wing from the end of January to the beginning of April; it was always about rather low shrubs, usually perching on the leaves, but occasionally sucking the flowers. The males, as a rule, perched higher than the females, keeping to the topmost sprigs; but they were not specially active, and their flights were very short. Colonel Bowker has taken this species in July and August.

Localities of *Hypolycæna Philippus.*

I. South Africa.
  B. Cape Colony.
    b. Eastern Districts.—King William's Town (*W. S. M. D'Urban* and *J. H. Bowker*).
  D. Kaffraria Proper.—Bashee River (*J. H. Bowker*).
  E. Natal.
    a. Coast Districts.—D'Urban. Umvoti. Mouth of Tugela River (*J. H. Bowker*). " Lower Umkomazi."—J. H. Bowker.
  F. Zululand.—St. Lucia Bay (*Colonel H. Tower*).
II. Other African Regions.
  A. South Tropical.
    a. Western Coast.—"Angola (*J. J. Monteiro*)."—Druce. "Chinchoxo (*Falkenstein*)."—Dewitz.
    b1. Eastern Interior —"Tette, Zambesi River."—Hopffer.
  B. North Tropical.
    a. Western Coast.—Sierra Leone.—Hope Mus. Oxon. and Brit. Mus.
    b1. Eastern Interior.—" Atbara."—Butler.

180. (3.) Hypolycæna Buxtoni, Hewitson.

♀ *Hypolycæna Buxtoni,* Hewits., Ent. M. Mag., x. p. 206 (1874).
♂ ♀ *Hypolycæna Seamani,* Trim.,[1] Trans. Ent. Soc. Lond., 1874, p. 332, pl. ii. ff. 3, 4.

Exp. al., (♂) 1 in. $1\frac{1}{2}$ lin.; (♀) 1 in. 3 lin.

♂ *Rich violaceous-purple. Hind-wing:* a hind-marginal black line from second median nervule to anal angle, immediately preceded by a

---
[1] I named this butterfly in memory of its discoverer, the late Dr. J. F. Seaman; but Hewitson's description was published in February 1874, six months before my own, and consequently his name, *Buxtoni,* has priority.

concurrent pure white line, the latter widening into a white space on anal-angular lobe; two very indistinct dark spots just before the white line, one above, the other below, first median nervule; a third spot, black, densely scaled with silvery-bluish and golden scales, on anal-angular lobe, edged interiorly and exteriorly with pure white; tails at extremities of first median nervule and submedian nervure respectively (of which the latter is nearly twice as long as the former), thin, black, conspicuously fringed and tipped with pure white. *Cilia* of fore-wing apparently greyish, *of hind-wing pure white*, both on hind and inner margin. UNDER SIDE.—*White, with thin yellow-ochreous striæ;* in both wings a short stria closing discoidal cell, and two transverse striæ (convergent downward, the outer one thinner and fainter than the inner) beyond middle. *Fore-wing:* the striæ beyond middle commence on costa, but do not reach inner margin, ending a little below submedian nervure. *Hind-wing: elongate spot near base,* below precostal nervure, *red;* the outer and inner striæ meet below third median nervule, but are thence independently deflected to inner margin; the outer stria becomes fuscous near the point of meeting with the inner, and is thence black; the usual hind-marginal spots between second and first median nervules and on anal-angular lobe respectively, the former black, inwardly bordered rather conspicuously with fulvous-yellow, the latter as on upper side; faint traces of a dusky line just before hind-margin, which is itself very finely edged with black.

♀ *White, with broad fuscous clouding and borders.* Fore-wing: fuscous basal clouding fills discoidal cell for about three-fourths of its length, and extends below it to inner-margin, but does not reach beyond middle; fuscous border extends from base to anal angle, and is very broad in apical region; disco-cellular stria indistinctly marked, and traces visible of the longer striæ beyond middle. *Hind-wing:* fuscous clouding in basal region fills cell, and extends irregularly beyond, above, and below it about to middle; the two under-side striæ strongly marked, fuscous, suffused, *not* meeting, but widely separated, between first median nervule and submedian nervure; some fuscous scaling near apex; hind-marginal and sub-marginal streaks and spots well marked. UNDER SIDE.—Quite as in ♂, except that in the *hind-wing* the two striæ beyond middle, though approximating much more nearly than on upper side, do not meet.

This butterfly is a close ally of *H. Philippus*, Fab. In the ♂ it is distinguished by the *more purple*, less cupreous *colour of the upper side*, and the *conspicuous white cilia* of the hind-wings; and in both sexes by the *whiteness of the under side*, with its thinner, much straighter striæ, and by the longer tails of the hind-wings.

The upper side of the ♀ is most strikingly different from the brownish-grey colouring of that of the ♀ *Philippus*, and the disparity is almost as remarkable as that between the ♂s of *Deudorix Antalus* (Hopff.), and *D. Dioeles*, Hewits., the ♀s of which can scarcely be distinguished, except by one or two slight characters that would escape a cursory comparison.

A single ♀ specimen was sent me from Pinetown, Natal, by the late Dr. J. E. Seaman, and a ♂ by Mr. Walter Morant, towards the end of 1869. Dr. Seaman noted the ♀ as having been taken in July, "at an opening in the bush;" and Mr. Morant described the ♂ as occurring in June "on small trees by the waterside," and further observed, with reference to the ♀ sent by Dr. Seaman (which he did not recognise as of the same species as the ♂ sent by himself), "This is very scarce; I have a single specimen in my collection, taken *on a low tree near water*" (see note on ♂ above) "about two years ago, since which time I have seen but one other." The late Mr. E. C. Buxton sent me the photograph of a ♀, taken by him in the Amaswazi country.

Since 1874 Colonel Bowker has forwarded a ♀ from the mouth of the Kei River, and five ♂s and three ♀s from Natal, the latter taken chiefly in the neighbourhood of Pinetown. The first-mentioned specimen was taken in March 1875; the Natalian examples at intervals from 1879 to January 1885.[1] In habits *H. Buxtoni* does not appear to present any peculiarities, but it is evidently very much rarer than its nearest congener, *H. Philippus*.

Localities of *Hypolycæna Buxtoni*.

I. South Africa.
   B. Cape Colony.
      b. Eastern Districts.—Kei River Mouth (west bank) (*J. H. Bowker*).
   E. Natal.
      a. Coast Districts.—Isipingo. Pinetown (also *J. E. Seaman* and *W. Morant*); and Inanda (*J. H. Bowker*).
   G. Swaziland.—" Usutu River."—E. C. Buxton.

181. (4.) Hypolycæna Hirundo, (Wallengren).

Thecla Hirundo, Wallgrn., K. Sv. Vet.-Akad. Handl., 1857, p. 35, n. 4.
Amblypodia Hirundo, Trim., Rhop. Afr. Aust., ii. p. 230, n. 133, pl. 4, f. 11 [♀] (1866).
Hypolycæna Hirundo, Hewits., Ill. Diurn. Lep., Suppl., p. 12 (1869).

Exp. al., 9 lin.—1 in. 1 lin.

Dark ash-grey, irrorated from bases with very pale bluish-grey; anal angle of hind-wing produced and lobed, and ending in a *long twisted, black, broadly white-edged and fringed tail*, on submedian nervure. *Fore-wing:* on hind-margin, close to posterior angle, a black, indistinctly white-ringed spot. *Hind-wing:* paler nearer inner margin; along hind-margin a row of dark white-ringed spots, very indistinct, excepting the three last, which are black and well-marked (that on anal-angular lobe being the largest); before this, a row of indistinct whitish lunules. *Cilia* of fore-wing grey, white at anal angle; of hind-wing wholly white. UNDER SIDE.—*Whitish-grey; with white-bordered yellow-ochreous striæ:* common to both wings—a

[1] The fine pair taken at this last date was presented by Colonel Bowker to the British Museum.

transverse, irregular streak, interrupted on nervules, beyond middle,—a submarginal lunulate streak,—a very indistinct row of dusky spots just beyond this,—and a line just within hind-marginal edge; in each wing a double disco-cellular striola, with two costal spots above it. *Fore-wing:* ocellus at anal angle conspicuous, bounded internally by last lunule of submarginal streak. *Hind-wing:* spot on lobe and that between second and third median nervules black, inwardly edged by a faint-yellow lunule,—the space between the two spots fuscous; streak beyond middle strongly recurved below first median nervule, and widely interrupted on submedian nervure.

In the ♀, the hind-marginal and submarginal white markings are more distinct on the upper side; and there is usually an additional, imperfect, smaller, white-ringed spot, immediately above that at the posterior angle of the fore-wing on the upper side.

This curious little species comes nearest to *H. Rabe* (Boisd.),[1] of Madagascar, but is smaller, and has a much darker upper side (that of *Rabe* being all pale-grey except near bases, and a broad costal, apical, and hind-marginal fuscous border of the fore-wing), while the underside striæ on disc are much more regular, and in their colour and arrangement more resemble those of *H. Philippus* (Fab.) The forewing ocellus and single long white tail of the hind-wings give it a very peculiar aspect.

I only once met with *H. hirundo*, and then with a single individual only. I was collecting on 22d February 1870 in the dense prickly scrub at the back of the village of Uitenhage, and noticed this butterfly flutter from a low bush and settle on the ground. Mr. W. S. M. D'Urban found the species abundant in the King William's Town district in the months of March, June, and October.

Localities of *Hypolycæna Hirundo*.

I. South Africa.
   B. Cape Colony.
      b. Eastern Districts.—Uitenhage. Keiskamma River near Bodiam; King William's Town (*W. S. M. D'Urban, M. E. Barber,* and *J. H. Bowker*).
   D. Kaffraria Proper.—Bashee River (*J. H. Bowker*).
   E. Natal.
      a. Coast Districts.—Esidumbeni (*M. J. M'Ken*). D'Urban (*J. H. Bowker*).
II. Delagoa Bay.—Lourenço Marques (*Mrs. Monteiro*).

---

[1] Mr. Butler (*Cat. Fab. D. Lep.*, p. 181) identifies *Rabe*—from a comparison of Jones's unpublished "Icones"—with the *Hesperia Phidias* of Fabricius (*Ent. Syst.*, iii. 1, p. 286, n. 99).

182. (5.) Hypolycæna Lara, (Linnæus).

Papilio Lara, Linn., Mus. Lud. Ulr. Reg., p. 320, n. 138 (1764); and Syst. Nat., i. 2, p. 791, n. 328 (1767).[1]
♀ *Papilio Iolaus*, Cram., Pap. Exot., iii. pl. cclxx. ff. F, G (1782).
♂ *Papilio Gorgias*, Stoll, Suppl. Cram. Pap. Exot., pl. xxxiii. ff. 5, 5D (1791).
♂ ♀ *Polyommatus Lara*, Godt., Enc. Meth., ix. p. 675, n. 179 (1819).
Thecla Iolaus and *Thecla Lara*, Wallgrn., K. Sv. Vet.-Akad. Handl., 1857; Lep. Rhop. Caffr., pp. 34, 35.
♂ ♀ *Chrysophanus Lara*, Trim., Rhop. Afr. Aust., ii. p. 260, n. 116 (1866).
Hypolycæna Lara, Hewits., Ill. Diurn. Lep., Suppl., p. 13 (1869).

Exp. al., (♂) 10 lin.—1 in. 2½ lin.; (♀) 1 in. 1–6 lin.

Glistening pale-ferruginous, darker on margins, with a brilliant-pearly basal lustre. *Fore-wing*: at posterior angle, a good-sized white-ringed black spot, often surmounted by one or two indistinct white rings, of which the lower is sometimes distinct and filled with black. *Hind-wing*: at anal angle two spots like those of fore-wing, but smaller; above them, along hind-margin, a series of whitish rings, becoming obsolete towards costa; beyond middle, occasionally an indistinct transverse row of whitish lunules. Cilia white, interrupted with fuscous at extremities of nervules. UNDER SIDE.—*Whitish-grey*. *Fore-wing*: tinged with brownish, except on costa and hind-margin; posterior-angular spots distinct, whitish rings above suffused, all interiorly edged by a brownish line; a pale-edged disco-cellular terminal streak; beyond middle, a macular, brownish, transverse, outwardly white-edged streak, sharply curved at costa. *Hind-wing*: disco-cellular streak and streak about middle (much sinuated) usually indistinct; no spots at anal angle or on hind-margin; an irregular, submarginal, suffused brownish fascia, broadest on discoidal nervule; near base two or three indistinct spots.

The male has the fore-wing at apex and the hind-wing at anal angle more acuminate than in the female, but the difference is not striking in the typical smaller form prevailing about Cape Town and the neighbouring districts, whereas in some of the larger examples from Kaffraria, Natal, and the Transvaal, it is very pronounced—almost as much as in Stoll's figure above quoted. This acute-winged larger ♂ also has the basal lustre usually more developed and slightly bluer in tint, and as a rule shows with tolerable distinctness the whitish lunules across the disc of the hind-wing. The ♀ accompanying this ♂ (the paired sexes taken near Potchefstroom on 28th February 1872 were sent me by Mr. W. Morant) is generally larger and less rufous in tint than the South-Western examples,—in two individuals

---

[1] As I have pointed out (*Trans. Ent. Soc. Lond.*, 1868, p. 287), the *Lara* of Donovan (*Nat. Repos.*, ii. pl. 71, 1824) has nothing to do with Linné's species; it is a small Satyride, and placed by Westwood and Butler in the genus *Ypthima*, near the Oriental *Y. Lisandra* (Cram.).

from the Transvaal, viz., that above mentioned from near Potchefstroom, and one from the Lydenburg district taken by Mr. T. Ayres, the expanse of wings reaches 1½ inches. If it were not for some intermediate specimens from the Eastern districts, I should have been disposed to treat Stoll's *Gorgias* as distinct; and it is perhaps entitled to rank as a Variety, especially as it exhibits a decided tendency to a constant greater development of the hind-marginal white rings. This is most remarkably shown in an aberrant ♀ taken near Pinetown in 1879 by Colonel Bowker, where each wing presents three perfect white rings with black centres, besides two superior imperfect white rings; the whole upper-side colouring of this specimen is paler and more metallic than usual. An abnormally small ♂, which I took near Cape Town in April 1860, is only 8 lines across the expanded wings.

Specimens of both sexes, taken in Basutoland by Colonel Bowker, are rather darker than usual, and with the under-side markings strongly developed. They are of moderate size, and the wings of the ♂ are rather blunt. In two examples (♂ and ♀) the ocelli of both wings are on the upper side ill defined, the white rings being very imperfect; and in one of them (the ♂) the upper ocellus is wanting in the fore-wing.

I found the butterfly uncommon in Natal. The specimens I took inland were like the Western Cape Colony form, though larger; but the solitary male I met with on the coast had the acute wings and other peculiarities of Stoll's *Gorgias* very pronounced. The latter form of ♂ has, however, been sent me from Estcourt (far inland) by Mr. J. M. Hutchinson.

LARVA.—Pale-green; head, front edge of second (?) segment, and a median dorsal line lake-red; on each side a row of very small black spots. Feeds on *Cotyledon cuneatum*.

PUPA.—Rather darker green than larva; a faint median line of red along hinder half of back of abdomen. Attached head downward to under side of leaf of *Cotyledon cuneatum*.

Larva and pupa described from a drawing of specimens observed near Grahamstown by Mrs. Barber, which is reproduced in Plate II. ff. 1, 1a. (Mrs. Barber gave me the name of the larva's food-plant; and it is interesting to observe in her drawing how closely the green, red-edged colouring of the insects accords with that of the *Cotyledon*.)

This delicately-marked little species is very easily recognised by its peculiar colouring and the conspicuous ocelli at the posterior angles of the wings. It frequents broken rocky ground at the foot or on the ascent of hills, and often occurs at considerable elevations. I have taken it in every month of the year, but it is scarce in the winter months. Its flight is rapid and frequent, but never far from the ground; on the wing it is particularly indistinct. Besides perching on twigs of low plants, it is fond of settling on stones, keeping the wings half open. I have not very often noticed it on flowers.

I have not found any record of *Lara's* occurrence within the tropical parts of the continent, except near Bamangwato (Shoshong), where Mr. H. Barber took an acute-winged ♂ in 1878, and the very remote locality of Shoa in Abyssinia.

Localities of *Hypolycæna Lara*.

I. South Africa.
 B. Cape Colony.
  a. Western Districts.—Cape Town. Malmesbury. Kalk Bay. Stellenbosch. Caledon (*J. H. Merriman*). Robertson.
  b. Eastern Districts.—Port Elizabeth. Grahamstown. King William's Town (*W. S. M. D'Urban*).
  c. Griqualand West.—Klipdrift, Vaal River (*J. H. Bowker*).
  d. Basutoland.—Maseru and Koro-Koro (*J. H. Bowker*).
 C. Orange Free State.—Special locality not noted (*C. Hart*).
 D. Kaffraria Proper.—Butterworth and Bashee River (*J. H. Bowker*).
 E. Natal.
  a. Coast Districts.—D'Urban.
  b. Upper Districts.—Great Noodsberg. Greytown. Estcourt (*J. M. Hutchinson*).
 F. Zululand.—Napoleon Valley (*J. H. Bowker*).
 H. Delagoa Bay.—Lourenço Marques (*Mrs. Monteiro*).
 K. Transvaal.—Potchefstroom and Lydenburg Districts (*W. Morant and T. Ayres*).

II. Other African Regions.
 A. South Tropical.
  b1. Eastern Interior.—Damangwato, Kama's Country (*H. Barber*).
 B. North Tropical.
  b. Eastern Coast.—"Shoa, Abyssinia (*Antinori*)."—Oberthür.

GENUS IOLAUS.

Iolaus, Hübn., Verz. Bek. Schmett., p. 81 (1816); Westw., Gen. Diurn. Lep., ii. p. 480 (1852); Hewits., Illustr. D. Lep., p. 40 (1865); Trim., Rhop. Afr. Aust., ii. p. 222 (1866).

IMAGO.—*Head* of moderate size; *eyes* smooth; *palpi* rather long, slender, separated throughout, densely scaled,—the second joint very long, rising to about level of summit of eyes,—terminal joint slender, of moderate length (rather shorter in ♂); *antennæ* rather thick, of moderate length (longer than in *Myrina*), very gradually clavate.

Thorax of moderate size, clothed with silky down in front, on sides, and at back. *Fore-wings* rather broad; costa rather strongly convex near base; apex usually rather prominent; hind-margin almost straight, or but slightly convex not far below apex; inner margin always somewhat projecting in the ♂ (in some species very strongly convex) before middle, and usually bearing on under side a tuft of long stiff hairs surmounted by a semicircular polished space; costal nervure short, ending a little before middle; subcostal nervures as in *Myrina*; upper radial nervule joined to subcostal nervure at upper end of extremity of discoidal cell; middle and lower disco-cellular nervules of about equal length, slightly curved inwardly,—the lower joining third median nervule about latter's origin. *Hind-wings* rather truncate, but somewhat produced in anal-angular portion; costa convex; hind-margin bearing two slender tails (the longer on submedian nervure, the other on first median nervule), and a dentation (rarely a short third tail) on second median nervule; anal angle more or less prominently lobed; costal nervure not reaching apex, but first subcostal nervule terminating there; in ♂, near base and costa (partly overlapped by inner margin of fore-wings), a circular polished space, large or very large, blackish or greyish, with a paler shining centre.

Legs as in *Myrina*, but very much more slender; the femora not hairy, and the first tarsal joint of the hind pair not swollen.

PUPA. (*I. Silas*, Westw.) Very thick, rounded; the back very convex, with a slight prominence on the thorax and second abdominal segment; posterior region rather suddenly narrowed; anal extremity truncate and slightly expanded. Attached with silk by the tail only, horizontally on the under side of a leaf.

Iolaus, as characterised above, seems a tolerably distinct genus, distinguished from *Myrina* by the wide separation of the radial nervules of the fore-wings at their origins, in addition to its much more slender structure in all respects, and constant possession of more than one tail on each hind-wing. The curious and conspicuous sexual badges in the ♂ appear not to exist in eight Indian and Malayan species catalogued by Hewitson (*loc. cit.*), but are constant in the African species, with the exception of *I. Pallene* (Wallengr.)

To the thirty-five species on record by Messrs. Hewitson and Kirby, I consider that *Myrina Creta*, Hewits., from Congo, and *M. Pallene*, Wallengr., should be added. Nearly all of these lovely butterflies are blue above, intense as a rule in the ♂, but duller and mixed with whitish in the ♀, with the apical part of the fore-wings black, while beneath they are shining-white or yellowish-white, sometimes tinged with grey, with one or two transverse streaks of black, crimson-red, ferruginous-red, or orange-ochreous, usually common to both fore and hind wings.

There are eight known South-African species. The finest is *I. Silas*, which is characterised by the very deep metallic-blue upper side of the ♂, and brilliant white under side, with a single crimson and black line, of both sexes. The rather larger but paler *I. Trimeni* is known only from the Transvaal. *I. Sidus*, a smaller butterfly, is of remarkable beauty, azure-blue above and beneath greyish-white, with two very pronounced crimson streaks. In *I. Bowkeri* both sexes are white-spotted on the disc of the fore-wings above; *I. Mimosæ* is of a rather dull blue, and on the under side is grey; *I. Aphnæoides* has the under side banded with orange-ochreous edged with black; and the aberrant *I. Pallene* is all cream-colour with a black edging, and beneath with two transverse black streaks.

The perfect insects rest on the twigs and leaves of shrubs or small trees, quite in the manner of *Thecla*, taking brief but rapid flights—usually in chase one of another—and occasionally visiting flowers. The only species as yet met with in the western districts of the Cape Colony is *I. Bowkeri*, of which I captured several specimens at Robertson in 1875 and 1876; but in the eastern districts *I. Silas, Sidus, Mimosæ*, and *Aphnæoides* occur,—the last-named being apparently of extreme rarity. *I. Pallene* is a tropical species, and has hitherto not been found in South Africa proper, except in Swaziland.

Of the thirty-seven species recorded, no fewer than eighteen are African; and of the rest, one is Arabian, ten are Indian, seven from the Malayan Archipelago, and one from some unregistered locality.

183. (1.) Iolaus Silas, Westwood.

♀ *Iolaus Silas*, Westw., Gen. Diurn. Lep., ii. p. 481 n. pl. 74, f. 5 [1] (1852).
♂ *Thecla Nega*, Herr.-Schäff., Aussereur. Schmett., ff. 51, 52 (1853 ?).
♂ ♀ *Iolaus Silas*, Trim., Rhop. Afr. Aust., ii. p. 222, n. 128 (1866).

Exp. al., (♂) 1 in. 5–7 lin.; (♀) 1 in. 6–9 lin.

♂ *Metallic pure-blue, without any purplish tinge. Fore-wing:* blue forming a large semicircle; a moderately-wide black border on costa becomes very broad at apex, extending thence rather broadly to anal angle; inner-marginal tuft of hairs ochrey-yellow. *Hind-wing:* a black costal and hind-marginal border, rather variable in width, but becoming very narrow towards anal angle; a more or less complete black streak parallel to it, between discoidal nervule and submedian nervure, interiorly edges two crimson-red spots between submedian and second median nervule; on anal-angular lobe, a crimson, blue-dusted spot; inner margin rather widely dark-grey; tails black, white-tipped, that on submedian white-edged; shining costal circular space black, filling nearly whole of cell. *Cilia* of *fore-wing* grey; of *hind-wing* white. UNDER SIDE.—*Glistening white; a common, transverse, dull, crimson-red line beyond middle. Fore-wing:* line commencing near costa, interrupted on each nervule, not reaching inner margin, often indistinctly marked. *Hind-wing:* streak well marked, straight from costa to second median nervule, where it bends outwardly, forming a spot (corresponding in position with the superior of the two spots on upper side), beyond which it is black, interrupted, and inclining inwards to inner margin; immediately before this part of line, a similar interrupted black one, not reaching beyond submedian; spot on lobe conspicuous, including a black spot; an indistinct greyish clouding between line and hind-margin, which is black-edged towards anal angle.

♀ *Blue pale, faint, inclining to violet*, not metallic, occupying a smaller space, so that the dark margins (which are dull-blackish) are broader, especially in hind-wing. *Fore-wing:* blue much paler on median nervules, occasionally almost white. *Hind-wing:* black streak

---

[1] In this figure the dentation at the extremity of the second median nervule of the hind-wings has been erroneously lengthened into a linear tail.

broader, usually more or less dentate, extending to join apical blackish; instead of two crimson spots, an *orange-ochreous broad band*, formed of from three to five large spots, between second subcostal and submedian.
UNDER SIDE.—Quite like that of ♂.
VAR. A. ♂ and ♀.

♂ Blue much duller, inclining to violaceous; black border narrower, especially at posterior angle of fore-wing, and throughout in hind-wing. *Hind-wing*: upper crimson-red spot obsolete, and the others small and dull.

♀ Blue almost obsolete, only indicated by bluish-grey scaling over a strong dull-fuscous suffusion from base.

Under side in both sexes normal.

Hab.—Delagoa Bay (*Mrs. Monteiro*). In the Collection of Mr. H. Grose Smith.

LARVA.—Dull-green, rather paler laterally; a pale-reddish dorsal median line. About 7 lines in length and 2 lines in width, across middle of back. Anterior extremity blunt and rounded; posterior extremity tapering and terminating bifidly; central portion very thick and convex superiorly.

(Described from a drawing by Captain H. C. Harford of a specimen found near D'Urban, Natal, on 20th September 1868. The figures—No. 8 on Plate I.—are from drawings of King William's Town specimens by Mr. J. P. Mansel Weale.)

PUPA.—Bright-green, paler on under side. Along median line of back a row of five sub-rhomboidal, creamy, ferruginous-edged spots, viz., one apart from the rest and more rounded on posterior part of thorax, and four along abdomen; on each side of abdomen a row of three similar, smaller, rounded spots.[1] Length ½ inch; width (greatest across anterior part of abdomen) ¼ inch.

(Described from two specimens received alive from Miss F. Bowker, of Pembroke, near King William's Town, in February 1873. A third specimen had produced the imago *en route*, and I obtained perfect butterflies from the two pupæ described. The figure on Plate I. is from a drawing by Captain Harford, who noted that the species remained twenty-five days in the pupal state.)

This lovely *Iolaus* haunts wooded places, and, like most of the larger Lycænids, is fond of perching on leaves at the summit of some tall shrub, thence taking short jerky flights, and returning often to the same seat, or to one close to the first. I noticed in Natal that the brilliant white of the under side, which looks so extremely conspicuous in the cabinet, was really protective to the insect when sitting among exceedingly glossy leaves in the full sunshine. Colonel Bowker found that, on the Bashee River, the flowers of mistletoe (*Loranthus*) were the favourite resort of this butterfly. I met with *Silas* but sparingly during my visit to Natal in the months of **February**

---

[1] These spots are variable; in one specimen the first, second, and third on back of abdomen are large and contiguous, while in another they are small and widely separate. In the latter, too, the lateral abdominal spots are altogether wanting.

and March; but it is numerous in some seasons, the late Mr. M. J. M'Ken having sent long series to the South-African Museum. In the Botanic Gardens at D'Urban I saw it most frequently on the leaves of the orange. Colonel Bowker notes its appearing in Kaffraria from November to July, and also met with specimens near D'Urban in August.

Localities of Iolaus Silas.

I. South Africa.
 B. Cape Colony.
  b. Eastern Districts.—" Grahamstown (Pluto's Vale)."—W. S. M. D'Urban. King William's Town (*Miss F. Bowker*). Kei River (*J. H. Bowker*).
 D. Kaffraria Proper.—Bashee River (*J. H. Bowker*).
 E. Natal.
  a. Coast Districts.—D'Urban. " Lower Umkomazi."—J. H. Bowker.
  b. Upper Districts.—Tunjumbili, Tugela River.
 F. "Zululand."—Coll. Brit. Mus.
II. Delagoa Bay.—Lourenço Marques (*Mrs. Monteiro*).

184. (2.) Iolaus Trimeni, Wallengren.

PLATE VII. fig. 4 (♀).

Iolaus Trimeni, Wallgrn. Öfv. K. Vet.-Akad. Förh., 1875, p. 87, n. 29.

Exp. al., (♂) 1 in. 8 lin. ; (♀) 1 in. 9½ lin.

♂ *Pure pale-blue. Fore-wing*: a black border, beginning about middle of costa, extremely broad apically, and thence narrowing to posterior angle, where it is tolerably broad; this border is curved and crenelated with ground-colour on its inner edge; costa from base to middle narrowly fuscous, the edge itself fulvescent close to base. *Hind-wing*: base and costa bordered with black; at middle of costa a rather large rounded black spot extending partly into discoidal cell; hind-margin slenderly, inner margin more widely bordered with black; near anal angle two black spots, of which the inner one is oblong and the outer rounded; some of the nervules black. UNDER SIDE.— *Silvery-white. Fore-wing*: without markings. *Hind-wing*: considerably beyond middle, a very slender transverse black line, interrupted on nervules, and towards anal angle strongly angulated; beyond it a still more slender fulvous line, ending in a rounded fulvous spot, between first and second median nervules ; at anal angle another larger fulvous spot, inferiorly black-bordered, and superiorly blue-edged; hind-margin edged with a thin black line ; tail on first median nervule black tipped with white ; tail on submedian nervure black, edged on both sides with white.

Cilia of fore-wings greyish above, white beneath ; of hind-wings white on both sides.

Forehead white, but fulvous mesially ; palpi white, with black tips;

antennæ black, ringed beneath with white. Thorax black above, here and there tinged with blue; breast and legs white. Abdomen black above, white beneath.

♀ *Rather dull pale-blue; borders fuscous in fore-wing, fuscous-grey in hind-wing. Hind-wing:* a discal transverse interrupted fuscous streak extending from second subcostal nervule to first median nervule; two black spots near anal angle enlarged and with large orange-fulvous centres; a similar brighter spot, inferiorly blue-scaled on anal angular lobe. UNDER SIDE.—*Hind-wing:* anal-angular spot more crimson than fulvous.

The above description of the ♂ is adapted from Wallengren's, as I have not seen an example of that sex.[1] The ♀ agrees with it in all respects except those just noted.

In tint of blue, *I. Trimeni* ♂ appears to resemble *I. Sidus*, to which species, as regards form, Wallengren compares it. In its large size, however, and very feebly-marked under side, *Trimeni* is much nearer to *Silas*, Westw.; but the ♂ appears to want the crimson upper-side spots near the anal angle of the hind-wing, while the ♀'s orange spots are very much less developed; and on the under side the red transverse discal streak is wanting in both wings.

This appears to be a rare species. The single specimen received by Wallengren is noted as having been taken in December on the crest of a mountain on Schoman's Farm near the Vaal River. The ♀ here described and figured was sent to me in 1873 by Mr. H. Barber, who took it in some part of the Transvaal, but made no note of the particular locality.

Localities of *Iolaus Trimeni*.

I. South Africa.
  K. Transvaal.—" Schoman's Farm, near Vaal River (*N. Person*)."—Wallengren. Some locality not noted (*H. Barber*).

185. (3.) Iolaus Sidus, Trimen.

♂ ♀ *Iolaus Sidus*, Trim., Trans. Ent. Soc. Lond., 3rd Ser., ii. p. 176 (1864); and Rhop. Afr. Aust., ii. p. 224, n. 129, pl. 4, ff. 5, 6 (1866).
♀ *Iolaus Sidus*, Hewits., Ill. Diurn. Lep., p. 41, pl. 20, f. 25 (1865).

♂ *Soft pale blue; apical half of fore-wing black;* circular costal patch on hind-wing large, conspicuous, glistening-whitish, broadly ringed with blackish. *Fore-wing:* costa grey, edged with pale-reddish near base, abruptly widening and deepening into apical black before extremity of discoidal cell; blue forming an imperfect semicircle, having for its base the inner margin of wing, but not reaching anal angle, to which the apical black (covering the entire hind-margin) narrowly extends. *Hind-wing:* hind-margin edged with a black line; three tails precisely as in *Silas*, but less twisted; inner-marginal groove dark-grey, clothed and fringed with whitish hairs; lobe of anal angle

---

[1] Though "♀" is printed in Wallengren's description, it is certain, from the mention of the circular costal spot of the hind-wings, that the ♂ was intended.

marked with a pale crimson-red spot, outwardly black dotted, and marked with a few bluish scales; a small, indistinct, similar spot immediately above it. Cilia greyish in fore-wing, white in hind-wing (tinged with yellow near anal angle). UNDER SIDE.—*White, tinged with greyish;* one transverse ferruginous-red streak in fore-wing, two in hind-wing; costa and hind-margin of fore-wing and hind-margin of hind-wing with an orange edging. *Fore-wing:* ferruginous-red streak rather broad, straight from costa beyond cell, in direction of anal angle to second median nervule, where it abruptly terminates; a semicircular patch of grey on inner margin near base, and lying over it a curious row of long black hairs springing from edge of inner margin; near anal angle, above submedian nervure, an elongate blackish mark. *Hind-wing:* first red streak as in fore-wing, from costal nervure (before middle) as far as first median nervule, where it narrows, and is thence a biangulated thin *black* streak to inner margin a little beyond middle; outer red streak not far from and almost parallel to hind-margin, narrowing and becoming almost obsolete after passing discoidal nervure; space between streak and hind-margin greyish, enclosing an additional red spot between second and first median nervules similar to that at anal angle, which is larger than on upper side, merged with the smaller red spot, and outwardly edged with a bluish line; from discoidal nervure the hind-marginal edging is black.

♀ *Violaceous-whitish, suffused with blue from bases. Fore-wing:* border narrower and dusky-blackish. *Hind-wing:* costa with a dusky border, widest at apex; from it run two diminishing dusky streaks along hind-margin, the outer as far as second median nervule, the inner indistinctly to inner margin; spot on anal angle, with that above it (which is large) centred with white; a third, conspicuous, subquadrate, orange, black-dotted spot between second and third median nervule. UNDER SIDE.—Quite like that of ♂, save for the absence of any grey patch on inner margin of fore-wing; and the spots on hind-wing being less conspicuous. (Described from single specimen.)

The ♂ is readily distinguished from ♂ *I. Silas*, Westw., by its smaller size, less glittering blue surface, strongly and doubly streaked under surface, with its orange edging, and more conspicuous circular patch on upper side of hind-wing. The front of the head and bases of pterygodes, which are both black (the former being white striped) in *Silas*, are ferruginous-red in *Sidus*. The fringe of hairs on inner margin of fore-wing, which is yellow in *Silas*, is black in the latter species. The ♀ is very different in the two species.

This very handsome *Iolaus* was discovered by Colonel Bowker in Kaffraria in the year 1862, and in the following year Mr. H. J. Atherstone met with it on the coast of the Bathurst District in Cape Colony. The former noted it as rare and local, occurring from December to April,[1] and took most of his specimens on a Sneezewood tree (*Pteroxylon utile*) at the edge of a forest. I

---

[1] He subsequently wrote that he had observed it throughout the year.

did not find this butterfly during my stay on the Natal Coast in 1867, but a good many examples taken at D'Urban have reached me at different times, and Colonel Bowker has fallen in with it both at Pinetown and on the Lower Umkomazi.

Localities of *Iolaus Silus*.

I. South Africa.
  B. Cape Colony.
    b. Eastern Districts.—Kleinemond River, Bathurst (*H. J. Atherstone*).
  D. Kaffraria Proper.—Bashee River (*J. H. Bowker*).
  E. Natal.
    a. Coast Districts.—D'Urban (*W. Guienzius, M. J. M'Ken, T. Ayres, J. H. Bowker*). Pinetown (*J. H. Bowker*). "Lower Umkomazi."—*J. H. Bowker.*
  F. Zululand.—St. Lucia Bay (*Colonel H. Tower*).

II. Other African Regions.
  A. South Tropical.
    *b*1. Eastern Interior.—"Lake Nyassa."—W. F. Kirby, Cat. Hewits. Coll.

186. (4.) Iolaus Bowkeri, Trimen.

♀ *Iolaus Bowkeri*, Trim., Trans. Ent. Soc. Lond., 3rd Ser., ii. p. 176 (1864); and Rhop. Afr. Aust., ii. p. 225, n. 130, pl. 4, f. 4 (1866).

Exp. al., (♂) 1 in. 2½–4 lin.; (♀) 1 in. 4–5½ lin.

♂ *Very pale blue; apical half of fore-wing and costal border of hind-wing greyish-fuscous, with white spots.* Fore-wing: blue extends from base, not quite reaching costal edge, to extremity of discoidal cell and along inner margin to beyond middle; a fuscous lunule closing cell, sometimes preceded by a more or less distinct small fuscous spot; immediately beyond lunule a small whitish space shot with blue, crossed by radial nervule; a submarginal row of six white spots, of which the first and second are lanceolate, the third minute, the fourth very small, the fifth large and quadrate, and the sixth (between first median nervule and submedian nervure) also quadrate, much the largest of all, more or less suffused with blue, and internally confluent with the ground-colour inferiorly. *Hind-wing*: costal border rather broad, darker at apex; a submarginal row of white spots, of which only the first and second (rarely the third) are distinct, the rest being lost in the blue field; immediately preceding this row, a series of four or five black spots, of which the two lowermost (between radial and second median nervules) are usually very distinct; immediately beyond the row of white spots a lunulate fuscous streak, interrupted on nervules, extending from costal border to submedian nervure; a narrow distinct black edge to hind-margin, bounded internally by a white line; three good-sized hind-marginal black spots, preceded by ill-defined whitish lunules, between second median nervule and anal angle; an indistinct fuscous lunule at extremity of discoidal cell; inner-marginal fold grey, clouded with whitish; tails black, with white tips and edges. *Cilia* of fore-wing fuscous-grey mixed with white, the latter predominating near anal angle; of hind-wing white, with small black marks at extremities of nervules. UNDER SIDE.—*White, with ochreous- and rufous-brown irregular transverse striæ*; in each wing, commencing on costa, a stria before middle, an elongate sub-lunulate mark closing discoidal cell, a broad stria beyond middle, a submarginal linear streak, and a hind-marginal linear edging. *Fore-wing*: close to base, a short broad transverse marking, between sub-costal and submedian nervures; outer transverse stria more irregular and much dentate on both edges. *Hind-wing*: costa at base narrowly edged with brownish-ochreous; inner transverse stria much narrower than in fore-wing; both striæ very acutely angulated on submedian nervure, and thence diverted, the inner quite, the outer very nearly, to base; sub-marginal linear streak adjoins two black spots (the lower larger) close to costa, is there, as well as on submedian nervure, interrupted, and from the latter point diverted, edging inner margin to before middle; black spots at and near anal angle inwardly scaled with bluish-silvery—the middle one often obsolete, and represented by some fuscous and bluish-silvery scales.

♀ Similar; the blue duller and its area more restricted; all the white markings larger and almost wholly free from any blue suffusion, especially in hind-wing. UNDER SIDE.—As in ♂, but the striæ (especially the sub-marginal and hind-marginal linear ones) of a decidedly clearer, more fulvous tint.

Aberration (or *Variety?*).

A ♀ from Springbokfontein, in Little Namaqualand, has in the fore-wing the white mark just beyond disco-cellular lunule obsolete, the two upper white spots of discal row quadrate, and the third and fourth entirely wanting; while in the hind-wing the white markings are even fainter than is usual in the ♂, and the sub-marginal lunulate blackish streak is obsolete. On the under side the principal striæ are greatly widened and of a duller brown; the space between the sub-marginal and hind-marginal streaks is filled with paler-brown mixed with grey, and in the hind-wing the space before middle is similarly but more thinly obscured. (In the South-African Museum, presented by Mr. G. A. Reynolds, who took the specimen at Matje's Kloof in 1873.)[1]

[1] In 1867 I made a note of an apparent aberration of *I. Burcheri* in the remains of the Burchell Collection at the Oxford Museum, which seems closely to resemble that described in the text. In Burchell's MS. list, kindly lent to me by Professor Westwood, I found the locality of this specimen, which was taken in the year 1814, given as "Chue Spring, on Maadje Mountains; lat. 26° 18′ 11″, long. about 24°,"—in the territory now known as Bechuanaland.

Hewitson (*Illust. Diurn. Lep.*, Suppl., p. 11) thought *I. marmoreus* (Butl.), from the White Nile, to be a variety of *I. Bowkeri*; but I have not been able to ascertain if this is the case.

The strongly and intricately marked under side strikingly distinguishes this fine *Iolaus* from its congeners. *I. aphnæoides*, Trim., resembles it in the width of the under-side striæ; but these are regular and even, and disposed quite differently from the corresponding markings in *Bowkeri*.

Colonel Bowker, to whom I have dedicated this very beautiful species, first met with it in Kaffraria in the year 1862, and afterwards took a good many examples in that territory. I have also received a specimen taken in the Albany District of the Cape Colony; and in 1867 I found two or three about stunted *Acacia horrida* near Greytown, Natal. It was not until 1876 that I discovered the butterfly as far to the south and west as Robertson in the Cape Colony; but 1 had received the ♀ *Variety* or *Aberration* from Namaqualand above described as early as 1873. Colonel Bowker's original specimen was taken on a sprig of mistletoe (*Loranthus*); and in October 1864 he noted that other specimens had the habit (which I subsequently observed in Natal and at Robertson) of settling on the dry stems and twigs of acacias and other thorny shrubs. At Robertson I found the few specimens I saw very fond of the yellow flowers of a tall straggling *Senecio* which was growing through a clump of rigid thorny bush. The flight of *I. Bowkeri* is exceedingly short, and, apart from the twigs and thorns about its favourite resting-places, it is very easily captured.

Localities of *Iolaus Bowkeri*.

I. South Africa.
 B. Cape Colony.
  a. Western Districts.—Springbokfontein, Little Namaqualand (*G. A. Reynolds.*—Var. or Aberr.) Robertson.
  b. Eastern Districts.—Rockdale, near Grahamstown (*H. I. Atherstone*).
 D. Kaffraria Proper.—Tsomo and Bashee Rivers (*J. H. Bowker*).
 E. Natal.
  b. Upper Districts.—Greytown.
 L. Bechuanaland.—"Chue Spring" (*W. Burchell*). [Var. or Aberr.]

II. Other African Regions.
 A. South Tropical.
  a. Western Coast.—"Angola (*J. J. Monteiro*)."—Druce. "Kinsembo, Congo (H. Ansell)."—Butler.
  b1. Eastern Interior.—Tauwani River (*F. C. Selous*).

187. (5.) Iolaus Ceres, (Hewitson).

♂ *Myrina Ceres*, Hewits., Ill. D. Lep., p. 39, n. 42, pl. 17, f. 63 (1865).

Exp. al., (♂) "$1\frac{4}{9}$ inch."

♂ "Lilac-blue. Anterior wing with the apex and outer margin broadly dark-brown. Posterior wing with the apex rufous-brown; two slender tails; a small black spot at the base of the outer tail; the apical" [evidently *anal-angular* is meant] "spot gold and carmine bordered below with pale-blue, on the left side with black. UNDER

SIDE.—Grey-white, clouded near the base of both wings with pale rufous-brown; both with a linear band at the end of the cell. Anterior wing with a lunular spot within the cell, a linear zigzag band a little beyond the middle (bordered inwardly with rufous-brown), a sub-marginal indistinct linear band, and the outer margin all carmine. Posterior wing crossed near the base by a curved broken band of five lunular spots, and by a very angular, linear, broken band, all carmine; a sub-marginal band indistinct and carmine from the costal margin to the middle, black above the anal spots; the spots black bordered above with orange, below with pale blue; the outer margin carmine; the apex pale rufous brown."—Hewitson, *loc. cit.*

♀ Pale-blue over discal inner-marginal area of both wings. *Hind-wing*: three fuscous spots near anal angle; hind-margin with a black bounding line edged with whitish on each side; both tails orange at base, the upper (shorter) one orange throughout except its black and white tip. UNDER SIDE.—Redder than in ♀ (judging from Hewitson's figure), especially the striæ themselves; hind-margin in both wings much clouded with brownish.

I have not seen the male of this species, which was described and figured by Hewitson from a Zululand example in the collection of Boisduval. The female above noted is in the Hewitson Collection at the British Museum, and came from Delagoa Bay; it was in very poor condition when I examined it in 1881.

Localities of *Iolaus Ceres*.

I. South Africa.
 F. "Zululand."—Hewitson.
 II. Delagoa Bay.—(Hewitson Collection).

188. (6.) Iolaus Mimosæ, Trimen.

Iolaus Mimosæ, Trim., Trans. Ent. Soc. Lond., 1874, p. 330, pl. ii. ff. 1, 2.

Exp. al., (♂) 1 in. 2⅔ lin.; (♀) 1 in. 3–4⅓ lin.

♂ *Glossy pale-blue, with fuscous-grey borders. Fore-wing*: costa from base rather widely bordered with pale-grey, diminishing to a point about middle; extreme costal margin very narrowly edged with ochreous; apical border very broad, commencing just at extremity of discoidal cell, but somewhat abruptly narrowing on second median nervule, whence the hind-marginal border to anal angle is not wide; sexual tuft on inner margin thin, dark-grey; inner margin rather prominently lobed in basal half. *Hind-wing*: costal and apical margin narrowly, hind-margin very narrowly edged with fuscous-grey; two usual hind-marginal spots black, that on anal angular lobe superiorly edged with greenish-silvery, and partly encircled with dull white; between the two spots a third lunulate one, immediately preceded by a whitish mark, which is itself preceded by a faint fuscous one; inner margin rather widely pale-grey, hoary towards base; sexual badge near base conspicuous, consisting of a fuscous spot in a shining grey ring. UNDER SIDE.—*Soft pale-grey, with thin ferruginous-ochreous striæ;* in both wings a short stria closing discoidal cell, and two long ones beyond middle, of which the inner is continuous and well defined, the outer sub-lunulate and rather faintly marked; *between these two striæ some white suffusion,* in *fore-wing* only towards inner margin, but in *hind-wing* from costa to inner margin. *Hind-wing:* before middle a third stria, irregular and angulated, well defined, extending from precostal to submedian nervure; first stria beyond middle very irregular, almost meeting the outer one between first median nervule and submedian nervure, where it is sharply deflected; a little before its inner-marginal extremity a small detached marking of the same ferruginous-ochreous; between the two hind-marginal spots some rather conspicuous greenish-silvery scaling.

♀ *Very similar to* ♂; *the blue scarcely duller; the fuscous bordering rather darker, and in parts broader or narrower. Fore-wing:* costal grey less pronounced, mixed with fuscous; apical border not so wide, not reaching to extremity of cell; hind-marginal border rather wider, especially at anal angle. *Hind-wing:* costal and apical border darker and considerably broader; a sub-marginal and hind-marginal row of faintly marked fuscous spots, the latter row in line with the usual three black spots, which are more strongly marked than in the ♂. UNDER SIDE.—Quite as in ♂, but slightly duller in tint, inclining to brownish, and with the white clouding beyond middle less distinct.

This species should be placed next to *I. Ceres* (Hewits.) On the under side it differs from that species in being wholly devoid of any rufous tinge or brown basal clouding; in having the transverse striæ beyond middle more regular and closer together; in wanting altogether the conspicuous lunular streak *in* the discoidal cell of the fore-wings; in possessing a continuous transverse stria before the middle of the hind-wings, instead of one broken into six or seven portions; and in wanting the conspicuous orange lunule which adjoins the upper hind-marginal spot of the hind-wings.

Mr. Henry I. Atherstone sent me two females of this butterfly as long ago as the end of 1863, having taken them at Rockdale and New Year's River, near Grahamstown, in August and November of that year. From the circumstance of finding one of them in company with *I. Bowkeri,* mihi, Mr. Atherstone imagined the two to be sexes of one species. In 1865 Mr. J. H. Bowker sent a male from the neighbourhood of the Tsomo River, in Kaffraria Proper, and noted it frequenting *Acacia* trees, and, like *I. Bowkeri,* having the habit of lighting in among the branches and settling on dry twigs, where it was easily taken with the fingers. This is the only ♂ of the insect that I have seen,[1] but three others, ♀ s, have reached me from Mrs.

---

[1] I have since received another ♂, taken in 1884 by Mr. J. M. Hutchinson near Estcourt, Natal. This example differs from Colonel Bowker's in being of a brighter blue above. I have also seen a third, from "Kaffraria," belonging to Mr. H. Grose Smith, and a fourth, taken by Mr. F. C. Selous, a little N. of Bamangwato (River Tauwani).

Barber, one taken near King William's Town by Miss Fanny Bowker in 1869, and the others by Mrs. Barber herself while travelling through the north-eastern portion of the Colony in 1872. Mrs. Barber confirms her brother's account of the habits of *I. Mimosæ*, and adds that both it and *I. Bowkeri* chiefly haunt the mistletoe (*Loranthus* sp.), which so generally infests the mimosa trees.

Mr. J. P. Mansel Weale sent me an excellent drawing of a ♀ that he captured at Cradock in December 1866 on a "thorn tree" (*Acacia horrida*).

Localities of *Iolaus Mimosæ*.

I. South Africa.
   B. Cape Colony.
      b. Eastern Districts.—Grahamstown (*H. I. Atherstone*). King William's Town (*Miss F. Bowker*). Cradock (*J. P. Mansel Weale*).
   D. Caffraria Proper.—Tsomo River (*J. H. Bowker*).
   E. Natal.
      b. Upper Districts.—Estcourt (*J. M. Hutchinson*).
   K. Transvaal.—Limpopo River (*F. C. Selous*).

II. Other African Regions.
   A. South Tropical.
      b1. Eastern Interior.—Tauwani River (*F. C. Selous*).

189. (7.) Iolaus Aphnæoides, Trimen.

♂ ♀ *Iolaus Aphnæoides*, Trim., Trans. Ent. Soc. Lond., 1873, p. 110.
Iolaus Canissus, Hewits., Ent. M. Mag., x. p. 123 (1873).
♀ *Iolaus Aphnæoides*, Hewits., Ill D. Lep., Suppl., pl. iv. *a*, ff. 50, 51 (1878).

Exp. al., (♂) 1 in. 2 lin.; (♀) 1 in. 2½ lin.

♂ *Pale-blue; the fore-wings broadly bordered with blackish. Fore-wing:* blackish border tolerably broad from base along costa, very wide in apical region, and narrowing to anal angle. *Hind-wing:* sexual patch on costa not strongly marked, dull-greyish, glistening; beyond middle, traces of two sub-oblique blackish streaks running to anal angle; on hind-margin, a sharp projecting point at end of second median nervule, and tails at ends of first median nervule and submedian nervure moderately long; on hind-margin a blackish spot on each side of first median nervule, that on the lower side edged with pale-yellowish both anteriorly and posteriorly. UNDER SIDE.—*White, with orange-ochreous, blackish-edged, rather broad, transverse stripes; common to both wings are* (1) a basal stripe, which in *hind-wing* runs parallel to and very near inner margin to a point a little before anal angle; (2) a stripe before middle, which from costa of *fore-wing* extends as far as first median nervule rather beyond middle of *hind-wing*; (3) a stripe about middle, which, after leaving costa of *fore-wing*, is abruptly interrupted from first median nervule as far as inner margin, but in *hind-wing* extends from costa straight to extremity of basal stripe before anal angle; (4) a row of small black spots (six in each wing) a little before hind-margin; and (5) a rather wide, orange-ochreous, hind-marginal edging, becoming obsolete at end of second median nervule. *Fore-wing:* beyond middle, an additional stripe from costa extending in the direction of anal angle, but becoming obsolete just beyond first median nervule. *Hind-wing:* at anal angle a black spot, from which runs a narrow black streak for a little way along inner margin.

♀ *Hind-wing and outer portion of fore-wing white;* in both wings, basal region to a little beyond middle clouded with pale-blue, and nervules clouded with blue and blackish mixed. *Fore-wing:* a rather strongly-marked fuscous streak closing cell. *Hind-wing:* a broad blackish streak corresponding to third transverse stripe of under side; a sub-marginal row of spots corresponding to those of under side, but smaller. UNDER SIDE.—Quite as in ♂.

The forehead is orange-red in both sexes.

Described from a single specimen of each sex, taken on a small tree at the edge of scattered bush about the base of Woest Hill, near Grahamstown, by Mr. James, in October or November. Both examples are considerably worn.

It is very singular that so few examples of this very distinct *Iolaus* should have been recorded, and those few from two such very widely distant localities as Grahamstown and Lake Nyassa. I much fear that the typical examples, of which I made a description in 1870, must have been lost, as they could not be traced in the Grahamstown Museum when I was last there in 1883. It is thus very fortunate that two specimens exist in the Hewitson Collection, and that Mr. Hewitson published figures of the female.

Localities of *Iolaus Aphnæoides*.

I. South Africa.
 B. Cape Colony.
  b. Eastern Districts.—Grahamstown (*E. James*).

II. Other African Regions.
 A. South Tropical.
  b1. Eastern Interior.—"Lake Nyassa (*Thelwall*)."—Hewitson.

190. (8.) Iolaus Pallene, (Wallengren).

Myrina Pallene, Wallengr., K. Sv. Vet.-Akad. Handl., 1857,—Lep. Rhop. Caffr., p. 36.

Exp. al., 1 in. 9–10 lin.

Cream-colour, the hind-wing rather yellower; a linear black hind-marginal edging. Fore-wing: a very small black spot at upper end of extremity of discoidal cell; at apex a rather broad but thinly-suffused dull-brownish patch, ill-defined inwardly, commencing on costa about termination of second subcostal nervule, and not extending along hind-margin below third median nervule; short transverse disco-cellular dusky stria and long transverse dusky discal stria faintly representing the markings of the *under side*. *Hind-wing*: an orange-fulvous ill-defined anal-angular stain, marked exteriorly by one or two small black spots; a median oblique dusky stria from costa towards anal angle, representing part of the *under-side* marking. *Cilia* blackish, slightly mixed with white at posterior angle of fore-wing and apex of hind-wing. UNDER SIDE.—*Tinged with ochre-yellow near bases and margins; black edging and cilia as on upper side*. *Fore-wing*: a very conspicuous short black transverse striola at extremity of discoidal cell; a long black transverse stria beyond middle, from costa almost to submedian nervure, inclining a little outward, interrupted on nervules, and slightly irregular. *Hind-wing*: anal-angular orange-fulvous deeper than on upper side, more suffused inwardly, its outer black spots better marked; an oblique black stria, from costa before middle to third median nervule beyond middle, where it is widely interrupted, but from submedian nervure continued, at an upward angle, almost to inner margin. Tails black.

Head ochre-yellow in front; palpi and antennae black,—the former creamy-white beneath to end of middle joint. Thorax and abdomen fuscous above, ochre-yellow beneath. Legs black, the femora of the middle and hind pair white beneath.

The sexes do not seem to differ in appearance, the male in my possession agreeing very well with Wallengren's description of the female.

This butterfly, which, from its pale-creamy colour and very well-defined jet-black streaks, should be a conspicuous member of its tribe, appears to be exceedingly rare. The late Mr. E. C. Buxton presented me with an injured specimen, one of three which he took in Swaziland; but I have seen no other South-African examples. Two *Pallene* from Lake Nyassa in the Hewitson Collection quite agree with Wallengren's description and with my solitary individual. The general aspect of the insect is very peculiar, and reminds one more of a small Pieride than of a Lycaenide. Mr. Buxton wrote that his three examples were taken on flowers in the month of August.

Localities of *Iolaus Pallene*.

I. South Africa.
   G. Swaziland (*E. C. Buxton*).

II. Other African Regions.
   A. South Tropical.
      *b*1. Eastern Interior.—"Lake Nyassa (*Thelwall*)."—In Hewitson Collection. "Lake Victoria Nyanza."—Butler, August 1883.

Genus MYRINA.

Myrina, Fab., "Illiger's Mag., vi. p. 286 (1807);" Latreille, Encyc.
Meth., ix. pp. 11 and 592 (1819-23) [Sect. C.]
? *Loxura*, Horsfield, "Cat. Lep. Mus. E. Ind. Comp., p. 119 (1829);"
Westwood, Gen. Diurn. Lep. ii. p. 474,—and *Myrina* [part], *op. cit.*,
p. 475 (1852); Trim., Rhop. Afr. Aust., ii. p. 219 (1866).

IMAGO. — *Head* rather broad; *eyes* smooth; *forehead* densely clothed with scales and short hair; *palpi* separate, parallel, very long, densely clothed with scales and with very short closely appressed hairs,—the second joint long, rather thick, only very slightly ascendant, not rising above middle line of eyes,—third joint long (rather longer and stouter in ♀), porrected horizontally or slightly deflected; *antennæ* short, stout, very gradually thickened from base to extremity, which is blunt.

Thorax long, very stout, with long silky hair laterally and posteriorly; pterygodes long and tufted. *Fore-wings:* short, broad, sub-truncate; costa very convex at base, but thence almost straight to apex; hind-margin more or less bluntly prominent about lower radial and third median nervules; subcostal nervure with four branches—the first and second originating considerably apart, far before extremity of discoidal cell, the third and fourth short, diverging not far from apex, at which the latter terminates; radial nervules with a common origin at upper corner of extremity of cell, so that middle disco-cellular nervule is obsolete; lower disco-cellular nervule rather long, curved inwardly, joining third median nervule not far beyond latter's origin; space between first median nervule and submedian nervure unusually broad. *Hind-wings:* rounded and blunted, except at anal angle, which is elongated, prominently lobed, and produced (on submedian nervure) into a long, or very long, broad, twisted tail; costa convex throughout, but especially at base; apex rounded; hind-margin very slightly dentate on nervules; inner margins convex and meeting, so as to conceal all but tip of abdomen beneath; costal nervure terminating at apex; discoidal cell short, truncate, closed by nervules, of which the lower is very slender and oblique, and joins third median nervule at latter's origin. *Fore-legs* of ♂ thick, rather large, scaly,—femur robust, closely hairy beneath,—tibia shorter than femur,—tarsus about two-thirds the length of tibia, obtuse, without apparent articulation, finely spinulose beneath and with a few longer inferior spines at the tip; of the ♀ about the same size and proportions, but the tarsus with distinct articulations and minute terminal claws. *Middle* and *hind legs* stout, densely scaly, the femur densely hairy beneath (especially in ♂), tibia a little shorter than femur, with strong terminal spurs, tarsus longer than tibia, its first joint very long (as long as the other four joints together), and thick (especially in hind-legs), very spinose beneath, broad at tip, with terminal claws short, strong, and widely apart.

Abdomen short, thick, but terminally rather acute.

LARVA.—Broad, very convex dorsally; each segment from second to ninth inclusive with a dorsal hump, most prominent on fourth, eighth, and ninth segments. Lateral margins of body widened so as to completely conceal head and legs from above. Food-plant, species of *Ficus*.

PUPA.—Stout and broad (not unlike a contracted larva in general form), constricted about middle; thorax very bluntly ridged on the back; abdomen very broad and globose. Attached to leaves or bark horizontally, by the tail only.

I follow Mr. W. F. Kirby in restoring the West-African *Silenus*, Fab. (= *Alcides*, Cram.), to its original position as the type of the Fabrician genus *Myrina*, placing with it the closely-allied African species *M. ficedula*, Trim., and *dermaptera*, Wallengr.; but I think it very questionable whether these butterflies can be identified generically with the Indian species (*Atymnus*, Cram., and allies) typical of Horsfield's genus *Loxura*. The latter are not only of much slenderer structure throughout, but have the palpi much thinner and longer, and present a very different neuration in part of the fore-wings, the radial nervules not originating together, but far apart, so that while the upper discocellular nervule is very short, the middle one (separating the radials) is of a good length, indeed as long as the lower one.

The two species found in South Africa are very strongly-made little butterflies, with robust bodies and thick legs and wings, *M. ficedula* more so than *M. dermaptera*. The former is readily distinguished by its larger size and chestnut-red or ferruginous apical and hind-marginal space in the fore-wings; it is also much more widely distributed over South Africa, *M. dermaptera* being only hitherto known from the coast of Natal and Zululand. Though capable of rapid flight for short distances, these *Myrinæ* do not seem to use their wings much, sitting very closely to their favourite perches among the wild and cultivated fig-trees, but occasionally visiting other plants. I have found *M. ficedula* sucking the fruit of the cultivated fig, and also the moisture exuding from wounds on a large kind of Acacia.

191. (1.) Myrina ficedula, Trimen.

♂ *Loxura Alcides*, Boisd., Sp. Gen. Lep., i. pl. 22, f. 3 (1836).
    „   „   Wallengr., K. Sv. Vet.-Akad. Handl., 1857,—Lep. Rhop. Caffr., p. 34.
♂ ♀ „   „   Trim., [part], Rhop. Afr. Aust., ii. p. 219, n. 125 (1866).
♂ ♀ *Myrina ficedula*, Trim., Trans. Ent. Soc. Lond., 1879, p. 324.

Exp. al., (♂) 1 in. $2\frac{1}{2}$–7 lin.; (♀) 1 in. 6–$7\frac{1}{2}$ lin.

Black, with very large basi-discal space of intense metallic ultramarine-blue in both wings; fore-wing with an apical hind-marginal ferruginous patch. Fore-wing: blue occupies inner margin and discoidal cell, but is rather widely bordered with black costally (most widely beyond extremity of discoidal cell), and outwardly from third median nervule to anal angle; ferruginous patch occupying hind-margin from apex to submedian nervure, irregularly convex inwardly, very broad on discoidal nervules, narrowing abruptly at apex, but more gradually to its lower extremity. *Hind-wing:* blue fills entire disc and discoidal cell, and is bordered to a moderate width with black along costa and hind-margin (the apical portion being broadest, and the hind-marginal narrowest); inner-marginal border broadly fuscous up to third median nervule, but blue-scaled along submedian nervure; anal-angular lobe and tail ferruginous, densely grey-scaled; edging base of lobe superiorly, and sometimes extending for a little distance along hind-margin, a fine streak of blue scales; on lobe a spot of unirrorated ferruginous. UNDER SIDE.—*Ferruginous-brown*, in most parts very finely and densely irrorated with yellowish and grey scales; in both wings (better defined in hind-wing) an ochreous-yellow line closing discoidal cell, and a transverse streak of the same colour beyond middle. *Fore-wing:* the streak only extends from near costal edge to between third and second median nervules; costa thinly edged with yellow; a conspicuous cloud of yellow scales occupies costa between streak and apex; inner margin pale-grey, gradually fading into ground-colour superiorly; hind-marginal border darker ferruginous, not (or very thinly) irrorated. *Hind-wing:* darker ferruginous before transverse streak; the streak itself well defined, continuous from costa to below first median nervule, where it is slightly broken and abruptly angulated, and is thence *white* to inner margin and along its edge to base; between this portion of the streak and anal angle, dense greyish-white irroration; a thin greyish line along hind-margin, indistinct superiorly; spot on lobe conspicuous, tinged with crimson.

♀ *Blue much less brilliant ; its area much smaller,* so that its black-bordering is broader, especially in *hind-wing,* where, in apical region, this is wider than inner-marginal fuscous. *Hind-wing:* hind-marginal streak of blue scales more marked, extending to third median nervule or a little farther. UNDER SIDE.—Quite as in ♂.

This *Myrina* is recognised at once from its congener, the West-African *M. Silenus,* Fab. (*Alcides,* Cram.), by (1) *the greatly-enlarged field of blue,* and (2) *the ferruginous bar along hind-margin of fore-wings from apex.* Judging from Cramer's figure (*Pap. Exot.* i. t. 96, D, E), *M. Silenus* has only a slightly paler fascia near the hind-margin of the fore-wings, with no trace of ferruginous; and the under side is generally much darker than in *M. ficedula,* and without yellowish clouding. Boisduval's figure purports to represent *Loxura Alcides* from "Guinée;" it is evidently that of rather a small ♂. In pattern and colouring (only the upper side being depicted) it agrees fairly with the Southern form, except that along hind-margin the ferruginous border has a narrow black-edging, and that the outline of the hind-margin of the fore-wing is not elbowed. If the locality of Boisduval's example be correctly recorded, it would appear that *M. ficedula*, or a very near ally, inhabits Western Africa in company with *M. Silenus*. That this is probably the case is further indicated by some specimens from Ambriz (Congo) in Mrs. Monteiro's collection, which do not differ from the Southern species except that the ♂ s have less ferruginous in the apical hind-marginal area of the fore-wing.[1]

LARVA.—Length, 9 to 11 lin.; greatest width, $3\frac{3}{4}$ lin. Very strongly convex on back, but flattened ventrally; lateral margin throughout produced into a fleshy border, concealing head and legs; each segment from second to ninth (both inclusive) with a dorsal hump, most prominent on fourth, eighth, and ninth segments. Yellowish-green; first and last segments, as well as a more or less interrupted median dorsal stripe, pale-ashy; the stripe irregularly varied with ferruginous-ochreous, slightly narrowing from first segment to dorsal hump (ferruginous-ochreous) on fourth segment, but widening much from sixth segment so as to cover all dorsal portion of seventh and eighth segments except a large and conspicuous pure-white spot on eighth segment terminating in white hump. On ninth segment, dorsal hump pale-ashy, and a good-sized pure-white spot on each lateral margin. All darker parts distinctly speckled with black; the green parts only very thinly so. Head and legs black. Spiracles and pro-legs ferruginous-ochreous. Feeds on *Ficus natalensis* and on the cultivated fig (*F. carica*).—Plate I., fl. 7 (from my own drawings).

PUPA.—Stout and broad, not unlike the contracted larva in general form, constricted about middle; back of thorax bluntly ridged; abdomen very wide and rotund, varying from greenish-brown to dull brownish-ochreous, the wing-covers and under side generally, and a narrow median dorsal stripe dull dark-brown. Frontal region of thorax varied with paler brown, which also forms a broad border on each side of dorsal stripe on abdomen. In some specimens the hinder part of the thorax and the sides of the abdomen are on the back varied with white. Attached by the tail only, but in a horizontal position, to leaves of the food-plant—usually on the under side, or to its twigs. —Plate I., ff. 7*a* (from my own drawings).

This very handsome *Myrina* first makes its appearance at the end of January, and worn individuals occur as late as the end of April. Though very swift when it does take flight, the butterfly is disinclined to move except in chase of other individuals of its species, and is perpetually resettling on the twigs and leaves of its food-plant. In these brief excursions it soon gets worn, and is apt to lose its long tails. It is fond of sucking the ripe figs split open by birds, and when so engaged may with caution be taken by hand.

The larvæ found by me in March 1859 at Knysna (see *Rhop. Afr. Aust.*, ii. p. 220, note), feeding on the cultivated fig, were proved to belong to this

---

[1] I have since seen examples of *M. ficedula* marked "West Africa" in the Collection of the British Museum.

species by the subsequent discovery and rearing to their perfect state of quite similar ones taken on *Ficus Natalensis* by Mrs. Barber near Grahamstown. In 1870 I had the pleasure of visiting Mrs. Barber at Highlands, and gladly took the opportunity of studying and rearing the larvæ, the foodplants of which were growing against the *stoep* of the house. There is no doubt that their peculiar colouring is highly protective, agreeing very thoroughly with that of the terminal green shoots, the bract and occasional withered portions of which are ferruginous; while the conspicuous white spots most completely resemble the drops of milk-like sap that exude from the stems and leaves on the slightest wound. Moreover, both the larva (when in a slightly contracted position) and the pupa bear a very strong resemblance to the small, rough, ashy-varied fruits of their food-plant. I found that those pupæ which were disclosed at large on the plants were much greener and more like the little figs than those which resulted from larvæ kept captive in a dimly-lighted breeding-cage.

M. ficedula does not seem to be often met with in Natal. I took a few in the inland districts, and Colonel Bowker has forwarded two or three from the coast, and two from quite the northern extremity of that Colony.

Localities of *Myrina ficedula*.

1. South Africa.
    B. Cape Colony.
        a. Western Districts.—Knysna. Oudtshoorn (*Adams*).
        b. Eastern Districts.—Grahamstown. "King William's Town." —W. S. M. D'Urban. Fort Warden, Kei River (*J. H. Bowker*).
    D. Kaffraria Proper.—Bashee River (*J. H. Bowker*).
    E. Natal.
        a. Coast Districts.—D'Urban (*J. H. Bowker*). "Lower Umkomazi."—J. H. Bowker. Udlands Mission Station. Great Noodsberg.
        b. Upper Districts.—Biggarsberg (*J. H. Bowker*). Pietermaritzburg (*Windham*).
    K. Transvaal.—Lydenburg District (*T. Ayres*).

II. Other African Regions.
    A. South Tropical.
        a. Western Coast.—Congo: Ambriz (*Mrs. Monteiro*).

192. (2.) Myrina dermaptera, (Wallengren).

♀ *Myrina n. sp.*, Angas, Kafirs. Illustr., pl. xxx. f. 9 (1849).
♀ *Loxura dermaptera*, Wallengr., K. S. Vet.-Akad. Handl., 1857,—Lep. Rhop. Caffr., p. 34.
♀ (as ♂) *Loxura dermaptera*, Hewits., Ill. D. Lep., pl. 4, ff. 3, 4 (1863).
♀ *Loxura dermaptera*, Trim., Rhop. Afr. Aust., ii. p. 220, n. 126 (1866).

Exp. al., (♂) 1 in. 2–3½ lin. ; (♀) 1 in. 2–5¾ lin.

♂ *Black, with very large basi-discal space of intense metallic-blue (with a greenish surface lustre in some lights) in both wings. Fore-wing:* blue occupies inner margin from base to beyond middle, forming a large semicircle, the upper part of which almost fills discoidal cell. *Hindwing:* blue occupies entire disc and discoidal cell, leaving costa and apex broadly bordered, but hind-margin only lineally edged with black; inner-marginal border broadly blackish as far as submedian nervure; longitudinal fold between median and submedian nervures marked with a broad silky black ray; anal-angular lobe marked with a dull-crimson spot speckled with a few bluish scales; tail black, with a white tip and a white central streak, more or less stained with orange at base. *Cilia* greyish, in hind-wing mixed with whitish. UNDER SIDE.—*Soft brownish grey; costa and hind-margin of fore-wing and hind and inner margins of hind-wing edged with ochre-yellow. Fore-wing:* paler about inner-marginal area. *Hind-wing:* dull-crimson spot on anal-angular lobe conspicuous, and accompanied by two similar smaller (sometimes contiguous) spots between it and second median nervule; before these spots, between the same nervule and inner margin, two short, irregular, subangulated transverse blackish lines, interiorly edged with white, of which the outer line is often and the inner occasionally indistinct; tail black, with a yellowish median streak in its basal half.

♀ *Fuscous or fuscous-brown; blue very variable in brightness and extent, and in some examples wholly wanting; when present, always paler and duller than in ♂. Hind-wing:* blue at its greatest development occupying a smaller area than in ♂, so that all the margins are more broadly fuscous.

In both sexes the top and front of head are dark-red mixed with black, and marked with the following white spots, viz., two on forehead, one at base of each antenna, and one on vertex; eyes edged with white; palpi black mixed with red, their middle joint externally white. Breast white and ferruginous mixed; legs ferruginous mixed with black, the femora with white hairs, the tibiæ and tarsi conspicuously barred with white.

From *M. ficedula* this species is readily known by its want on the upper side of the apical hind-marginal ferruginous, and by presenting a brownish-grey instead of a ferruginous-brown under side; it has, too, the tails of the hind-wing considerably narrower and shorter. The two latter characters also separate it from *M. Silenus*, Fab., of Western Africa, to which on the upper side some of the duller ♀ s of *M. dermaptera* bear considerable resemblance. There is in these ♀ s a complete gradation from individuals with the field of blue in the fore-wing quite, and in the hind-wing almost as much developed as in the ♂, to those in which even the few sprinkled scales of blue found in others are totally absent.

PUPA.—Resembling that of *M. ficedula*. Dull-brown, paler along middle of back. Under side, including head and wing covers, dark olivaceous-brown.—Described from a drawing by Mr. (now Captain) H. C. Harford, who wrote in 1869 that he had found the larvæ on a fig-tree near D'Urban.

This curious *Myrina* seems to be extremely local, and but few specimens are seen in collections. During my stay in Natal I saw only one example, a fine ♀, which I captured in the Botanic Garden at D'Urban, on 14th February 1867;—it was flitting about the top of a tall shrub, and settling on the leaves. The late Mr. M. J. M'Ken took several specimens in the same locality, but at long intervals. Colonel Bowker sent three *Dermaptera* taken near D'Urban in August 1873, and in the course of 1879 forwarded in all nine others; but it was not until July 1880 that he discovered a little metropolis of the species at Claremont, near D'Urban, in the shape of a large wild fig-tree.[1] He wrote at that time: "My attention was attracted to a fine specimen sitting with closed wings on the bark of the tree. He was soon boxed, and I then looked round for others. You may guess my surprise at finding them in great numbers, in all degrees of development, from the little twisted-up lump creeping out of the pupa skin to the fully expanded butterfly. I secured about fifty, and an equal number must have got away. They were most numerous at about a foot from the ground, and the pupæ were collected together in the hollows of the bark and suspended to a mass of web. I send some of this web, which you will see is full of imperfect specimens, bits of wings, &c." Colonel Bowker thought that this web was the work of the congregated larvæ of the butterfly, but it appeared to me to be certainly that of a spider; and he himself added that he found a spider in one part of the mass. On the 30th July he further noted that the butterflies were still coming out, but not so numerously, and estimated that over a thousand must have appeared from the same tree between that date and the 15th of the same month.

Localities of *Myrina dermaptera*.

I. South Africa.
   E. Natal.
      a. Coast Districts.—D'Urban.
   F. "Zululand."—Coll. Brit. Mus.

GENUS APHNÆUS.

Aphnæus, Hübn., Verz. Bek. Schmett., p. 81 (1816); Hewits., Illust. Diurn. Lep., p. 60 (1865).
Amblypodia [part], Westw., Gen. Diurn. Lep., ii. p. 477 (1852); Trim., Rhop. Afr. Aust., ii. p. 226 (1866).
Spindasis, Wallengr., Lep. Rhop. Caffr., in K. Sv. Vet.-Akad. Handl., p. 45 (1857).

IMAGO.—*Head* of moderate size; *eyes* smooth; *palpi* moderately long, separated throughout, divergent, ascendent, densely and compactly clothed with scales,—the second joint not rising quite to level of summit of eyes,—terminal joint shorter than in *Iolaus* or *Hypolycæna*, not very slender; *antennæ* of moderate length, rather thick (less so than in *Iolaus*), very gradually incrassated from rather before their middle.

Thorax robust, proportionally more so than in *Iolaus*, well clothed with silky down. *Fore-wings* apically acute and subapically somewhat convex in the ♂, but blunter and sub-truncate in the ♀; costa very slightly arched near base, and thence almost straight; costal nervure

---

[1] Mr. T. Ayres, in the list furnished by him of a collection of South-African insects, mentions his having reared *Dermaptera* from the pupa found near D'Urban, "at the foot of a banyan fig-tree."

very strong, terminating about middle of costa; subcostal nervure four-branched,[1]—the first and second nervules originating (widely apart) considerably before extremity of discoidal cell, third rather nearer apex than to extremity of cell, and fourth terminating at apex; upper radial nervule springing from subcostal nervure a little distance beyond extremity of cell; middle and lower disco-cellular nervules about equal in length, slightly curved; no tuft on inner margin in ♂. *Hind-wings* prominently produced in anal-angular portion; costa rather strongly arched; costal nervure much arched, terminating at a little distance before apex; radial and disco-cellular nervules as in *Hypolycæna*; two rather short, linear tails, respectively on submedian nervure and first median nervule, of which the former is the longer; no badge in ♂. *Fore-legs* of ♂ rather large, scaly,—femur with some thin fine hair beneath; tibia with three or four pairs of rather long spines beneath; tarsus closely spinulose beneath, and not perceptibly articulate, but with pairs of longer spines indicating the articulations;—of the ♀ but a little larger and thicker,—tarsus considerably thicker, distinctly articulate, and with a pair of claws at extremity. *Middle* and *hind legs* stout,—tibiæ with well-developed terminal spurs; tarsi strong, thickly spinulose beneath.

LARVA.—Rather elongate, broad anteriorly and about middle, but narrowing considerably posteriorly; segment next head apparently projecting as a short hood; anal segment produced, and bristly. " On *Convolvulaceæ* (Thwaites)."—F. Moore.

PUPA.—Rather slender, elongate; thorax angulated laterally.

(These characters of the larva and pupa are taken from the figures of those of the Cingalese *A. lazularia* in Moore's *Lepidoptera of Ceylon*, pl. 41, ff. 1c.)

Aphnæus is strictly an Old-World genus, but ranges widely through Africa and Southern Asia, from Sierra Leone (*A. Orcas*, Drury[2]) to the Philippine Islands (*A. Syama*, Hewits.) Two species are Arabian, and one of them (*A. Acamas*, Klug) is recorded from Asia Minor. Of the twenty species known, nine are African and nine from India and the Indo-Malayan Islands, four of the latter inhabiting Ceylon. Seven of the African species are found in Southern Africa, and four of them seem to be peculiar to the sub-region. I exclude from the genus the Syrian and North-African species *Cilissa*, *Zohra*, and *Siphax*,

---

[1] In *A. Hutchinsonii*, Trim. (and I believe in *A. Orcas*, Drury), this nervure has five branches, the third nervule arising rather nearer end of cell, and the fourth and fifth nervules (of which the former ends at apex and the latter a little below it) originating about midway between the origin of the third and the apex.

[2] Hewitson (*loc. cit.*) describes an example of what he believes to be the female of Drury's *Orcas* from the Hope Collection at Oxford. Through the kindness of Professor Westwood, I had the opportunity, in 1867, of examining and noting the characters of this very specimen. It appeared to me to be a ♂, and certainly not identical with *Orcas*,—the under side being ochreous-yellow, with all the silvery markings edged with purplish-ferruginous, while that of *Orcas* is both described and figured by Drury as chocolate-brown generally, without mention or delineation of any edging to the silvery markings.

included in it by Hewitson and Kirby, but which are rightly referable to the genus *Zeritis*.

The *Aphnæi* are on the upper side for the most part of rather dull-brownish colouring varied with ochre-yellow, but several (especially the ♂ s) have a vivid purple or violaceous-blue gloss extending from the bases over a considerable area of the wings. On the under side their beauty is very remarkable, the pale-creamy or yellowish ground-colour being crossed by numerous well-defined bands of orange or ochre-yellow, brown or purple, containing brilliant silvery or very pale-golden stripes. In the *Orcas* group (noted above as having five branches to the subcostal nervure of the fore-wings), the silvery marks, though very largely developed, are broken up into separate spots, the rest of the bands being either merged in the ground-colour or represented by a dark edging to the spots. A tendency in the same direction is exhibited by the under-side pattern of *A. Iza*, Hewits., a small species from the Gaboon.

Of the South-African species, only one—the largest and most beautiful, viz., the newly-discovered *A. Hutchinsonii*, Trim.—belongs to the *Orcas* group; five represent the most numerous group, viz., that of *Etolus*, Cram.; and the remaining one, *A. Pseudo-zeritis*, Trim., has a very distinct aspect, and in the under-side characters shows a resemblance to some species of *Zeritis*.

These butterflies are very swift on the wing, but settle very frequently on twigs and leaves of shrubs, more rarely on flowers, and sometimes on the ground. Of the four species I have seen in nature, I think the very handsome *A. Masilikazi* is the most easily captured, though it usually frequents higher twigs and flowers than the others. This species is most prevalent on the Natal coast, but several of the genus haunt exceedingly dry or almost desert country, notably *A. Namaquus*, Trim.,[1] which I discovered in Little Namaqualand.

193. (1.) **Aphnæus Hutchinsonii**, *sp. nov.*

Exp. al., (♂) 1 in. 6 lin.

♂ *Fuscous, with disco-inner-marginal area from base widely suffused with violaceous-blue in both wings; six conspicuous white spots in sub-apical area of fore-wing.* Fore-wing: blue occupies basal two-thirds of discoidal cell, and all the space between first median nervule and inner margin, except a moderately wide border on hind-margin; immediately beyond extremity of cell a rather large subquadrate pure white spot; a submarginal row of five similar but rather smaller spots, of which the third (between lower radial and third median nervules) is out of line

---

[1] The North-African and Arabian *A. Acamas* and *A. Tamaniba* are also dwellers in desert tracts.

with and beyond the rest, the upper two almost united, and the lower two (between third and first median nervules) a little apart; costa yellow-ochreous at and near base, with a reddish stain. *Hind-wing:* blue occupies entire cellular and discal area, being bounded superiorly by subcostal nervure and its second nervule, inferiorly by submedian nervure, and externally by a hind-marginal border rapidly narrowing inferiorly; lower part of hind-marginal border traversed by an ill-defined ochre-red streak widening on anal-angular projection. *Under-side: dull, pale ochreous-yellow, with numerous (mostly large) silvery-white spots, distinctly outlined with black, and further less regularly edged with ferruginous. Fore-wing:* the following silvery markings, viz., along costal border and superiorly touching its edge, a moderate-sized longitudinally ovate spot close to base,—two large transversely-lying spots (or short bands), one crossing discoidal cell, the other just beyond its extremity,—a smaller and shorter but similarly-shaped spot beyond that last named, and a still smaller one just before apex; two sub-marginal spots, one small and round, between lower radial and third median nervules, the other large and elongate on the median nervules; another spot (the smallest on the wing) in discoidal cell below and slightly beyond basal spot on costa; between first median nervule and submedian nervure, a very large superiorly arched and black-edged but inferiorly and interiorly ill-defined white patch, slightly silvery in its upper part; traces of a ferruginous line very near and parallel to hind-margin, which is narrowly edged with black. *Hind-wing:* the following silvery spots, viz., at base, one in size and shape like that in fore-wing, but placed transversely; before middle, a transverse row of four, of which the first is large and very round, between costal and subcostal nervures,—the second in discoidal cell, also rounded, but very much smaller,—the third very small, and with the fourth (smaller, quadrate, and farther from base) situate on inner margin; and about and beyond middle an irregular series of four very large differently-shaped spots (the central one, just at extremity of cell, the largest and roundest, and the lowest, on inner margin, very elongate and crossed by a black line), and two very small ones, situate respectively below upper and middle large spots; two very small similar spots respectively at the beginning and end of a submarginal ferruginous streak; a blackish spot on anal-angular projection; hind-margin narrowly edged with black.

Head reddish-brown; a conspicuous white spot on vertex and two smaller ones in front; palpi white with black tips; antennæ dark reddish-brown with cream-coloured tips; eyes narrowly edged with white; thorax brown beneath, spotted with white; legs reddish-brown, barred here and with white, and with tufts of white hair on coxæ; abdomen black, with white segmental stripes on the sides and beneath.

(Described from a single specimen in worn condition, in which the tails and part of the anal-angular lobe of the hind-wings are wanting.)

This *Aphnæus* is evidently related to *A. Orcas*, Drury, judging from the figures given on pl. xxxiv. of the third volume of that author's *Illustrations*, but is at once distinguishable by its conspicuous white spots on the upper side in the fore-wing and its yellowish instead of chocolate-brown under side. The large and brilliant silvery spots of the under side (which altogether separate it from any other South-African species known) are arranged much as in *Orcas*, but those in and beyond middle are perfectly separate, instead of being confluent into transverse bands.

The only example I have seen of this exceedingly beautiful butterfly was captured by Mr. J. M. Hutchinson, who liberally presented it to the South-African Museum, and after whom, in recognition of his services to Entomology, I have great pleasure in naming the species. Mr. Hutchinson wrote in February 1886, that he met with only this single specimen, on the summit of a hill near Estcourt, Natal. It was flying round a tree with much rapidity, but settled several times; when captured, it was resting on the upper side of a leaf. A second example, believed by Mr. Hutchinson to be also a ♂, was taken by a Mr. Morrison at a spot about twelve miles distant. Colonel Bowker, who saw both specimens at Estcourt, informs me that Mr. Morrison's was also without tails on the hind-wings. When in perfect condition, this *Aphnæus* must rival in beauty the most brilliant of the family *Lycænidæ*.

Locality of Aphnæus Hutchinsonii.

I. South Africa.
  E. Natal.
    b. Upper Districts.—Estcourt (*J. M. Hutchinson*).

194. (2.) Aphnæus Natalensis, (Westwood).

♀ (?) *Amblypodia Natalensis*, Westw., Gen. Diurn. Lep., ii. p. 479, pl. lxxv. f. 4 [*Aphnæus N.*] (1852).
♂ ♀ *Aphnæus Natalensis*, Hewits., Ill. Diurn. Lep., p. 62, pl. 25, ff. 1, 2 (1865).
♂ ♀ *Aphnæus Caffer*, Trim., Trans. Ent. Soc. Lond., 1868, p. 88, and 1870, p. 368.

Expl. al., 1 in.—1 in. 4 lin.

♂ *Fuscous-brown, with glistening violaceous-blue discs; fore-wing with three transverse stripes of ochre-yellow. Fore-wing:* blue occupies inner-marginal area to beyond middle, but enters discoidal cell only just at base; a short ochre-yellow stripe, crossing discoidal cell towards extremity, from subcostal nervure to base of first median nervule,—another, beyond middle, longer, from near costa rather obliquely to first median nervule or a little below it,—the third one, submarginal, sometimes ends on third median nervule and sometimes joins second stripe just above second median nervule. *Hind-wing:* blue covers all but a wide costal and apical, and a narrow inner-marginal dull-grey border; a very indistinct, darker, rather oblique, short stripe from costa beyond middle, and a similar but longer submarginal one; at anal angle a large, conspicuous orange-yellow spot, marked inferiorly by a silvery-spangled small black spot; a well-marked black edging along hind-margin; tails black, white-tipped, the lower one orange at base. *Cilia* ochre-yellow. UNDER SIDE.—*Pale whitish-yellow, with orange-ochreous or orange, mesially golden-spangled transverse stripes, finely but distinctly black-edged on both sides. Fore-wing:* three stripes from costal edge, viz., first short, quite transverse, before middle to median nervure or a little below it,—second long, oblique, fuscous below first median nervule, extending to submedian nervure not far before posterior angle,—third beyond middle, quite transverse, abruptly terminating (with rounded, black-edged extremity) just below third median nervule; between second and third stripes a costal spot of the same colouring; a similar sub-basal spot in discoidal cell; two submarginal black streaks, of which the inner is thicker and more irregular inferiorly; hind-margin with a very well-defined linear black edging. *Hind-wing:* base and inner margin to beyond middle rather widely bordered with orange-ochreous, edged outwardly by a series of five small black spots; two stripes from costal edge of a brighter orange than those of fore-wing, viz., first very long, oblique, from before middle to below first median nervule, where it is sharply angulated, and whence its black edges only are continued to inner margin,—the second almost parallel from a little before apex to third median nervule, where (like the first at its angulation) it joins an irregular, inferiorly-widened, golden-spangled orange inner submarginal streak terminating at anal angle; outer submarginal black streak and hind-marginal linear edging as in fore-wing; anal-angular black spot larger than on upper side. *Cilia* orange. *Collar* rufous.

♀ Like ♂, but blue in fore-wing occupying more of discoidal cell, and usually more or less obscuring first transverse ochre-yellow stripe; two outer stripes of fore-wing broader, and usually more widely confluent at lower extremities. UNDER SIDE.—Quite as in ♂.

Numerous specimens of both sexes collected in Basutoland by Colonel Bowker, as well as one which I took in Griqualand West, are rather smaller than the Natal examples, and the orange of the under-side stripes, &c., is replaced by pale creamy-ochreous with a slight ferruginous tinge; the tails of the hind-wings are also rather shorter.

I have had great difficulty in deciding whether Westwood's *Natalensis* is the butterfly above described, or the species immediately following. There is no description given in the *Genera of Diurnal Lepidoptera*, and only the upper side of an apparently worn individual (? female) is figured. The prevalent form on the coast of Natal is the species (*Masilikazi* of Wallengren, *Natalensis* of Hopffer and myself) with purple fasciæ on the under side; and in the British Museum Collection this was the species labelled "*Natalensis*," although associated with it was the single Sierra Leone individual mentioned in my *Rhopalocera Africæ Australis* (p. 228) with orange fasciæ. When in 1868 (*loc. cit.*) I described the orange-banded form as a distinct species (*A. caffer*), I found that Hewitson (*op. cit.*) had figured it as *Natalensis*, Westw., giving the upper side of a ♂ and the under side of a ♀. Re-examination of Hewitson's figure of the upper side in the *Genera*, in comparison with a large number of specimens, has led me to conclude—especially in view of the large size of the orange anal-angular marking in the hind-wing, and the small development of the adjacent hind-marginal lunulate whitish streak—that it represents the species with orange-banded underside; but it must be admitted that much uncertainty attends this conclusion. My *A. caffer* being thus sunk in *Natalensis*, Westw., the reputed *Natalensis* (purple-banded beneath) must take Wallengren's name of *Masilikazi*, proposed in 1857.

From the latter, the form under notice is separated on the upper side by its paler, less purplish blue, by the much more constant and well-developed ochreous-yellow stripes of the fore-wing, and by the brighter and wider anal-angular orange of the hind-wing; while on the under side the stripes are orange or ochreous instead of dull-purple; the fore-wing has no spot on costa at base or at origin of inner submarginal streak; and the hind-wing has the base and inner margin orange-ochreous, the band beyond middle inclining outward more than inward, and always joining the inner submarginal streak, which latter is orange instead of dull-purplish.

A. Natalensis is nearer than *A. Masilikazi* to the Indian *A. Etolus*, Cram., having the under-side stripes of the same colour; but both the African species want the hind-marginal and sub-marginal orange stripes of the fore-wing, and the additional sub-basal orange stripe of the hind-wing which *Etolus* displays.

Remarkably diminutive specimens of *Natalensis* occur; one ♂, taken in Natal by Mr. T. Ayres, expands only 10 lines, and a ♀ from the Transvaal country barely 11 lines.

I found this butterfly not uncommon in the upland grassy hills in the interior of Natal.[1] It was swift and active in flight, but settled very frequently on the leaves and flowers of low plants. I captured the paired sexes in the Noodsberg on the 16th March 1867. In Basutoland the slight variety found abundantly by Colonel Bowker was described by him as usually sitting on the ground or on stones, keeping much in pairs, and only flying for a very few yards at a time, so as to be easily captured.

Localities of *Aphnæus Natalensis*.

I. South Africa.
   B. Cape Colony.
      c. Griqualand West.—Kimberley.
      d. Basutoland.—Maseru (*J. H. Bowker*).
   E. Natal.
      b. Upper Districts.—Great Noodsberg. Udland's Mission Station.
   K. Transvaal.—Potchefstroom District (*T. Ayres*).

195. (3.) Aphnæus Masilikazi, (Wallengren).

♂ ♀ *Spindasis Masilikazi*, Wallengr., K. Sv. Vet.-Akad. Handl., 1857, Lep. Rhop. Caffr., p. 45.

Amblypodia Natalensis, Hopff., Peters' Reise n. Mossamb.,—Ins., p. 399 (1862).

♂ ♀ „   „   Trim., Rhop. Afr. Aust., ii. p. 227, n. 131 (1866).

Exp. al., 1 in. 2–7 lin.

♂ *Dull violaceous-blue shot with rich purple, with broad brownish-black borders. Fore-wing:* blue forms a semicircle on inner margin

---
[1] Colonel Bowker has this month (August 1885) sent specimens taken at Malvern, ten miles from D'Urban.

from base to beyond middle, sometimes extending upwards into cell; just beyond end of cell an *orange* spot, sometimes very indistinct; in some specimens a faint trace of a second orange spot a little beyond the other; more rarely these two markings are developed into short stripes (of which the inner is oblique and longer), and in one specimen they are confluent above second median nervule; occasionally a third yellow mark in discoidal cell. *Hind-wing:* blue leaves a tolerably broad costal, hind-marginal and inner-marginal dark border; hind-marginal border containing a bluish lunular line near anal-angular projection, which is *orange-yellow*, marked with two silvery-dotted black spots; *tails* black, orange at base and white at tip. UNDER-SIDE.— *Pale whitish-yellow, with gold-lined, purple-bordered, black-edged transverse fasciæ;* common to both wings, a fascia from costa before middle (confluent on median nervure with a black basal patch), extends across hind-wing to before anal angle, whence it is angulated to about middle of inner margin,—a fascia from costa about middle (becoming dull and gradually obsolete below second median) to third median nervule of hind-wing beyond middle,—two submarginal purplish striæ (of which the inner commences with a costal gold spot) becoming confluent and bright-orange, with a gold streak before anal angle of hind-wing,—and a hind-marginal, black, edging line. *Fore-wing:* at base, two purple spots, and a short transverse streak a little beyond them, *not* gilt; beyond second fascia, a costal spot and a fascia from costa to third median nervule, coloured like other fasciæ. *Hind-wing:* seven spots, of which the larger have gilt centres, in basal area; anal-angular projection of ground-colour, its black spots very conspicuous; *a remarkable longitudinal fold,* clothed with silky pale-brown hairs, runs from base between median and submedian nervures as far as orange colouring before anal angle.

♀ *Paler; blue much duller and paler, without purple gloss. Fore-wing:* orange spots beyond cell larger, sometimes expanded into *transverse bands,* sometimes confluent on median nervules; a short dull-whitish band before them, crossing cell and joining blue. *Hind-wing:* a dusky band from costa crosses a dull-whitish space before hind-marginal border; anal-angular orange pale and dull; lunular marginal line *white.* UNDER SIDE.—As in ♂.

·In this beautiful *Aphnæus,* while the under-side pattern is most constant, the upper side of the fore-wing is remarkably variable, especially in the ♂, as regards the ochre-yellow markings. In a series of ♂s before me there is every gradation, from uniform fuscous beyond the blue area to a development of the yellow markings as full as in the allied *A. Natalensis,* Westw. In the ♀ the markings in question are broader and always to some extent represented, but they are very often incomplete, and sometimes inclining to whitish and suffused.

I have noted under *A. Natalensis* the differences existing between that species and *A. Masilikazi.*

On the coast of Natal this butterfly is by no means rare. I always found it in wooded spots, keeping about rather high bushes, and often settling on their twigs or flowers. It was on the wing in February, March, and April; and Colonel Bowker writes that he has taken it also in July. Though hitherto not recorded from Kaffraria proper, the insect probably occurs there, as Colonel Bowker sent an example from British Kaffraria, and in 1870 Miss M. Barber gave me two ♂ s captured by her as far south as the coast of the Bathurst district of the Cape Colony.

Localities of *Aphnæus Masilikazi*.

I. South Africa.
  B. Cape Colony.
    b. Eastern Districts.—Mouth of Kleinemond River (*Miss M. Barber*). British Kaffraria (*J. H. Bowker*).
    c. Griqualand West.—Vaal River (*J. H. Bowker*).
  E. Natal.
    a. Coast Districts.—D'Urban. Verulam. Pinetown and Mouth of Tugela River (*J. H. Bowker*).
    b. Upper Districts.—Rorke's Drift, Buffalo River (*J. H. Bowker*).
  F. Zululand.—St. Lucia Bay (*the late Colonel H. Tower*).
  K. Transvaal.—Potchefstroom (*T. Ayres*).

II. Other African Regions.
  A. South Tropical.
    a. Western Coast.—Damaraland (*C. J. Andersson*).
    b. Eastern Coast.—"Querimba."—Hopffer.
    b1. Eastern Interior.—Shashani and Makloutze Rivers (*F. C. Selous*). "Near Victoria Falls, Zambesi River (F. Oates)."—Westwood.

196. (4.) Aphnæus Ella, Hewitson.

Aphnæus Ella, Hewits., Ill. Diurn. Lep., p. 63, n. 1c, pl. 25, f. 6 (1865).
Aphnæus Chaka, Wallengr., Öfv. K. Vet.-Akad. Förh., 1875, p. 89, n. 46.

Exp. al., 1 in.—1 in. 3 lin.

♂ *Colouring and general pattern quite like those of A. Natalensis*, Westw. *Fore-wing:* middle ochre-yellow stripe not oblique outwardly, but quite transverse, irregularly dentate on its outer edge, narrowed inferiorly, and sometimes confluent on median nervure (between first and second nervules), with the preceding short ochre-yellow stripe crossing discoidal cell ; outer ochre-yellow stripe completely separated from, being parallel to, middle stripe, very variable in form and width, —its superior extremity very narrow and often indistinct ; its middle portion, on third median nervule, partly or sometimes wholly separated by the fuscous-brown ground-colour. *Hind-wing:* two indistinct oblique discal darker stripes, and a submarginal one ; bluish-white streak close along lower half of hind-margin well developed ; no orange at anal angle, but tails of that colour at their base. *Cilia* whitish mixed with brown in fore-wing. UNDER SIDE.—The stripes not orange or orange-ochreous, as in *Natalensis*, but *of a darker purple than in Masilikazi, and differing in form and arrangement from those of both* the species named. *Fore-wing*: two basal and two sub-basal small darkpurple spots, two on costa and the other two in discoidal cell; first stripe from costa as usual; second, about middle, wholly different, being transverse instead of outwardly oblique, and not longer than the first, terminating abruptly (with a dark-purple edging) just below median nervure; costal spot as usual; third stripe from costa very much longer, roughly parallel with hind-margin as far as submedian nervure, irregular, in some specimens even interrupted on third median nervule; inner submarginal streak commencing with a costal spot and widening into a similar spot on third median nervule, when it almost touches a projection of the third stripe; outer streak linear, close to hind-margin. *Hind-wing*: sub-basal row of spots as in *Masilikazi*, but the spots larger; three smaller basal spots better marked; long central oblique angulated stripe having same direction as usual, but irregular and more or less interrupted on second subcostal and third median nervules; second stripe in usual position, but curved inwardly, and touching, or even occasionally confluent with, first stripe at the latter's points of interruption; inner submarginal streak replaced by a stripe similar to the others, which (as in *Etolus*, Cram.) is widened about the middle, and there meets both the extremity of the second stripe and the angulation of the first, and from the latter point is itself narrowed and angulated to inner margin; outer submarginal streak obsolete except near apex and anal angle; at the latter two black spots, of which the upper one is silvery-spangled; immediately bordering outer edge of lower part of submarginal stripe some bright-orange scaling.

♀ *Blue much paler, without purple lustre. Fore-wing*: first short ochre-yellow stripe much reduced, second and third more even and wider than in ♂. *Hind-wing*: inclining to whitish on upper part of disc, rendering the dusky stripes less indistinct; submarginal bluish streak whiter and broader. UNDER SIDE.—As in ♂.

The specimen, of which the under side is figured by Hewitson, has all the stripes of a faded greyish ochre-yellow instead of dark-purple; he gives Natal as its locality. Two similar ♂s (one with the stripes rather darker) were sent to me by Colonel Bowker from Griqualand West.

This species, most like *Natalensis* on the upper side, is nearer to *Masilikazi* on the under side, but may at once be distinguished from both by the shortness of the second and length of the third stripes in the fore-wings (just the reverse of the corresponding markings in those species, and indeed in the other South-African *Aphnæi*), and by the irregularity and crowding together of the stripes of the hind-wing.

This is by no means a common species, the Transvaal territory seeming to be its metropolis. It occurs as far north as the Makalaka country.

Mr. W. Morant captured specimens near Potchefstroom at the end of February, and Mr. F. C. Selous in the North-West Transvaal in February and March. Wallengren's full description is taken from a single individual captured in the Transvaal by Mr. N. Person. Mr. H. L. Feltham notes it as occurring in plenty at Barkly West (= Klipdrift) at about the end of December 1885; and four specimens have been presented by him to the South-African Museum.

Localities of *Aphnæus Ella*.

I. South Africa.
  B. Cape Colony.
    c. Griqualand West.—Klipdrift, Vaal River (*J. H. Bowker* and *H. L. Feltham*).
  E. "Natal."—Hewitson.
  K. Transvaal.—Potchefstroom (*W. Morant* and *T. Ayres*). Marico and Limpopo Rivers (*F. C. Selous*).

II. Other African Regions.
  A. South Tropical.
    *b*1. Eastern Interior.—Makloutze and Tati Rivers (*F. C. Selous*).

197. (5.) Aphnæus Phanes, Trimen.

♂ ♀ *Aphnæus Phanes*, Trim., Trans. Ent. Soc. Lond., 1873, p. 111, pl. i. ff. 4, 5.

Exp. al., (♂) 1 in. 1½ lin. ; (♀) 1 in. 3½–5 lin.

♂ *Fuscous, shot with rich purple; fore-wing with yellow-ochreous markings. Fore-wing:* the purple gloss covers inner-marginal region, but extends upward only as far as median nervure and its second nervule; costa rather broadly marked with dull ochreous at and near base; an almost square marking in discoidal cell near extremity; beyond cell, near costa, an irregular roughly V-shaped marking; a narrow, submarginal, irregular stripe commencing close to apex, more or less distinctly interrupted on third median nervule, and ending on first median nervule. *Hind-wing:* purple gloss does not extend above subcostal or below submedian nervure, but covers the space between those nervures from base to hind-marginal edge; a hind-marginal whitish streak from radial to anal angle; on either side of submedian nervure a hind-marginal black spot dotted with silvery; just before the space between these two spots, an indistinct yellow-ochreous mark; tails black, ochreous at base and white at tip. *Cilia* white. UNDER SIDE.— *Metallic silvery-white, with dull ochreous, mesially silvery-streaked, narrowly black-edged, broad transverse bands. Fore-wing:* base narrowly suffused with yellow-ochreous, inner margin widely with pale grey; three transverse bands commence on costal edge; the first, before middle, short, straight, ending a little below median nervure; the second, long, oblique, commencing about middle, extending in direction of anal angle, but becoming obsolete above submedian nervure; the third, near apex, rather short, ending abruptly between third and second median nervules, where it touches the second band; between second and third bands a costal spot of the same colouring; a fourth band (not marked with silvery) occupies hind-margin, and is traversed by an interrupted black line, which is inwardly bordered by white sub-lunulate marks, both line and marks being strongest at submedian nervure where they abruptly terminate. *Hind-wing:* a basal and inner-marginal band, irregularly dentate on its edges, leaving a very narrow inner-marginal edging of white; a second band, continuous of first band of fore-wing, crossing obliquely from costa about middle to a little distance before anal-angular lobe, where it narrows and coalesces with extremity of inner-marginal band; a third band, running parallel to the second, is confluent at apex and about middle of hind-margin, with a hind-marginal band similar to that of fore-wing, and near anal angle with the second band; no black traversing line in hind-marginal band, but the white markings more continuous than in fore-wing.

♀ *Without purple gloss; yellow-ochreous markings in both wings;* basal region of wings irrorated with light-bluish scales. *Fore-wing:* all the markings much more developed than in ♂; the quadrate cellular marking more or less indistinctly produced into an oblique band, which, on submedian nervure beyond middle, meets the extremity of the variable but uninterrupted submarginal stripe; the much-enlarged V-shaped subcostal marking beyond middle is prolonged to join submarginal stripe on second median nervule (in two specimens it is confluent with the stripe). *Hind-wing:* inner-marginal region from base clothed with light-bluish hairs; a narrow submarginal yellow-ochreous stripe, commencing indistinctly about first subcostal nervule, is joined between third and second median nervules by an oblique wider stripe of the same colour, commencing on first subcostal nervule about middle. UNDER SIDE.—As in ♂, but the transverse bands paler, and inclining to an orange-ochreous tint; whitish lunulate marks in hind-marginal border much enlarged and suffused. *Fore-wing:* extremity of third band does not meet the second band. *Hind-wing:* the inner-marginal band is externally more irregularly dentate than in ♂, a small portion at origin of first median nervule forming a separate spot.

This *Aphnæus* is distinguished from all its congeners except *A. Namaquus* by *the silvery-white ground-colour of the under side* of the wings. The arrangement of the bands of the under side comes nearest to that presented by *A. Natalensis,* Westw. On the upper side the ♂ is further remarkable for its rich purple gloss and want of discal blue, and the ♀ for the great development of the yellow-ochreous markings, particularly in the hind-wings.

Mr. J. H. Bowker sent me a ♀ of this beautiful species from the Vaal River, Griqualand West, in July 1871, and has since forwarded three ♂ s and three ♀ s from Klipdrift, on the same river. He notes the habits of the insect as closely resembling those of *A. Natalensis.*

On the 18th September 1872 I captured a ♂ at Klipdrift; it was settling on a high bush.

I have not seen any other examples;[1] and the only record I have met

---
[1] Mr. H. L. Feltham has lately (March 1886) presented three specimens to the South-African Museum, which were taken at Barkly (Klipdrift). He describes the butterfly as very plentiful in that locality about the end of December 1885.

with of the occurrence of the species elsewhere is by Mr P. Aurivillius (*Öfv. K. Vet.-Akad. Förh.*, 1879, p 44), who notes a single ♂ taken by G. de Vylder in Damaraland on February 4th, and a ♀ in the Stockholm Museum, found by Wahlberg in "Kaffraria."

Localities of *Aphnæus Phanes*.

I. South Africa.
 B. Cape Colony.
  c. Griqualand West.—Klipdrift, Vaal River (*J. H. Bowker* and *H. L. Feltham*).
II. Other African Regions.
 A. South Tropical.
  a. West Coast.—"Damaraland (*De Vylder*)."—Aurivillius.

198. (6.) Aphnæus Namaquus, Trimen.

♂ ♀ *Aphnæus namaquus*, Trim., Trans. Ent. Soc. Lond., 1874, p. 334, pl. ii. ff. 5, 6.

Exp. al., (♂) 1 in. 2–3 lin.; (♀) 1 in. 4½ lin.

♂ *Fuscous, with a vivid purple gloss. Fore-wing*: three rather narrow yellow-ochreous markings, viz., a short, quadrate, cellular one, adjoining median nervure between origins of first and second nervules, an elongate, curved, irregular streak beyond middle, from close to costa, as far as first median nervule, and a small sub-lunulate spot near apex, between upper radial and third median nervules; the purple gloss covering inner-marginal region from base to hind-margin, but not rising above median nervure or its second nervule. *Hind-wing*: purple extends from base to hind-margin, between subcostal nervure and its first nervule and submedian nervure; on anal-angular lobe two blackish dots, scaled and ringed with silvery and whitish. *Cilia* white. UNDER SIDE.—*Silvery-white, with broad, dull ochreous-brown, centrally silver-streaked, black-edged bars and spots. Fore-wing*: a quadrate spot at base, leaving costal edge and median nervure very narrowly silvery-white; before middle, a short bar from costa to submedian nervure; from about middle of costa a long oblique bar extending towards anal angle, but becoming obsolete just above submedian nervure; a round spot on costa beyond middle; from costa, near apex, a much-curved bar extending to between third and second median nervules, where it abruptly and bluntly ends; a hind-marginal border of the same colour as the bars, commencing at apex and becoming obsolete about sub-median nervure, interiorly black-edged and rather sharply dentated, mesially traversed by an interrupted black line edged with whitish on both sides. *Hind-wing*: a basal and inner-marginal submacular bar, externally presenting five blunt projections, extending to about middle; an irregular, oblique central bar (continuous of first bar of fore-wing) extending to a little before anal angle, where it is angulated backward to inner margin by an elongate usually separate portion; a large costal spot just before apex, in contact or confluent with a very strongly angulated bar, which, commencing just below apex, *touches or is confluent with the central bar* just beyond extremity of discoidal cell, and between third and second median nervules becomes confluent with a hind marginal border similar to that of fore-wing; the traversing white-edged line of the hind-marginal border is more continuous than in fore-wing; black dots on anal-angular lobe more conspicuous than on the upper side.

♀ *Without purple gloss, only presenting a slight violaceous suffusion from bases; yellow-ochreous markings strongly developed in both wings.* *Fore-wing:* the yellow-ochreous markings much enlarged and prolonged inferiorly, so that the first and second are widely confluent below first median nervule, and the second and third narrowly so between third and second median nervules; a faint yellow mark in cell, near base. *Hind-wing:* the position of the silvery-white parts of the under side is roughly indicated by suffused markings of *yellow-ochreous*, viz., one in cell near base, one on costa near apex, one on hind-margin below apex, and one (largest) on median nervules; a whitish line close and parallel to hind-margin scaled with silvery on anal-angular lobe. UNDER SIDE.—As in ♂, but the bars and spots proportionally narrower, leaving more of the silvery-white ground-colour unoccupied.

(Described from ten ♂ and one ♀ specimens.)[1]

This *Aphnaeus* is a near ally of *A. Phanes*, mihi (*Trans. Ent. Soc.*, 1873, p. 111, pl. i. figs. 4, 5), resembling the latter particularly in the silvery-white ground colour of the under side, and the great development of the yellow-ochreous bands on the upper side of the ♀. The chief difference of importance is presented by the *under side of the hind-wing*, in which, instead of being rather even and almost parallel, the oblique bars are irregular and almost submacular, *and the outer one is so strongly angulated as to be confluent with the inner one* near the end of the discoidal cell. This arrangement breaks the silvery ground-colour beyond the middle into three irregular markings, and gives the under side an appearance quite different from that of other *Aphnaei*. Other distinctions from *A. Phanes* are (in the ♂) the very undeveloped state of the ochreous fore-wing upper-side markings, which in one example are very small and dull, and in another all but obsolete; and (in the ♀) the different arrangement of the hind-wing upper-side markings, which in both species follow or correspond with the silvery-white portions of the under side. In both sexes, *the very dark colouring of the spots and bars of the under side* is a marked distinguishing feature.

---

[1] Four ♀ examples, taken in Namaqualand by Mr. Péringuey during November 1885, closely resemble the one here described, only varying in the development of the outermost yellow bar, which in two of them is in both wings much narrowed and interrupted.

I first met with this species under a thorn-tree (known as the "One Tree," in a wide expanse of country) a few miles from Annenous, on the line of railway laid down by the Cape Copper Mining Company, and afterwards on the road between Elboogfontein and Kockfontein; near the Komaggas Mission Station; and at Oograbies; but it was numerous at the latter place only. It has the short active flight of its congeners, but is less wary when settled. It usually rests on the bare twigs of some low shrub, with its head downward, and when disturbed will sometimes return to the same perch. The ♂ has a very dark, almost black appearance on the wing; but the only ♀ I met with had in flight more the look of *A. Natalensis*, Westw.

Localities of *Aphnæus Namaquus*.

I. South Africa.
 B. Cape Colony.
  a. Western Districts.—Annenous, Oograbies, &c., District of Namaqualand. Garries, Spectakel, Ookiep, and Klipfontein, District of Namaqualand (*L. Péringuey*).

199. (7.) Aphnæus pseudo-zeritis, Trimen.

♂ *Aphnæus pseudo-zeritis*, Trim., Trans. Ent. Soc. Lond., 1873, p. 113, pl. i. f. 6.

Exp. al., 11 lin.

♂ *Fuscous, glossed with metallic blue. Fore-wing :* blue forms a patch on inner margin, rising only very little above first median nervule, and not entering discoidal cell or extending much beyond middle. *Hind-wing :* blue occupies greater part of wing from base to hind-margin, leaving a broad costal and narrower inner-marginal fuscous border; a good-sized fulvous-ochreous spot at anal angle, marked exteriorly by two black dots, and interiorly by a few minute brassy scales. UNDER SIDE.—*Dull, pale greyish-ochreous, with sub-quadrate, darker spots centred with glittering brassy scales. Fore-wing :* five spots in discoidal cell, irregularly placed, the elongate one at extremity and the spot nearest to it being considerably larger than the other three; a row of three minute brassy dots along costal edge near base; beyond the latter, four small elongate spots form a curved row, near costal edge, from second cellular spot to a little beyond middle; an irregular transverse row of about six spots beyond middle, of which the first three are distinct and well-separated, but the lower ones con-fluent, larger, and indistinct; two parallel rows of six spots each along hind-margin, the outer row on hind-marginal edge, and consisting of smaller spots than those of the inner row; a short, oblique, apical streak is formed by the confluence of the first spots of the two rows; cellular region and costa adjacent suffused with fulvous-yellow from base; between median nervure and inner margin an irregular, elongate, fuscous marking. *Hind-wing :* spots arranged similarly to those in fore-wing, but more suffused and indistinct, especially in basal region and beyond middle towards lower part of hind-margin, where there is a clouding of brownish-fuscous; anal-angular fulvous-ochreous spot smaller than on upper side, and edged inferiorly with black.

This curious little species appears to constitute a passage between the genera *Aphnæus* and *Zeritis*; the upper side of the wings, with a blue gloss and fulvous anal-angular spot (but without the usual yellow-ochreous markings), resembling that prevalent in the former genus, while the metallic-centred spots of the under side are so similar in arrangement and appearance to the characteristic spotting of *Zeritis* (and particularly to that of the little *Z. Phosphor*, mihi) that, until I detected the remains of a second tail on each hind-wing, I was strongly disposed to place the butterfly in the latter genus.

I am indebted to Miss Fanny Bowker, of Pembroke, near King William's Town, for the first specimen that I have seen; it was taken by her on a low shrub (a species of *Euclea*), on the border of a wood at Tharfield, in the Division of Bathurst.[1]

A second example occurred in a collection formed by Mr. J. M. Hutchinson, shown to me in the year 1881; he informed me that he captured it on the Bushman River in Natal.

Localities of *Aphnæus pseudo-zeritis*.

I. South Africa.
   B. Cape Colony.
      b. Eastern Districts. — Tharfield, Bathurst District (*Miss F. Bowker*).
   E. Natal.
      b. Upper Districts.—Bushman River (*J. M. Hutchinson*).

GENUS CHRYSORYCHIA.

Chrysorychia, Wallengr., Kongl. Svensk. Vetensk.-Akad. Handl., 1857,— Lep. Rhop. Caffr., p. 44.
Axiocerses, Hübn., Verz. Bek. Schmett., p. 71 (1816).
Zeritis [part], Trim., Rhop. Afr. Aust., ii. p. 261 (1866).

IMAGO.—*Head* rather small, rather roughly hairy in front; *eyes* smooth; *palpi* short,—second joint rather roughly hirsute and scaly, especially towards extremity,—terminal joint short, scaly, moderately slender, acuminate, obliquely ascendant; *antennæ* rather long, moderately thick, with an elongated, cylindrical, but very pronounced club.

Thorax robust, clothed above frontally, laterally, and posteriorly, with close silky hair, and beneath with dense woolly hair. *Forewings* rather produced apically and elbowed hind-marginally, especially in ♂; costa slightly hollowed about middle; costal nervure ending about middle; subcostal nervure with only three nervules,— the first arising midway between base and extremity of discoidal cell,—the second about midway between the first and extremity of cell,—the third at a little beyond extremity of cell (having a com-

---

[1] About the same bush were several *Zeritis Chrysaor*, Trimen; and Mrs. Barber informs me that *Ebenaceæ*, of the genus *Euclea*, are the plants most frequented by the species of *Zeritis* in the eastern districts of the Colony.

mon origin with upper radial nervule) and terminating at apex; upper disco-cellular nervule much shorter than lower one. *Hind-wings* narrowly somewhat produced at anal angle, which is prominently but bluntly lobed inferiorly; hind-margin denticulate, and bearing at extremity of submedian nervure a short or moderately long rather hirsute tail, wider at its base; costal nervure basally strongly arched, terminating at apex; subcostal nervure branched a little before middle. *Fore-legs* of ♂ very robust,—femur and tibia about equal in length, both densely hairy (especially the latter, which has a terminal external spur, and two internal terminal spines),—tarsus thick, scaly, with three pairs of strong spinules beneath, several lateral spinules, and a rather long terminal slightly-curved claw;—of ♀ less hirsute,—the tarsus longer, complete, with short curved terminal claws. *Middle* and *hind legs* very robust,—femora and tibiæ moderately hirsute (the latter much shorter and with short terminal spurs),—tarsi long and thick (especially first joint), strongly spinulose beneath, with terminal claws short.

This genus, which includes only the *P. Harpax*, Fab., *Perion*, Cram., and *Amanga*, Westw., is nearly related to *Deudorix*, *Capys*, and *Zeritis*, but differs from all in having only three branches to the subcostal nervure of the fore-wings. In colouring and pattern, especially on the under side, these butterflies nearly approach *Zeritis*, but are further structurally distinguished from that genus by their longer, more slender, distinctly clavate antennæ; shorter palpi; more hirsute head, body, and legs; and prominently lobate anal angle of hind-wings.

Hübner gave the generic name of *Axiocerses* to Cramer's *Perion*, a West-African species; but I give preference to *Chrysorychia*, seeing that Wallengren was the first to diagnose the group, taking his *Tjoane* (= *Harpax*, Fab.) as the type.

All the three species are known to inhabit Tropical Africa, and *Perion*, Cram., has not been found out of that region. *C. Harpax* has the widest distribution, ranging from the North-Tropical West Coast to the eastern districts of Cape Colony; while *C. Amanga* appears to be proper to the South-Tropical belt, but has occurred in Natal. Both the South-African species are orange-red above with dark-brown borders, but *Harpax* is darker in tint and has blackish spots; while beneath *Amanga* presents only a few silvery markings near the base; in contrast to the numerous golden or brassy spots of *Harpax*.

200. (1.) Chrysorychia Harpax, (Fabricius).

♀ *Papilio Harpax*, Fab., Syst. Ent., App., p. 829, n. 327–328 (1775).
♂ *Chrysorychia Tjoane*, ♀ *Chr. Thyra*, Wallengr., K. Sv. Vet.-Akad. Handl., 1857,—Lep. Rhop. Caffr., p. 44.[1]

---

[1] I pointed out that Wallengren's *Thyra* was quite distinct from Linné's insect, and suggested (*Trans. Ent. Soc. Lond.*, 1870, p. 372) that it might be a well-marked ♀ *Z. Chrysaor*, Trim.; but Aurivillius (*K. Sc. Vet.-Akad. Handl.*, 1882, p. 117) has satisfactorily determined it to be the ♀ of *Tjoane*, Wallengr.

♂ ♀ *Zeritis Crœsus*, Trim., Trans. Ent. Soc. Lond., 3rd Ser., i. p. 283 (1862).
♂ ♀ *Chrysophanus Perion*, Hopff., Peters' Reise n. Mossamb.,—Ins., p. 403, pl. xxvi. ff. 1-3 (1862).
♂ ♀ *Zeritis Perion*, Trim. [part], Rhop. Afr. Aust., ii. p. 267, n. 166 (1866).

Exp. al., (♂) 1 in.—1 in 3 lin. ; (♀) 1 in. 2-3 lin.

♂ Red, inclining to orange, not brilliant ; with shining brown-blackish borders. *Fore-wing*: base suffused with brown-blackish ; border wide along costa and hind-margin, and very broad in apical region (almost reaching extremity of discoidal cell) ; two spots in cell, and a quadrate one closing it, all touching costal border, and dark-brown ; two similar spots between second and first median nervules, one just beneath extremity of cell, the other more or less incorporate with hind-marginal border (a spot above, and rarely another below, this latter spot, touching it, sometimes separable from border) ; cilia narrow, white, interrupted with brown. *Hind-wing*: costa from base widely, hind-margin very narrowly, bordered ; a fuscous streak closing cell ; a sub-marginal row of blackish lunular marks, more or less distinct ;[1] *anal angle prominently lobed*, ferruginous-red, marked with a gilded dot, *bearing a short, slightly twisted, acute tail* of the same hue, white-tipped, on submedian nervure. *Cilia* mingled greyish and ferruginous. UNDER SIDE. — *Fore-wing*: dull orange-yellow, very pale on inner margin ; border varying from cinereous to ferruginous-brown ; spots in and bordering discoidal cell, transverse row of spots beyond middle, and row of four dots on costa, all with large, brilliant, greenish-golden centres ; below median nervure, a large, dull-black, whitish-centred, often gold-dotted spot ; a whitish, gold-dusted streak along bend of costal edge at base ; a sub-marginal row of golden dots, sometimes indistinct. *Hind-wing*: varies in tint like border of fore-wing ; three transverse rows of small golden spots, some of which are indistinct, the middle row including a golden streak closing cell ; a dark-brown transverse shade near hind-margin, ending on inner margin with a golden streak, and often marked externally with some indistinct golden dots ; a pale hind-marginal edging becomes golden near anal angle.

♀ Dull orange-yellow ; brown borders paler, narrower. *Fore-wing*: base more widely suffused with brown, mingled with ochreous ; apical border not half as wide as in ♂ ; spots in and about cell larger ; beyond middle, a zigzag row of small, quadrate, blackish spots across wing. *Hind-wing*: base dark-brown to extremity of discoidal cell ; costa broadly bordered ; hind-margin edged with a brown line ; sub-marginal lunular row well marked, the lunules contiguous ; a more or less incomplete row of small spots a little beyond middle, like that in fore-wing. UNDER SIDE.—Quite like that of ♂ ; *rather paler ;* spots larger, the gilding slightly paler but not less brilliant.

[1] In three Natalian specimens this marking is altogether wanting.

Antennæ, in both sexes, marked beneath with a conspicuous broad white bar, just at the base of the club.

There is considerable variety in the depth of colouring of the under side, especially in the ♂, the ferruginous in some being much paler and duller, and the metallic spots much reduced and mostly indistinct. The upper side of the ♀ varies much in the extent of the basal fuscous suffusion and the completeness of the discal row of spots in the hind-wing.

Hopffer (*loc. cit.*) points out that the Mozambique specimens are smaller and of a duller red, but have more brilliant metallic spots (especially in the hind-wings) than those from the Cape. His figures represent longer and straighter tails on the hind-wings than I have seen in any South-African examples.[1] Four ♂s and a ♀ from Sierra Leone, in the Hope Museum at Oxford (1867) also differed from the South-African specimens in their longer tails and more brilliant under-side spots, and the ♀ had the hind-wings uniformly fulvous to the hind-margin.

Though a near ally of the West-African *Perion*, Cram. (*Pap. Exot.*, t. ccclxxix, B, C), with which both Hopffer and myself associated it, this butterfly is really quite distinct, presenting a much less regular transverse series of spots on both surfaces, and a very much shorter and narrower tail in the hind-wings.

Mr. W. S. M. D'Urban found *Harpax* very abundant near King William's Town, taking it from October to December, and again in March; he noted that it frequented bushes with sweet-scented flowers, one of its favourites being the thorny *Arduina ferox*. Colonel Bowker noted the same habits in Kaffraria Proper; and the few individuals I met with in Natal were all taken on or about various shrubs in February and March. Its flight and motions quite resemble those of the species of *Aphnæus*. I met with single specimens at Uitenhage and at East London in the month of February.

This butterfly has a very wide range over Africa, but, as far as it is known, seems more prevalent to the south of the Equator.

Localities of *Chrysorychia Harpax*.

I. South Africa.
   B. Cape Colony.
      b. Eastern Districts.—Uitenhage. Grahamstown and Fish River (*M. E. Barber*). King William's Town (*W. D'Urban*). East London.
   D. Kaffraria Proper.—Tsomo and Bashee Rivers (*J. H. Bowker*).

---

[1] Two ♂s and a ♀ since received from Delagoa Bay quite agree with Hopffer's diagnosis, and the tails of the hind-wings (though not so straight as in his figures) are much longer and more linear than in Natalian and other more southern examples. The red of the ♂ is decidedly paler and more orange on the upper side; and in both sexes the under side is paler, and its metallic spots more brilliant and more numerous, especially in the hind-wing,—the sub-marginal spots in the fore-wing being also metallic, as well as a hind-marginal streak in the hind-wing.

Specimens from Matabeleland are remarkable in both sexes for their paler upper-side colouring, and in the ♂ for the narrower apical border; while in the ♀ (which is larger than usual) the discal spots, and, in the hind-wings, the sub-marginal spots are much reduced.

Two ♂s from Zumbo on the Zambesi, taken by Mr. Selous, agree closely with those from Querimba.

E. Natal.
   a. Coast Districts.—D'Urban and Mouth of Tugela River (*J. H. Bowker*). "Lower Umkomazi."—J. H. Bowker. Verulam. Mapumulo. Intzutze River.
   b. Upper Districts.—Estcourt (*J. M. Hutchinson*). Blue Krantz, near ↓Colenso (*W. Morant*). Junction of Tugela and Mooi Rivers (*J. H. Bowker*).
F. Zululand.—Napoleon Valley (*J. H. Bowker*).
H. Delagoa Bay.—Lourenço Marques (*Mrs. Monteiro*).
K. Transvaal.—Marico and Limpopo Rivers (*F. C. Selous*).

II. Other African Regions.
   A. South Tropical.
      a. Western Coast.—"Angola (*J. J. Monteiro*)."—Druce.
      b. Eastern Coast.—"Querimba."—Hopffer.
      b1. Eastern Interior.—Tauwani, Makloutze, Tati, and Zambesi Rivers (*F. C. Selous*). "Gubulewayo (*F. Oates*)."—Westwood.
   B. North Tropical.
      a. Western Coast.—Whydah.—Coll. Brit. Mus. Sierra Leone.— Coll. Hope, Oxon.

201. (2.) Chrysorychia Amanga, Westwood.

PLATE IX. fig. 1 (♂).

♂ ♀ *Zeritis Amanga*, Westw., in Oates' Matabele Land, &c., p. 351, n. 62 (1881).

Exp. al., 1 in. 3–4 lin.

♂ *Fuscous, with an orange-red transverse discal band, which in fore-wing commences rather acutely immediately below third median nervule, and thence widens so as to occupy nearly all inner margin, and in hind-wing occupies entire area except a fuscous space at base.* Fore-wing: costa for a little distance from base with a rather wide orange-ochreous border; first and second median nervules and submedian nervure more or less defined with fuscous where crossing the red band. Hind-wing: basal fuscous broader near costa; subcostal nervules fuscous; a fine linear fuscous hind-marginal edging; anal-angular lobe and tail ferruginous. Cilia of *fore-wing* dark-greyish, slightly mixed with ferruginous and whitish, but wholly whitish from apex to third median nervule; of *hind-wing* ferruginous. UNDER SIDE.—*Warm ferruginous; outer half of costal area in both wings clouded with lilac-grey.* Fore-wing: basi-costal border *widely brilliant-silvery*, tinged with pale-yellow on extreme edge; two small spots close together (sometimes confluent inferiorly) in discoidal cell, a rather larger spot immediately below cell between first and second median nervules, a macular thin streak at extremity of cell inferiorly joining last-named spot, and two or three dots close to costa,—brilliant-silvery edged with black; faint traces of a discal row of thin silvery-and-fuscous spots; inner-marginal area orange-yellow, much paler inferiorly. Hind-wing: traces of three indistinct trans-verse rows of darker lunulate marks, of which the outermost (submarginal) becomes silvery-scaled near anal angle; a silvery edging along excavation of inner margin before projection of anal-angular lobe.

♀ *Orange-red paler, in fore-wing much enlarged, forming a very broad discal band, which leaves only a narrow fuscous border along costa, apex, and hind-margin to just above submedian nervure. Fore-wing*: basi-costal orange-ochreous border paler, longer, and wider; fuscous of basal area much varied with orange-red, especially in discoidal cell. *Hind-wing*: basal fuscous narrower, not so dark; hind-marginal linear edging obsolete. UNDER SIDE.—*Very much paler than in ♂, of an almost uniform reddish-ochreous tint, without any lilac-whitish cloud-ing. Fore-wing*: basi-costal silvery border much narrower; cellular and adjacent metallic spots smaller, and steely rather than silvery; discal traces of spots wanting; inner-marginal orange-yellow fainter. *Hind-wing*: markings quite obsolete, except inner-marginal silvery edging before anal-angular lobe.

The ♀ here described is from the interior of Natal; that described by Westwood from Matabeleland appears to have been more like the ♂ on the under side.[1]

This ally of *Harpax*, Fab., is well distinguished by the very brilliant and conspicuous basi-costal silvery border on the under side of the fore-wings, and the small development or absence of the metallic spots, with the exception of those before the middle of the fore-wing. On the upper side *Amanga* has in both sexes a basi-costal ochreous-orange border not developed in *Harpax*; the red in the ♂ is of a much more orange tint and in the fore-wing of smaller extent, while in the ♀ it is unspotted in both wings.

I first saw a specimen of this butterfly in the year 1860; it was in the collection of the late Mr. C. J. Andersson, who took it during his exploration of Damaraland. Mr. J. A. Bell brought me a very tattered individual from the same region in 1862. It was not until 1867 that I saw a third example, viz., the Zambesi ♂, in the Hopeian Museum at Oxford, described by Professor Westwood (*loc. cit.*) The ♂ figured in the present work was taken in 1875 in the northern part of the Transvaal by Mr. H. Barber, and the Natal ♀ just described in 1884 by Mr. J. M. Hutchinson. Mr. F. C. Selous met with the species at several places on the road between Bamang-wato and the Zambesi River. Mr. Hutchinson wrote that the specimen he captured (on 22d May 1884) was on the summit of a very lofty hill, and settled several times on the same stone.

Localities of Zeritis Amanga.

I. South Africa.
 E. Natal.
  b. Upper Districts.—Weenen (*J. M. Hutchinson*).

---

[1] A ♀ received from Mr. T. Ayres, with the note "Between Limpopo and Zambesi Rivers," is coloured and marked on the under side quite similarly to the ♂, except in being paler and in the reduction of the basal silvery border of the fore-wings. Mr. Selous's specimens from the Tropical Interior (especially from the Tauwani River) exhibit in both sexes considerable variation both in depth of colouring and distinctness of markings on the under side.

II. Other African Regions.
A. South Tropical.
  a. Western Coast.—Damaraland (*C. J. Andersson* and *J. A. Bell*).
  b. Eastern Coast.—Zambesi (*Rev. H. Rowley*).
  b1. Eastern Interior.—Makalapisi and Tauwani Rivers, Tchakani Vley, Makloutze River, and Zumbo (Zambesi River).—(*F. C. Selous*). "Gwailo River (*F. Oates*)."—Westwood.

GENUS ZERITIS.

Zeritis, Boisd., Sp. Gen. Lep., t. 22, f. 6 (1836).
Phasis and *Alocides*, Hübn., Verz. Bek. Schmett., p. 73 (1816).
Nais, Swainson, Zool. Illustr., 2nd Series, iii. p. 136 (1833).
Zerythis, Blanchard, Hist. Nat. Ins., iii. p. 463 (1840).
Cigaritis, Lucas, Expl. Alger., Zool., iii. p. 362 (1849).
Zeritis, Westw., Gen. Diurn. Lep., ii. p. 500 (1852); and Trim. [part], Rhop. Afr. Aust., ii. p. 261 (1866).
Crudaria, Wallengr., Sv. Vetensk.-Akad. Förhandl., 1875, p. 86.

IMAGO.—*Head* rather broad, very hairy in front; *eyes* smooth; *palpi* long, separate throughout their length, thickly scaly,—second joint rather stout, long, ascendant, densely scaly (and sometimes rather hairy) beneath,—terminal joint porrected horizontally, slender, acuminate, smoothly scaled, usually rather long (longer in ♀ than in ♂); *antennæ* of moderate length, or rather short, straight, thick, very gradually incrassate from before their middle, or sometimes even from their base, the tip obtuse.

Thorax robust (especially in ♂), clothed as in *Chrysorychia*. *Fore-wings* usually more or less prolonged, especially in ♂, but blunt and with convex hind-margin in ♀; in some species, however, hind-margin in both sexes is angulated or elbowed at end of lower radial nervule, and slightly dentate throughout; subcostal nervure five-branched,[1]—the first and second nervules originating widely apart from each other well before extremity of discoidal cell,—the third and fourth similarly well beyond extremity of cell, the latter being short and terminating at apex,—the fifth terminating on hind-margin a little distance from apex; upper radial nervule united to superior extremity of upper disco-cellular nervule, lower radial at junction of two disco-cellulars. *Hind-wings* not produced in anal-angular portion, but roundly prominent outwardly about middle of hind-margin; costa but slightly arched or nearly straight; hind-margin more dentate than that of fore-wing, often with a more or less marked projection or short pointed tail at extremity of submedian nervure, and sometimes with also a similar shorter one at extremity of first median nervule; neuration as in

---

[1] In *Z. Chrysantas* (Trim.) and *Z. Leroma* (Wallengr.) the fifth subcostal is wanting, the third being given off very much nearer to apex, and the shorter fourth (which ends at apex) being somewhat sinuated. Wallengren has made *Leroma* the type of his new genus *Crudaria* (K. Sv. Vet.-Akad. Förhandl., 1875, p. 86), but in his short diagnosis I find no character of importance except that of the subcostal nervure just mentioned.

Chrysorychia. Legs longer and thinner than in *Chrysorychia*, and not, or very slightly, hairy; fore-legs of ♂ usually with more or less developed spur and spine at extremity of tibia;—of ♀ rarely similarly armed; *middle* and *hind-legs* with tibial spurs longer than in *Chrysorychia.*

PUPA.—Thick, rounded, with blunt head and dorso-thoracic prominence. Without silken attachment, lying quite free under stones (*Z. Thyra*).

Boisduval did not describe this genus, but merely figured as its type (*op. cit.*) *Zeritis Nericne,* a species said to be from Guinea. Blanchard's brief diagnosis gives the generic name as "*Zerythis,* Boisd.," with the obvious intention of preserving the earlier author's designation; his types are *Z. Thero* and *Z. Thysbe* of Linnæus. Hübner's much earlier nominal generic titles of *Phasis* (for the two species just named and *Z. Palmus* of Cramer) and *Alocides* (for *Z. Thyra* of Linnæus and *Z. Pierus* of Cramer) have never been diagnosed by any author,[1] and are therefore not adopted; while Swainson's *Nais,* as Westwood (*op. cit.*) remarks, is inadmissible from the fact that it is Cramer's species-name for the type (= *Thysbe,* Linn.) converted into the name of the genus.

There are five tolerably pronounced forms in this interesting genus, represented by the following species, viz., *Z. Nericne,* Boisd. (with which I provisionally—not having seen *Nericne* in nature—associate *Leroma,* Wallengr.); *Z. Zeuxo* (Linn.); *Z. Thysbe* (Linn.); *Z. Thero* (Linn.); and *Z. Thyra* (Linn.) The first section is characterised by a rather squarely chequered under surface of ochrey-yellow and cream-coloured spots separated by black lines; but this in *Leroma* is much obscurer in tint (though dotted here and there with silvery points), while the upper side is in both sexes uniform glossy dull-grey. The *Zeuxo* section includes the brilliant forms which, in their shining golden or coppery-red upper sides spotted with black, so nearly resemble the genus *Chrysophanus;* their under side is more or less ornamented with glittering steely or brassy spots. The third or *Thysbe* section is closely related to the second, but the fore-wings are angulated, the hind-wings with a distinct process at anal angle, and the under side adorned with remarkable silvery H- and W-like characters in the hind-wings. In the *Thero* section, the upper-side colouring is dark-brown with orange-red (not metallic or glossy) spots or patches; the under side is splendidly adorned with metallic silvery-white spots and other markings; the fore-wings are angulated or elbowed; and there is usually, besides the anal-angular projection, a small pointed process at the end of the first median nervule. The last section, represented by *Thyra,* is of a more robust type, with the upper side non-metallic, almost always orange-red bordered with blackish-brown and unspotted, while the under side is coloured with

---

[1] Hübner's own line and a half of description are, as usual, utterly insufficient for the purpose.

ochreous-brown, ferruginous, or vinaceous, varied with sub-metallic white or greyish spots.

This is an eminently South-African genus, only three of the twenty-eight species known being peculiar to other parts of Africa. The two North-African species (*Syphax*, Lucas, and *Zohra*, Donzel) belong to the *Zeuxo* group, which includes eight South-African species. The *Thysbe* group contains three, the *Thero* group four, and the *Thyra* group nine natives of South Africa. It is very noteworthy that out of the twenty-five recorded South-African species only two (*Taikosama*, Wallengr., and *Orthrus*, Trim.) have been found within the Southern Tropic, and not one elsewhere in the Ethiopian Region. *Z. Leroma* and the *Zeuxo* and *Thysbe* groups have much the same habits as *Chrysophanus*, actively flitting about bushes and flowering shrubs, but several of them are at the same time fond of settling on the ground, a practice which prevails also in the *Thero* group (though not in the case of *Thero* itself), and becomes quite constant in the *Thyra* group. Many inhabit the most arid and desolate tracts of country, and seem to delight in the intense heat of the parched sandy soil under the noon-day sunshine, seldom visiting flowers or seeming to need liquid nourishment of any kind. The under surface colouring of the *Thyra* group is generally highly protective, closely resembling the tints of the ground frequented by these butterflies. The flight of the ground-frequenting species is rapid but very short; and I have noticed a great difference among them in wariness and alacrity in evading attack. Thus, while it is easy to capture *Wallengrenii* and *Aylaspis* with finger and thumb, *Thyra* and *Barklyi* are extremely shy of the collector's advances,—the latter (except when settled on flowers) being on this account by no means easily taken.

Zeritis is very generally distributed over South Africa, fourteen species being found alike in the Western and Eastern districts; four others appearing to be limited to the former, and seven to the latter, but mostly possessing a considerable range towards the Interior.

202. (1.) Zeritis Leroma, (Wallengren).

♀ *Arhopala? Leroma*, Wallengr., K. Sv. Vet.-Akad. Handl. 1857, Lep. Rhop. Caffr., p. 42.

Amblypodia? Leroma, Trim., Rhop. Afr. Aust., ii. p. 231, n. 134 (1886).

♂ ♀ *Zeritis Leroma*, Trim., Trans. Ent. Soc. Lond., 1870, p. 375, pl. vi. f. 10 (♂).

♂ *Zeritis Zorites*, Hewits., Trans. Ent. Soc. Lond., 1874, p. 354.

Exp. al., (♂) 9½ lin.—1 in. 1¼ lin.; (♀) 1 in. 1–4 lin.

♂ *Pale brownish-grey, with a silky or sub-metallic lustre; a linear hind-marginal fuscous edging; cilia whitish with a shining-greyish gloss. Hind-wing:* a short black linear tail, slightly tipped with whitish, on submedian nervure; just below it, an anal-angular small indistinct fuscous spot bounded interiorly by a small shining-whitish or silvery spot; sometimes an indistinct hind-marginal ochreous-yellow spot just above submedian nervure. UNDER SIDE.—*Very pale dull brownish-grey, with more or less of an ochrey-yellow tinge; disco-cellular, discal, and submarginal spots paler (almost whitish), with dark edging on both sides, in fore-wing generally, in hind-wing more sparsely, scaled with silvery. Fore-wing:* three ordinary disco-cellular spots; discal row of six spots, of which the second is out of line, being before the first and third; a longitudinal row of three similar much smaller spots not far from costal edge; submarginal row of spots less distinct, usually only their outer black edging represented by blackish dots; a good-sized blackish basal mark between median and submedian nervures. *Hind-wing:* a spot between costal and subcostal nervures near base; two in discoidal cell; a sub-basal row of four, of which the second is at extremity of cell; eight spots in irregular discal row, their darker edging usually very indistinct; submarginal row regular but usually indistinct; anal-angular black spot better marked than on upper side, and really the outer edging of the last spot of submarginal row.

♀ *Slightly paler; in fore-wing usually an indistinct lunulate darker marking at extremity of discoidal cell.* UNDER SIDE.—Markings generally more defined, especially those of fore-wing, whose black edging is usually well developed.

Though the unicolorous upper side of this dull-coloured species presents little or no variation except in depth of tint, the under side is very variable, whether as regards the shade of the ground-colour, the distinctness of the markings, or the amount and distribution of the silvery scaling of the spots. The latter feature is best developed in a ♂ from Pinetown, and a ♂ and two ♀s from the Vaal River, Griqualand West. Three of the ♂s I took near Grahamstown have the black interior edges of the discal spots enlarged so as to form three inter-nervular rays as far as median nervure in the fore-wings. The Transvaal specimens are much paler than those taken in Cape Colony and Natal, the silvery scaling is almost obsolete, and there exists in three of them a small faint ochre-yellow mark on hind-margin just above submedian nervure (which is also very fully indicated on the upper side. The ♀ described by Wallengren is evidently of this local variation, in which that sex is considerably larger than elsewhere. The only specimen (a ♂) taken by Colonel Bowker in Basutoland was darker than usual, and the largest of that sex I have seen; the fore-wings, too, were acuter at the apex than in any other example.

Wallengren (*Öfv. K. Vet.-Akad. Förh.*, 1875, pp. 86, 87) has made this insect the type of a new genus, viz., *Crudaria;* but the characters given do not seem to me to warrant this course,—the only feature of moment being the subcostal nervure of the fore-wings, which is vaguely described as "*biramosa vel triramosa*," and *Leroma* having really a short third branch of that nervure ending on costa not far from apex. The palpi have the terminal joint long and slender in both sexes, but more so in the female.

I did not receive any examples of this butterfly until 1869, when two specimens reached me from Natal.[1] In January 1870 the Basutoland ♂ above-

---

[1] I had in 1867 seen and described a damaged specimen in the Burchell Collection at the Hopeian Museum, Oxford, but did not at the time identify it with *Leroma*.

mentioned arrived from Colonel Bowker, and during the same and the following months I had the pleasure of capturing numerous examples in the neighbourhood of Grahamstown. It is a very obscure little species, and would readily be passed over for one of the duller species of *Lycæna*. The first individual I noticed was sitting on a flower of *Acacia horrida*, but numerous others were taken flitting about near the ground, among herbage and low shrubs. Colonel Bowker's specimen was noted by him as taken "on the stony ground among short grass and flowers, 14th December 1869."

Localities of *Zeritis Lerona*.

I. South Africa.
   B. Cape Colony.
      b. Eastern Districts.—Grahamstown.
      c. Griqualand West.—Klipdrift, Vaal River (*J. H. Bowker*). Kimberley (*H. G. Smith*).
      d. Basutoland.—Vogel Vley, Jammerberg (*J. H. Bowker*).
   E. Natal.
      a. Coast Districts.—Pinetown (*W. Morant*). D'Urban (*the late M. J. M'Ken*).
      b. Upper Districts.—Estcourt, (*J. M. Hutchinson*). Colenso (*W. Morant*). Valley of Mooi and Tugela Rivers (*J. H. Bowker*).
   K. Transvaal.—Lydenburg District (*T. Ayres*).

II. Other African Regions.
   A. South Tropical.
      a. Western Coast.—" Damaraland (*De Vylder*)."—Aurivillius.

203. (2.) **Zeritis Zeuxo,** (Linnæus).

Papilio Zeuxo, Linn., Mus. Lud. Ulr. Reg., p. 331, n. 149 (1764); and Syst. Nat., i. 2, p. 789, n. 231 (1767).
♂ ♀ *Zeritis Zeuxo*, Trim., Rhop. Afr. Aust., ii. p. 262, n. 162, pl. 5, f. 2 [♂] (1866).

Exp. al., 1 in.—1 in. 2 lin.

♂ *Metallic golden-orange, with a slight basal fuscous suffusion, and spotted with black ; a hind-marginal blackish border*, broad and even in *fore-wing*, narrow and dentated in *hind-wing*. *Fore-wing:* a small spot in cell; an elongate spot closing cell, with a small costal spot a little above and beyond it; an irregular discal transverse row of seven spots, of which the first three are usually confluent, and the two last (just above submedian nervure) rarely so; costa slightly fuscous-clouded; usually a dot before middle below cell. *Hind-wing:* spots similarly arranged, but none in cell or just below it; a submarginal row of lunular spots, near costa confluent with hind-marginal border; costa and inner margin fuscous-clouded. UNDER SIDE.—*Hind-wing and border of fore-wing* (except inner margin) *brownish-grey. Fore-wing:* yellowish-orange, not metallic ; an additional dot on costa before middle, and another in cell at base; all spots above median nervure and its third branch centred with glistening-steely; spot below cell large; a very faint submarginal row of spots, the two about its middle dimly steely-scaled. *Hind-wing*: spots indistinct, but little paler than ground-colour, indicated by their dark edges; a row of three before middle; a discal row of eight sub-lunulate spots; two indistinct lunular submarginal streaks, of which the outer one is more strongly marked and interrupted.

♀ Like the ♂, but paler and not quite so metallic; basal fuscous suffusion rather wider.

Three specimens which I took at Blaauwberg in the Cape District in October 1878 are smaller and duller than those found in the more immediate vicinity of Cape Town, and have *a costal* blackish border in the fore-wing, besides a broader basal suffusion and hind-marginal border in both wings.

The fore-wings of this bright little species are both in colouring and pattern very like those of the abundant European *Chrysophanus Phlæas*, the "Small Copper" of English collectors. Its range seems to be extremely limited, and I am not aware of its occurrence beyond the Cape District. It is, moreover, very local, haunting almost exclusively the leaves and flowers of a tall, shrubby, thick-leaved *Senecio*, which flourishes about the rough broken slopes and rocky "kopjes" near Cape Town, but grows in rather detached groups in certain spots only. I feel pretty sure that the larva must feed on this plant, but much searching has not resulted in its discovery. The butterfly is usually numerous where it occurs, and easily captured. October and November are its favourite months, but I have met with it from September to January.

Localities of *Zeritis Zeuxo*.

I. South Africa.
   B. Cape Colony.—Cape Town. Blaauwberg, Cape District.

204. (3.) Zeritis Chrysaor, Trimen.

PLATE IX. fig. 2 (♂).

Zeritis Chrysaor, Trim., Trans. Ent. Soc. Lond., 3rd Ser., ii. p. 177 (1864); and Rhop. Afr. Aust., ii. p. 263, n. 163 (1866).

Exp. al., 10 lin.—1 in. 2 lin.

Glittering golden-orange; spots arranged much as in *Zeuxo*, but *usually smaller and more distinct* (the submarginal row of hind-wing wholly wanting); *hind-marginal border much narrower* (in *fore-wing* widest at apex, in *hind-wing* macular or nearly so, being sharply indented interiorly on nervules). *Hind-wing:* a dot or short linear mark at end of cell; no costal clouding. UNDER SIDE.—*Hind-wing, and costa at base with apical area of fore-wing, varying from pale creamy-ochreous to pale ferruginous-ochreous. Fore-wing:* spots arranged as in *Zeuxo*, but filled with more glittering silvery, the whole (except dots of submarginal row) usually metallic-centred, but sometimes only those near costa. *Hind-wing:* spots small, slightly glistening, arranged as in *Zeuxo*, but less distinct; on hind-margin of paler specimens (usually ♀ s) some ferruginous clouding; anal-angular projection more acute than in *Zeuxo*.

This species further differs from *Zeuxo* in having the lowest spot of the discal row (below first median nervule) in both wings situated *beyond*, instead of before, the fifth spot, and in wanting on the upper side the isolated spots in and below the cell. It exhibits considerable variation in the size of the spots of the discal row. In a ♂ from Basutoland the lowest spot in both wings is wanting, and those of the hind-wing are minute; while in three ♀ s from Port Elizabeth and Uitenhage the spots in the hind-wing are so much enlarged as to be for the greater part contiguous. In a very small ♀ taken at D'Urban, Natal, by Colonel Bowker, this enlargement of the spots is carried still farther, those of the fore-wing also uniting to form a continuous irregular stripe. Both this last-named example[1] and a ♂ taken by myself in the same locality present a rather conspicuous ferruginous submarginal suffused streak in the hind-wing, which is also more faintly represented in two ♂ s sent by Colonel Bowker from the Bashee River and near Somerset East respectively. The spots on the underside of the hind-wings are usually less indistinct in the ♀ ; they are in both sexes more apparent in specimens from the eastern side of South Africa, especially in the few I have seen from Natal, and in one ♂ that I captured at Port Elizabeth they are sub-metallic. The anal-angular projection of the hind-wing is little developed in specimens found near Cape Town, and not much more so in Western examples generally, but farther eastward it is marked, and in the Natalian examples becomes very prominent and widened, forming quite a "tail." In both sexes from Basutoland the silvery spots of the fore-wings are very faintly developed.

This is the most metallic of the South-African members of the genus, rivalling the European *Chrysophani*. It is rather scarce near Cape Town, but numerous at Malmesbury and other places in the Western Districts, frequenting rocky "kopjes" and the stony sides of hills. On Table Mountain I have usually found it at a considerable elevation, and always singly. It is an active and conspicuous insect on the wing, and settles very frequently on low shrubs. Near Malmesbury it specially affected a species of *Cotyledon*, and at Lady Grey, in the Robertson District, I observed it on the flowers of *Mesembryanthemum*. It seems to occur throughout the year, but I have no record of it as appearing in December.

Localities of *Zeritis Chrysaor*.

1. South Africa.
    B. Cape Colony.
        a. Western Districts.—Cape Town. Kalk Bay and Blaauwberg, Cape District. Malmesbury. Waagenmaaker's Kraal, Beaufort District. Robertson and Lady Grey. Swellendam (*L. Taats*).
        b. Eastern Districts.—Port Elizabeth. Uitenhage (*S. D. Bairstow*). Zwaarte Ruggens, Uitenhage District (*J. H. Bowker*). Grahamstown. King William's Town (*J. H. Bowker*). Between Somerset East and Murraysburg (*J. H. Bowker*). "Bodiam, near Keiskamma River."—W. D'Urban. Summit of Gaika's Kop, Amatola Mountains (*J. H. Bowker*).
        c. Griqualand West.—Vaal River (*J. H. Bowker*).
        d. Basutoland.—Maseru and Koro-Koro (*J. H. Bowker*).
    D. Kaffraria Proper.
        Kei and Bashee Rivers (*J. H. Bowker*).
    E. Natal.
        a. Coast Districts.—D'Urban.

---

[1] A ♀ of ordinary size, found by Colonel Bowker at Malvern, near D'Urban, in August 1885, has just reached me. The upper-side spots are considerably enlarged, but not quite confluent, and the under side has the ferruginous streak on the hind-wing well marked.

205. (4.) Zeritis Lyncurium, Trimen.

PLATE IX. figs. 3 (♂), 3a (♀).

♂ ♀ *Zeritis Lyncurium*, Trim., Trans. Ent. Soc. Lond., 1868, p. 86.

Exp. al., (♂) 1 in.—1 in. 0½ lin.; (♀) 1 in. 1–1 lin.

♂ *Metallic golden-red, with narrow hind-marginal black borders. Fore-wing:* base slightly blackish; costa narrowly black to beyond middle, where the edging widens to join hind-marginal black border, which is broadest at apex, and the inner edge of which is irregularly dentated with the ground-colour on median nervules; a narrow black spot at extremity of discoidal cell; sometimes an indistinct smaller spot in cell, usually from one to four spots representing a discal transverse row. *Hind-wing:* base and costa broadly or very broadly clouded with blackish; in some specimens a faint, narrow, blackish mark at extremity of discoidal cell; hind-marginal border very sharply indented on its inner edge with the ground-colour on nervules. UNDER SIDE.—Hind-wing and a narrow costal and apical border of fore-wing pale creamy-ferruginous. *Fore-wing:* discal row of spots irregular, interrupted,—the sixth (lowest) spot situated, as in *Æthon*, Trim., not beyond the fifth; a submarginal row of five or six small spots; below median nervure a short basal blackish suffusion, succeeded by a black spot; the following spots more or less marked with glittering steely scales, viz., two in discoidal cell, one at extremity of cell, two minute costal ones, first of discal row, and first three of submarginal row. *Hind-wing:* the following inconspicuous darker markings, here and there edged interiorly by dull metallic scales, viz., a thin terminal disco-cellular striola; a spot in cell; two spots between costal and subcostal nervures, one before, the other about middle; and an irregular, interrupted discal macular row; some red suffusion along hind-margin, varied by a very indistinct submarginal row of paler lunulate marks. *Cilia* dull-grey, with nervular blackish interruptions.

♀ *Paler, less metallic. Hind-wing:* costal clouding considerably narrower and paler; red indentations of border more numerous and deeper, almost reaching hind-marginal edge. UNDER SIDE.—Hind-wing and costal-apical border of fore-wing paler, yellower. *Fore-wing:* steely scaling much duller and fainter, and in submarginal row of spots wanting entirely. *Hind-wing:* markings generally more distinct.

From both *Chrysaor*, Trim., and *Æthon*, Trim., this butterfly is distinguished by its smaller size, rather deeper red, comparatively broader hind-marginal border, great imperfection of discal row of spots in the fore-wing, and entire absence of that of hind-wing, and particularly by the *broad costal black or blackish clouding in hind-wing,* of which latter no trace is found in either of the species mentioned. On the under side, *Lyncurium* is decidedly nearer to *Æthon* than to *Chrysaor*, especially as regards the position of the lowest spot of the discal row in the fore-wing, the extent of red and the slight development of the metallic scaling in the same wing, and the nature and position of the markings of the hind-wing.

This species was discovered by Colonel Bowker near the River Tsomo in December 1864. In that month and in the following January he captured a good many specimens, but only in two spots, "flitting about stunted bushes growing between rocks upon a lofty hill ridge." No other examples have come under my notice.

Locality of *Zeritis Lyncurium*.

I. South Africa.
   D. Kaffraria Proper.—Tsomo River (*J. H. Bowker*).

206. (5.) Zeritis Lycegenes, Trimen.

♀ (?). *Zeritis Lycegenes*, Trim., Trans. Ent. Soc. Lond., 1874, p. 337, pl. ii. f. 7.

Exp. al., 11½ lin.

♀ (?). *Bright sub-metallic orange-red, hind-marginally edged with black; fore-wing only with black spots.* Fore-wing: an elongate, sub-ovate spot at extremity of discoidal cell; an indistinct smaller spot below cell, close to origin of first median nervule; beyond middle a row of six spots, of which the first three (between costa and third median nervule) are farther from the base than the other three (between third median nervule and submedian nervure); costa narrowly edged with blackish, widening at apex; hind-marginal edging narrow, with slight projections between the nervules. *Hind-wing*: no spots or other markings, except the hind-marginal edging, which emits prominent inter-nervular projections. *Cilia* long, dull brownish-grey, very slightly mixed with whitish. UNDER SIDE.—*Hind-wing and basal, costal and apical border of fore-wing, pale greyish-ochreous.* Fore-wing: besides markings of upper side (of which the spot below cell is conspicuous), there are two cellular spots, of which that nearer base is minute; two costal dots above and beyond the spot closing cell; and a submarginal row of four small sub-lunulate spots between lower radial nervule and submedian nervure (traces of two spots commencing this row are just visible in the apical ochreous). *Hind-wing*: five small indistinct round brown spots in basal region, viz., two close to costa (one near base, the other about middle), one in discoidal cell, one (elongate) closing cell, and one between first median nervule and submedian nervure; a little beyond middle, a transverse, irregular, sub-macular brown streak, *not* parallel to hind-margin, extending from near apex to submedian nervure; a submarginal, very indistinct, deeply-festooned, brownish streak, touching the submacular streak on sub-costal nervules.

This little *Zeritis* is in character intermediate between *Z. Chrysaor*, Trim., and *Z. Lyncurium*, Trim. From the former it is separable by its smaller size, less metallic upper side, different arrangement of the discal row of spots in the fore-wings, and *total want of spots* in the hind-wings, while on the under side the brilliant metallic spots of *Chrysaor* are scarcely indicated in the fore-wings, and the duller ones of the hind-wings wanting. From *Z. Lyncurium* it differs on the upper side in its paler colouring, well-marked discal row of spots, and narrower hind-marginal border in the fore-wings, *and entire want of the broad basal, and especially costal, fuscous clouding in the hind-wings;* but on the under side is very similar, differing chiefly in the discal row of spots in the fore-wings being much more regular.[1]

The above description is made from the only example that I have seen, which was taken by Mr. Walter Morant near the Mooi River, in Natal, on the 15th September 1870.

I believe it to be a ♀, but cannot decide, the first pair of legs being absent.

Locality of *Zeritis Lycegenes*.

1. South Africa.
   E. Natal.
       b. Upper Districts.—Mooi River (*W. Morant*).

207. (6.) Zeritis Æthon, sp. nov.

Plate IX. fig. 4 (♀).

Exp. al., (♂) 1 in. 2 lin.; (♀) 1 in. 3 lin.

♂ *Metallic golden-red, deeper than in Chrysaor, spotted and narrowly bordered with black.* Fore-wing: hind-marginal border rather broader than in *Chrysaor*, its inner edge more regularly and deeply indented with the ground-colour on nervules; terminal disco-cellular spot and spots of discal row larger and much broader; the latter arranged as in *Chrysaor*, with the exception of the sixth (lowest) spot, which is minute and *immediately below*, not beyond, the fifth. *Hind-wing:* hind-marginal border rather broader; discal row of spots very differently arranged,— the first (upper-most) spot being isolated from the others and considerably nearer origin of first subcostal nervule than in *Chrysaor*,—the second much nearer termination of that nervule, and with the third and fourth forming a nearly straight transverse row as far as third median nervule, —the fifth and seventh wanting, and the sixth minute but occupying the same position (between first and second median nervules) as in *Chrysaor*. Under side.—*Hind-wing and narrow apical border of fore-wing pale creamy-ferruginous.* Fore-wing:—Pale orange-yellow almost to costal edge (which has an almost linear border of pale creamy-ferruginous), and quite to hind-margin below second median nervule; a slender dark-red line on extreme edge of hind-margin; all the spots very much thinner than above; two spots in cell reduced to mere dots, and spot beneath cell very small; an incomplete submarginal row of very small

[1] In this latter feature, *Lycegenes* differs from both the species named as well as from th allied *Z. Æthon*, Trim., the three lower spots being almost in a straight line.

spots; only two minute costal spots, upper part of terminal discocellular spot, first and second spots of discal row, and upper three spots of submarginal row filled with metallic-golden scales. *Hind-wing*: the following slender dark-ferruginous markings interiorly rather conspicuously edged with subdued-silvery, viz., a spot near costa before middle, a terminal disco-cellular striola, and a spot and two separate striolæ representing and following the course of the disjointed discal row of spots on the upper side; a very indistinct submarginal greyish suffused streak, and a very indistinct small ferruginous spot in discoidal cell. *Cilia* grey, with nervular blackish interruptions.

♀ *Less metallic than* ♂. *Fore-wing*: last spot of discal row larger; a minute spot just above and slightly beyond terminal disco-cellular spot, and another just below origin of first median nervule. *Hind-wing*: spots of discal row larger, especially the sixth,—the fifth spot present. UNDER SIDE.—As in ♂, but all the markings more distinct. *Hind-wing*: the two discal striolæ almost united on first median nervule.

This close ally of *Z. Chrysaor*, Trim., is further distinguished by its larger size, and by the different outline of its fore-wings, which in the ♂ are not produced apically, and in the ♀ are not nearly so convex hind-marginally.

My description is made from a single example of each sex, acquired by the South-African Museum in 1879 from Mr. T. Ayres, who noted them as having been captured in the Lydenburg District of the Transvaal. I have not met with any other specimens.

Locality of *Zeritis Æthon*.

1. South Africa.
   B. Transvaal.—Lydenburg District (*T. Ayres*).

208. (7.) Zeritis Chrysantas, Trimen.

Zeritis Chrysantas, Trim., Trans. Ent. Soc. Lond., 1868, p. 85, pl. v. f. 6.

Exp. al., 1 in. 1–2 lin.

♀ *Pale-orange, scarcely sub-metallic, with black discal spots and narrow hind-marginal border; cilia broad, blackish, interrupted conspicuously with white between nervules*. *Fore-wing*: a terminal well-marked disco-cellular black lunulate spot; an irregular discal row of five small spots (of which the third and fifth are beyond the line of the other three) extending from close to costa as far as first median nervule; base narrowly greyish; hind-marginal border broadest at apex, narrowing to posterior angle, its inner edge dentated with ground-colour on median nervules; a rounded whitish spot at apex, touching white cilia; on costal edge, beyond middle, three very short sub-oblique white lines, separated by black. *Hind-wing*: base very narrowly greyish; in one specimen without any spots, in the other with a minute terminal disco-cellular one, and a discal row of three spots (of which first and third are minute) between second subcostal and first median nervules; hind-marginal border broadest near costa, its inner edge irregularly but deeply dentated with ground-colour on nervules. UNDER SIDE.—*Hind-wing, and costal edge and narrow apical and hind-marginal border of fore-wing, pale-grey varied with whitish; in both wings some glittering pale-golden spots.* Fore-wing: orange fading into pale-yellowish on inner margin; a submarginal row of six small black spots, of which the upper three are more or less filled with golden scaling; first spot of discal row, two small costal spots before it, terminal disco-cellular spot, and two additional spots in discoidal cell, all filled with pale-golden scaling; below median nervure a blackish spot; a hind-marginal row of indistinct whitish lunules, commencing with that at apex. *Hind-wing*: the following pale-golden brownish-edged spots, viz., three in discoidal cell—first (minute) at base, second (larger) midway, third (large) at extremity; two near base—one just below costal, the other just below median nervure; nine forming a regular submarginal row—the third and fourth conspicuously clouded with dark-brown; a rather irregular discal row of nine whitish-grey spots, of which the first, second, eighth, and ninth are more or less marked with pale-golden scales; several whitish subquadrate marks interspersed between the golden spots; hind-marginal lunules as in fore-wing, that next anal angle largest, interiorly edged with dark-brown. Dark part of cilia mixed with grey.

This very distinct species appears on the whole to be nearer to *Z. Pyroeis*, Trim., than to any other member of the genus, resembling it in the absence of metallic lustre in the orange of the upper side, the shape of the wings, and the arrangement of the under-side markings; but also differing conspicuously in its total want of any basal blue on the upper side, and possession of golden spots on the under side of the hind-wings, and of cilia black and white instead of almost uniform greyish. The defective or obsolete condition of the discal spots on the upper side of the hind-wings reminds one of *Z. Lyncurium*, Trim., but it is not at all like the latter in other respects.

I have seen only two examples—both ♀—of this beautiful little *Zeritis*. The first (on which I founded the species in 1868) was found at Murraysburg in the Cape Colony by Dr. and Mrs. Muskett, who sent it to me in July 1864. The other I had the good fortune to capture on August 20, 1873, at the very distant locality of Oograbies, in Little Namaqualand. It was very active and wary, settling sometimes on the ground, and sometimes on a shrubby *Mesembryanthemum*, with small white flowers, at the opening of a dry ravine. It for a long time eluded my pursuit, as I lost sight of it altogether for more than an hour, much to my disappointment at the time, as I had at the first glance recognised it as the hitherto unique *Chrysantas*. This Namaqualand example is a little smaller than the Murraysburg one, and has the under-side markings all less distinct, but presents an imperfect discal row of spots on the upper side of the hind-wings.

Localities of *Zeritis Chrysantas*.

I. South Africa.
  B. Cape Colony.
    a. Western Districts.—Oograbies, Namaqualand.
    b. Eastern Districts.—Murraysburg (*Mrs. Muskett*).

209. (8.) Zeritis Phosphor, Trimen.

♀ *Zeritis Phosphor*, Trim., Rhop. Afr. Aust., ii. p. 269, pl. 4, f. 12 (1866).

Exp. al., 10 lin.—1 in. 1 lin.

♀ *Shining golden-orange, with dark borders. Fore-wing:* costa dusky at base: a conspicuous brownish-black spot closing discoidal cell, and united to a border of the same colour, which, commencing on costa just above it, and very broad in apical portion, diminishes in width to anal angle, where it turns inward a little on inner margin. *Hind-wing:* costa, base, and inner margin broadly bordered or suffused with blackish-brown; a dark disco-cellular lunule; beyond middle, parallel to hind-margin (which is edged with a black line) a row of blackish spots, forming a macular stripe from costal to inner-marginal bordering; lobe and tail on anal angle reddish-brown, the former sparsely scaled with bluish-silvery, the latter white-tipped and slightly twisted. UNDER SIDE.—*Fore-wing:* costa, apex, and hind-margin bordered with pale greyish-ochreous, with a reddish tint on hind-margin; ground-colour very pale orange-yellow; three metallic, black-edged spots in cell,—below cell a black spot contiguous to middle cellular one; two metallic dots on costa; transverse stripe of six metallic, black-edged spots beyond cell (of which the three lower are confluent) turning inwards so as to appear almost continuous of spot closing discoidal cell; a submarginal metallic-dotted streak rather strongly marked. *Hind-wing:* greyish-ochreous, with a ferruginous hind-marginal tinge; metallic dark-edged spots arranged much as in *Harpax*, Fab.; two in cell, one closing it; row beyond middle more conspicuous, composed of seven spots, arranged in pairs, except the seventh,—which, with the sixth, is large and brilliant; submarginal streak as in fore-wing, but strongly metallic at anal angle.

The above description is that of the first example discovered by Colonel Bowker, a ♀ expanding only 10 lines across the wings, taken on the Bashee River in Kaffraria Proper. Two ♀ specimens subsequently captured on the Tsomo River, in the same territory, are considerably larger, expanding respectively 12 and 13 lines. Of these two, the larger has the outline of the hind-margin of the fore-wings elbowed about extremity of third median nervule, while in the other (as well as in the small Bashee River example) there is only a slight prominence in that part. The hind-marginal dark border of the fore-wings is very broad in the largest specimen, and the submarginal dark streak of the hind-wings suffused and almost continuous. I have not seen any but these three examples.

Though near *Harpax*, Fab., in outline of wings (including the form of the tail on the hind-wings) and in the pattern of the under side, this curious species is very different as regards the palpi, the length of which—together with the metallic-orange of the upper side of the wings—approximate it to *Zeuxo*, Linn., *Chrysaor*, Trim., and their allies.

Concerning the Bashee River specimen, Colonel Bowker wrote that it was the only one he saw, and was caught at the edge of a forest in the month of March. The two larger examples were also taken on the edge of a large forest (called the "Boolo") in December 1865, and were observed to fly down from some trees to drink at a small pool.

Localities of *Zeritis Phosphor*.

I. South Africa.
 D. Kaffraria Proper. Tsomo and Bashee Rivers (*J. H. Bowker*).

210. (9.) Zeritis Pyroeis, Trimen.

♂ ♀ *Zeritis Pyroeis*, Trim., Trans. Ent. Soc. Lond., 3rd Ser., ii. p. 178 (1864); and Rhop. Afr. Aust., ii. p. 264, n. 164, pl. 5, f. 1 [♂] (1866).

Exp. al., (♂) 1 in.—1 in. 3 lin.; (♀) 1 in. 3–4½ lin.

Orange-yellow, not metallic, with black spots; base of both wings broadly blackish and densely blue-scaled; hind-wing of ♂ richly shot with a shifting blue lustre. *Fore-wing:* costa and hind-margin bordered with black; costal border broadest near base, where it is powdered with fulvous-ochreous, but usually much narrowed beyond middle; hind-marginal border very regular, of even width, very slightly crenelated on inner edge; just beyond basal clouding a small round spot in discoidal cell, a longer quadrate spot closing cell; beyond middle an irregular row of six spots, of which the three first form a continuous costal stripe, the fourth is nearer base, the fifth in a line beneath the three first, and the sixth (above submedian nervure) in a line beneath the fourth. *Hind-wing:* a row of six spots as in fore-wing, except that the three first spots are more separate,—in ♂ the first spot is almost merged in costal blackish band, which is wider than in ♀; on hind-margin a row of black arches (sharper in ♂) intersecting the ground-colour between nervules, which carry the orange to edge; closing cell is a black streak usually merged with basal clouding. *Cilia* fuscous mixed with white; in hind-wing mingled with red. UNDER SIDE.—*Costa and apex of fore-wing and whole of hind-wing pale creamy-ferruginous.* *Fore-wing:* two cellular spots and first spot of transverse row silver-centred; a third black dot near base in cell, and another (silver-centred) on costa beyond cell; beneath central spot of cell a good-sized rounded spot below submedian nervure; close and parallel to hind-margin a row of fuscous spots, indistinct near costa. *Hind-wing:* often almost spotless; a pale glistening line closing cell; near it, towards inner margin, a pale dot; a submarginal row of pale dots, sometimes replaced by a faint reddish line; transverse discal row sometimes represented by spots barely distinguishable from ground-colour except by their fusco-ferruginous outlines.

From its allied congeners, *Z. Zeuxo* and *Z. Chrysaor*, mihi, this species is at once distinguished by its blue-glistening basal patches, while its entire outline of wings and plainly-tinted under surface preclude its being mistaken for the angulated *Z. Thysbe*, Linn. In the two species first named, also, the ground-colour is metallic, strikingly so in *Chrysaor*. The width and evenness of the hind-marginal border of the fore-wings resemble the corresponding features in *Zeuxo*, Linn.

Sandy spots in level country. October (c)—December (m)—February (b) —March (m).

I first met with this very beautiful *Zeritis*, interesting from its relations to the three species just mentioned, on the sand flats near the coast of False Bay, a few miles from Wynberg, in October 1861. I have since taken it in other parts of the Flats, usually settling on spaces of white sand, or on the low plants that fringe such arid spots. I once took a specimen at a little distance from the Flats, on a hillside at Wynberg, and subsequently took five specimens at Kalk Bay, in the Cape District, about the steep hillside above the village. I also captured a single ♀ on the summit of the hill behind Simon's Town. Mr. Péringuey, in 1882 and 1883, met with two or three examples (one an unusually large ♀) at a considerable elevation in the Worcester District, viz., at Hex River Mountain and Touws River.

It does not appear to be numerous in any locality, occurring singly or in pairs.

Localities of *Zeritis Pyrocis*.

1. South Africa.
   B. Cape Colony.
      a. Western Districts.—Cape Town, Wynberg, Kalk Bay, Simon's Town, Cape District. Hex River Mountain and Touws River, Worcester District (*L. Péringuey*).

211. (10.) Zeritis Thysbe, (Linnæus).

Papilio Thysbe, Linn., Mus. Lud. Ulr. Reg., p. 330, n. 148 (1764); and Syst. Nat., i. 2, p. 789, n. 228 (1767).
Papilio Nais, Cram., Pap. Exot., pl. xlvii. ff. D, E [♂ *aberr.*], (1779).
Polyommatus Thysbe, Godt. [part], Enc. Meth., ix. p. 663. n. 157 (1819).
Nais splendens, Swains., Zool. Illust., 2nd Ser., iii. pl. 136 [♂ and ♀ var.], (1833).
♂ ♀ *Zeritis Thysbe*, Trim. [part], Rhop. Afr. Aust., ii. p. 265, n. 165 (1866).
  „   „   Butl., Proc. Zool. Soc. Lond., 1868, p. 223, pl. xvii. f. 5 [♂ *aberr.*]

Exp. al., (♂) 1 1 lin.—1 in. 3 lin.; (♀) 1 in. 1½–3 lin.

♂ *Orange, slightly glistening but not metallic, with brilliant silvery-blue gloss from bases; spotted and margined with black. Fore-wing*: blue usually completely covering basal half of wing, extending obliquely from costa at end of cell nearly to anal angle; cellular spot and discal row as in *Pyrocis*, but the spots squarer; a rather wide black costal border from end of cell to apex, marked with three white dashes between nervules; joining it a hind-marginal border of variable width, sometimes entire up to cilia, but usually with an external lunulate orange edging. *Hind-wing*: blue not extending beyond discal row of spots, which are arranged as in *Pyrocis*; costal and hind-marginal border very much narrower than in fore-wing, its latter portion often reduced to a row of dots at ends of nervules; anal-angular projection long and rather acute. *Cilia* white, interrupted with blackish mixed with orange on dentations of margin. UNDER SIDE.—*Fore-wing*: orange-yellow; costa narrowly, apex broadly, hind-margin rather widely *bordered with pale creamy-ochreous*, the nervules crossing border being ferruginous, with short white dashes from margin between them; spots as in *Pyrois* (occasionally two first of discal row silvery-centred); submarginal macular streak as in *Pyrois*, of variable intensity. *Hind-wing:* pale brownish-ochreous, clouded with darker-brownish, and spotted with silvery *liturae* edged broadly with ferruginous; two elongate liturae between costal and subcostal nervures,—another (sometimes like a V reversed) closing and piercing far into cell,—occasionally a small silvery dot both above and below this mark,—and a discal row of three double ones, very singularly shaped (more or less resembling reversed W's); hind-margin varied as in fore-wing. No black in cilia.

♀ Wings more rounded, not so dentate. Similar to ♂, but *blue of very much less extent and duller*, being mixed with blackish; ground-colour rather paler and duller; spots larger. UNDER SIDE.—As in ♂, but paler and duller.

Aberration ♂ (*Hab.*—Cape Town). Blue suffusion of unusual brilliancy and extent, completely obliterating all orange of *fore-wing* as well as black spots, but leaving a very broad apical and hind-marginal black border, edged outwardly by the usual small orange lunules; while in the *hind-wing* the blue extends to beyond middle, but leaves a broad even hind-marginal border of orange. Under-side markings as usual, but very dark in tint. Fore-wing more acutely angulated than usual; hind-wing with unusually long anal-angular projection.

This remarkable and very beautiful "sport" of *Thysbe* was taken on Table Mountain by Herr Gross in 1865, and has been figured by Mr. A. G. Butler (*loc. cit.*)

Another somewhat similar ♂ example was captured near Cape Town by Mr. E. L. Layard (I believe in 1868), and is in the South-African Museum. It is small (exp. 11¼ lin.), and the blue is even more developed than in the specimen just described, leaving no trace whatever of orange in fore-wing, and reducing that of hind-wing to an imperfect hind-marginal row of narrow spots. On the under side the inner-marginal region of fore-wing is suffused with blackish, and metallic spots are broadly black-ringed, while hind-wing generally is darker than usual, with all the markings less distinct.

The extent of the basal blue in the ♂ *Thysbe* varies considerably in the Western Districts of the Cape Colony, specimens taken on the coast having it more developed than those found more inland, in which latter (from Berg River Bridge, Robertson, &c.) the blue nowhere reaches beyond middle, and in the fore-wing discoidal cell is bounded by a small black spot some way before the extremity. These inland ♂ specimens are also smaller than the coast ones, though I have not found the ♀s to be so; but the under side is in both sexes paler, with the ferruginous in the hind-wing much less developed. This latter feature in some Western examples (including one from Malmesbury) constitutes quite a broad basal space and wide hind-marginal border, leaving little more than a median band of the pale ground-colour.

In Kaffraria Proper a different variation prevails, three ♂s from the Bashee River being of smaller size than the ordinary Western ♂s, and having the blue very broad, reaching in *fore-wing* from extremity of discoidal cell obliquely quite up to hind-marginal black border at posterior angle, and in *hind-wing* quite up to spots of discal row. The hind-marginal border is broader and more even than usual in fore-wing (and with no orange lunules beyond it), and in hind-wing continuous, though narrow, to anal angle. On the under side the silvery lituræ are paler and thicker than usual. A ♀ from the same locality differs similarly from Western individuals of that sex (except as regards the blue), and has the blackish border of both wings—though less distinctly in hind-wing—pierced exteriorly by acute white denticulations adjoining the white parts of the cilia.

This *Zeritis* is the most highly ornamented of its genus, and perhaps excels in beauty all other South-African *Lycænidæ*. On the wing its general hue seems to be silvery, but when basking in the sunshine with half-opened wings it looks like some brilliantly burnished jewel. Though local in its haunts, frequenting sandy hillsides and the dunes on the sea-coast, it is usually rather numerous where it occurs; it settles frequently on the ground or on low plants, and is partial to the fleshy leaves of the larger species of *Mesembryanthemum*. I have found it at various dates from the middle of September to the end of April, but it appears most numerously in October and January. The males are much oftener met with than the females. I have not taken it near Cape Town, where it is very scarce; but it inhabits Kalk Bay, and is occasionally almost abundant on the hills at the back of Simon's Bay. At Mossel Bay, where I first saw the species, it was very numerous on 20th September 1858, and I took many near Robertson in January 1876.

Localities of *Zeritis Thysbe.*

I. South Africa.
  B. Cape Colony.
    a. Western Districts.—Cape Town (*E. L. Layard* and — *Gross*). Kalk Bay and Simon's Town, Cape District. Mossel Bay. Knysna. Malmesbury (*J. H. Bowker*). Swellendam (*L. Taats*). Caledon (*J. X. Merriman*). Robertson and Lady Grey. Berg River Bridge (Piketberg side).
    b. Eastern Districts.—Port Elizabeth. Top of Gaika's Kop, Amatola Mountains (*J. H. Bowker*).
  D. Kaffraria Proper.—Bashee River (*J. H. Bowker*).
  F. Zululand.—Special locality not noted (*Dr. Andrew Smith*). In Coll. Brit. Mus.

212. (11.) Zeritis Osbecki (Aurivillius).

♂ *Phasis Osbecki*, Auriv., Lep. Mus. Lud. Ulr., in K. Sv. Vet.-Akad. Handl., Bd. 19, n. 5, p. 117 (1882).

Exp. al., 1 in.—1 in. 1 lin.

Very closely allied to *Thysbe*, Linn.

♂ *Pale-orange, with an irregular discal row of quadrate black spots; very broadly shot with silvery-blue.* Fore-wing: blue extends over the whole surface from base to extremity of discoidal cell and along inner margin to beyond middle, usually almost obliterating terminal disco-cellular black spot and sometimes partly the lower three spots of discal row,—in the latter case meeting hind-marginal blackish border at posterior angle; this border moderately wide, even throughout, usually marked externally with more or less defined orange lunules; spots of discal row arranged as in *Thysbe*. *Hind-wing:* discal row of spots as in *Thysbe*, but smaller, the fourth, fifth, and sixth sometimes obsolete; blue usually extending as far as discal row; hind-marginal blackish border linear, usually obsolete or fragmentary below third median nervule. *Cilia* only slightly mixed with whitish between nervules. UNDER SIDE.—*Hind-wing, and moderately-wide costal, apical, and hind-marginal border of fore-wing, dull greyish ochrey-yellow, without or with very faint markings. Fore-wing:* all the spots as in *Thysbe*, those above median nervure and its third nervule similarly centred with silvery; submarginal macular streak indistinct, ferruginous, with a broad blackish inferior termination above submedian nervure; hind-marginal border unvaried by any crossing streaks. *Hind-wing:* markings either absent altogether or very faintly exhibited in outlines indicating the characteristic darker clouding and silvery liturae of *Thysbe*.

♀ *Blue very much paler and duller, and not extending to middle in either wing;* spots of discal row larger, especially in hind-wing, where all six are well developed; cilia more whitish between nervules. UNDER SIDE.—As in ♂; but in hind-wing the darker cloudings and silvery liturae more apparent—especially the latter, which are faintly metallic.

Hind-margins not so dentate as in *Thysbe*; the angulation in fore-wing blunter, not so prominent.

It is doubtful whether this form is really more than a variety of *Thysbe*, especially as the only ♀ s (two) I have seen, and also one ♂ taken near Malmesbury and another at Port Nolloth, in the Cape Colony, indicate with some clearness on the under side of the hind-wings the characteristic markings of that species. At the same time, the smaller size, blunter outline of wings, usually more extended silvery-blue in the ♂, and (even in the examples just specified) obscureness of the under side, render *Osbecki* easily recognised. The character last named approximates this form to *Pyrois*, Trim., in which, however, the spots indicated are rounded.

Mr. Aurivillius sent me for inspection the type of this species from the Stockholm Museum; it is a ♂. He notes (*loc. cit.*) that it bore the undoubtedly wrong label of "India orientalis." It agreed very nearly with the ♂ mentioned beneath as having been taken by myself at Blaauwberg.

I took a ♂ and a very small ♀ (exp. only 10 lines) at Port Nolloth in August 1873, and did not meet with the form again until September 1879, when Colonel Bowker and myself each captured a ♂ near Malmesbury. At Blaauwberg, on the coast of the Cape District, I met with a fine pair in the following January; and Mr. L. Péringuey gave me four similar ♂ s which he captured rather farther to the north, at Paternoster Bay, Malmesbury District. Thus far, therefore, it would appear to be a form peculiar to the Western Coast Districts of the Cape Colony. There is nothing in its habits to distinguish it from its nearest allies.

Localities of *Zeritis Osbecki*.

I. South Africa.
    B. Cape Colony.—Port Nolloth, Little Namaqualand. Paternoster Bay, Malmesbury District (*L. Péringuey*). Malmesbury. Blaauwberg Beach, Cape District.

213. (12.) Zeritis Palmus (Cramer).

Papilio Palmus, Cram., Pap. Exot., iv. pl. cccxli. ff. F, G (1782).
♂ ♀ *Zeritis Thysbe*, *var.*, Trim., Rhop. Afr. Aust., ii. p. 265, n. 165 (1866).

Exp. al., (♂) 1 in.—1 in. 2½ lin.; (♀) 1 in.—1 in. 3½ lin.

♂ Orange-red, sub-metallic near bases; an irregular discal row of small black spots; and a narrow black hind-marginal border, becoming obsolete in lower half of hind-wing. *Fore-wing*: base (except on costa) with a narrow black suffusion; costa with a linear edging of black; a terminal disco-cellular spot, and a discal row of six spots, arranged quite as in *Thysbe*, Linn., but all considerably smaller and less quadrate; hind-marginal border and external orange-red lunules as in *Thysbe*. *Hind-wing*: a basal black suffusion, broader than in fore-wing and prolonged somewhat widely along inner margin to considerably beyond middle, where it terminates in a point; discal spots arranged as in *Thysbe*, but smaller and thinner, the sixth (last) spot of row minute or sometimes absent; hind-marginal linear border quite as in *Thysbe*.
UNDER SIDE.—As in *Thysbe*, except that the black discal spots of the fore-wing are smaller, and the ferruginous markings of hind-wing constantly well developed.

♀ Paler than ♂; less sub-metallic near bases; basal blackish duller, and slightly narrower. UNDER SIDE.—As in ♂.

Seven specimens collected at Maseru, in Basutoland, by Colonel Bowker are smaller than usual; the bases of the wings have a much wider though duller blackish suffusion, the hind-marginal border is wider, and the cilia strongly alternated with black and white. One ♂ has the costa of both wings (more especially that of the hind-wing) clouded with blackish; all the hind-wing discal spots prolonged interiorly, and a total want of the external lunulate orange edging between hind-marginal border and cilia. A ♀ of this small dark description has been sent from Burghersdorp by Dr. Kannemeyer.

Though so conspicuously different from *Thysbe* (Linn.) in the total want of the splendid silvery-blue basal gloss, *Palmus* is so exceedingly close to that species that I have long doubted whether to treat it as distinct. The under-side pattern is identically the same in the two butterflies; but the discal spots seem to be constantly smaller (except in the small northern examples just mentioned) on both surfaces; and on the upper side the deeper red ground-colour and the more pronounced inner-marginal black of the hind-wing appear to be constant distinctions.

As noted in *Rhop. Afr. Aust.*, ii. p. 266, I possess a single ♂, taken near Cape Town, which has a faint but distinct blue lustre over the basal region of the hind-wings, but in no other way differs from ordinary specimens.

Though not numerous in individuals, *Palmus* is more generally to be met with than *Thysbe*, especially near Cape Town. Its warm bright-red colouring makes it conspicuous during its short flights, or when perched on twigs or flowers of low plants. It appears from the middle of September to the end of April.

Localities of *Zeritis Palmus*.

I. South Africa.
   B. Cape Colony.
      a. Western Districts.—Cape Town. Eerste River, Noord Hoek, and Simon's Town, Cape District. Paarl. Robertson. Swellendam (*L. Taals*). Knysna. Plattenberg Bay.
      b. Eastern Districts.—Burghersdorp (*D. R. Kannemeyer*).
      d. Basutoland.—Maseru (*J. H. Bowker*).

214. (13.) Zeritis Thero, (Linnæus).

♀ [*Papilio Thero*, Linn., Mus. Lud. Ulr. Reg., p. 328, n. 146 (1764); and Syst. Nat., i. 2, p. 787, n. 219 (1767).
Papilio Rumina, Dru., Ill. Nat. Hist., i. pl. ii. ff. 1, 1 (1770).
♀ *Papilio Salmoneus*, Cram., Pap. Exot., iv. pl. cccxli. ff. D, E (1782).
Papilio Erosine, Fab., Mant. Ins., ii. p. 51, n. 506 (1787).
Papilio Pulsius, Herbst., Nat. Bek. Ins.,—Schmett., vi. pl. 156, ff. 6, 7 (1793).
♂ ♀ *Polyommatus Thero*, Godt., Enc. Meth., ix. p. 662, n. 154 (1819).
Zerythis Thero, Chenu, Enc. d'Hist. Nat., Pap., i. f. 351 (1852).
♂ ♀ *Zeritis Thero*, Trim., Rhop. Afr. Aust., ii. p. 276, n. 172 (1866).

Exp. al., (♂) 1 in. 4½–8 lin.; (♀) 1 in. 8 lin.—2 in.

♂ *Dark-brown, with orange-red spots;* bases rather widely scaled with orange-ochreous; *hind-wing with two tails*, a rather long acute one on submedian nervure, and a shorter one on first median nervule; *cilia* whitish, with wide brown interruptions at ends of nervules. *Fore-wing*: spots subquadrate; one in cell, and another just beyond it; a discal row of five parallel to hind-margin (of which the first three are smaller and sometimes indistinct) between fourth subcostal and third median nervules; below first median, rather before middle, the largest spot in wing. *Hind-wing*: a submarginal row of four broad lunular spots between discoidal nervule and submedian nervure (the first, and sometimes the second, and even the third also, occasionally obsolete), of which the last, as well as anal-angular lobe, is marked with a pale greyish-edged black spot. UNDER SIDE.—*Hind-wing and borders of fore-wing, soft-grey clouded with brown. Fore-wing: orange-red;* two silvery-white centred black spots in cell, and a third, larger, closing it; spots of discal row more or less confluent, the first four (there being an additional costal one) dimly white-centred; a spot on either side of origin of first median nervule; midway between discal row and hind-margin, a parallel submacular suffused fuscous stripe, outwardly mixed with ferruginous. *Hind-wing*: central area brown; the following silvery-white markings, sometimes with ferruginous edges, viz., a small spot near base, just below costal nervure; a transverse row of four before middle, of which the second (in cell) is large and elongate; *closing cell, a large irregular marking, emitting a broad ray along dis-*

coidal nervule to beyond middle, where it unites with the fourth spot of a very irregular discal row of seven or eight linear spots; immediately beyond this row, a dark-brown narrow fascia; hind-margin paler, with a submarginal row of more or less distinct blackish lunules.

♀ Similar to ♂; the orange spots larger, those of discal row of fore-wing confluent. *Fore-wing*: sometimes an additional orange spot between second and first median nervules, close to their origin. *Hind-wings*: some two or three small orange marks on disc before lunular row, of which the spots are larger and brighter. UNDER SIDE.—All markings more clearly defined.

VAR. A. (♂ and ♀).

Orange markings feebly developed,—in ♀ almost obsolete in one example, and in another broad and suffused in fore-wing only. UNDER SIDE.—*Grey of a hoarier tint, the brown clouding indistinct. Fore-wing*: dark band beyond discal row more suffused, and marked mesially with a row of seven small rather indistinct hoary-grey spots. *Hind-wing: silvery-white spots reduced in number, those present attenuated,* especially the large elongate marking from extremity of discoidal cell; all spots before middle (except lowest of row of four) obsolete,—the absence of the usually conspicuous spot in cell very noticeable. The subapical angulation of the fore-wing is much less marked, especially in the ♀. (*Hab.*—Namaqualand, Cape Colony.)

An approach to this variety is exhibited by a ♂ taken by Colonel Bowker between Somerset East and Murraysburg, Cape Colony. In this example, while the upper side is normal, there is almost as much failure of the silvery-white spots on the under side of the hind-wing, where, however, the dark-brown clouding remains.

Specimens of both sexes taken at Uitenhage by the same gentleman are larger than usual, and, while they resemble the variety as far as the hoary-grey and rather attenuated markings of the under side of the hind-wings are concerned, differ remarkably from it on the upper side, having the orange markings extremely well developed,—the ♂ s, in particular, exhibiting one four and the other *five* conspicuous spots in the submarginal row of the hind-wings, and having the lowest of them tinged with lake. These very handsome examples further differ from the variety in having the dark-brown clouding of the under side of the hind-wings well marked.

This very handsome and distinct species attains a wider expanse of wings than any other South-African Lycaenide. The large and conspicuous silvery-white marking (which, as Linné remarks (*Mus. Lud. Ulr. Reg.*, p. 328), is hooked at the extremity), is a central feature of the under side of the hind-wings quite peculiar to the species, and often catches the eye when the butterfly is at rest. *Thero* frequents shrubs in waste sandy places, and seems specially to delight in the hardly rigid plants that grow on sandhills close to the sea. Near Malmesbury I observed that it was very partial to the common *Melianthus major*. It is very wary and its short flights are very rapid, but it settles so frequently on exposed twigs or shoots, that, with a little patience, the collector will not fail to secure specimens. When flying it much resembles an *Erebia* or *Pseudonympha* in size and colour, but its motions are very different. September and October are the months in which it is most prevalent, but it is not uncommon in August and November, and I once took it early in January.

Localities of *Zeritis Thero.*

I. South Africa.
B. Cape Colony.
  a. Western Districts.—D'Urban Road, Hout Bay, and Kalk Bay, Cape District. Malmesbury, Riebeck's Kasteel, Kalbas Kraal, and Bridgetown, Malmesbury District. Piketberg. Stellenbosch. Paarl. Worcester, and Hex River Station (*L. Péringuey*), Worcester District. Robertson. Swellendam (*L. Taats*). Mossel Bay. Messkraal, Elboogfontein, and Oograbies, Namaqualand District [*Var*].
  b. Eastern District.—Uitenhage (*J. H. Bowker*). Between Somerset East and Murraysburg (*J. H. Bowker*).

215. (14.) Zeritis Sardonyx, Trimen.

♂ ♀ *Zeritis Sardonyx*, Trim., Trans. Ent. Soc. Lond., 1868, p. 83, pl. v. f. 5, pl. vi. ff. 6, 7.

Exp. al., (♂) 1 in. 7–9½ lin.; (♀) 1 in. 9 lin.—2 in. 1 lin.

♂ *Fulvous-orange, with fuscous markings and hind-marginal border. Fore-wing:* base with a slight fuscous-grey suffusion; costa rather narrowly bordered with creamy or vinous-yellow from base to a little before middle, thence with fuscous to apex; hind-marginal border broad, of nearly even width to about posterior angle, where it is considerably enlarged; a rather irregular submacular transverse discal band, of moderate width, from costal border, its lower extremity joining hind-marginal border on first median nervule; usually a small rounded black spot in discoidal cell; a larger, squarer, terminal disco-cellular spot confluent with costal border. *Hind-wing:* base next costa shining pale ochreous-grey, next inner margin narrowly suffused with fuscous-grey; apical area occupied broadly with brownish-black, which divides to form (1) a narrow hind-marginal edging, disappearing before attaining submedian nervure, and (2) a broader discal ray, ending rather abruptly on first median nervule; inner margin whitish; anal-angular lobe and caudal projection vinous-ochreous. *Cilia* conspicuously alternated with black and white. UNDER SIDE.—*Hind-wing and margins of fore-wing varying from dull vinous to dark ferruginous-ochreous. Fore-wing:* orange ground-colour paler than on upper side; the following silvery-white black-ringed spots, viz., three disco-cellular (the additional spot being near base), and an irregular discal band of six. *Hind-wing:* two conspicuous, short, transverse, pure-white streaks about middle, between costal and subcostal nervures, at a little distance apart; beyond middle a similar but much longer, slightly irregular, streak extending from first subcostal nervule to inner margin; adjoining this streak a hoary-grey inner-marginal space, extending narrowly along margin, outwardly to lobe at anal angle, and inwardly as far as a small narrow white mark; externally bounding upper part of discal white streak, a small purplish-fuscous cloud; usually some pruinose bloom over basal half of inner margin and middle of hind-margin. Cilia of the colour of the margins alternated with whitish.

♀ *Considerably paler, markings similar. Fore-wing*: hind-marginal border comparatively narrower, the space of ground-colour between it and discal band broader. *Hind-wing*: apical black not so extended, the discal part ending on third median nervule and the hind-marginal edging being narrower. UNDER SIDE.—As in ♂.

Considerable variation is shown in the width and extent of the discal band. A ♂ from Griqualand West, taken by Mrs. Barber, has the band in both wings much attenuated inferiorly; and a ♀ from the same country, captured by myself, has it in the fore-wing very narrow throughout, and in the hind-wing prolonged, as in the ♂, but broader. A very worn ♀, the locality of which is not recorded, in the South-African Museum exhibits the band in the fore-wing in a yet more reduced form, its upper part being very narrow and its lower part obsolete.

This very distinct species has no close ally known to science. In outline of wings and the character of the upper-side markings it has much of the aspect of *Chrysorychia Harpax* (Fab.), but, apart from its much larger size, the under side of the hind-wings has a totally different pattern, three pure-white streaks replacing the numerous metallic spots of that species. Structurally it has no near affinity to *Harpax*, the gradually clavate antennæ, long palpi, and five-branched subcostal nervure of fore-wing showing its close relationship to *Z. Thero* (Linn.) and *Argyraspis*, Trim.

Sardonyx inhabits the dry upland districts of the north and north-east of the Cape Colony. In 1864 single examples reached me simultaneously from Burghersdorp and Murraysburg, having been captured respectively by Mr. D. R. Kannemeyer and Mr. J. J. Muskett. The latter correspondent subsequently sent me several fine specimens, from which I described the species at the end of 1867. In December of that year Mr. J. P. Mansel Weale sent me a good drawing of a ♀ taken at Cradock, with the note that the species was "abundant about Cradock Flats on the ice-plant;[1] all the specimens varying slightly." The only living example I have seen is the ♀ above mentioned as having been captured in Griqualand West; I met with it at Kolberg on the 6th September 1872, flying about a rocky hillside.

Localities of *Zeritis Sardonyx*.

I. South Africa.
 B. Cape Colony.
  b. Eastern Districts.—Murraysburg (*J. J. Muskett*). Cradock (*J. P. Mansel Weale*). Burghersdorp (*D. R. Kannemeyer*).
  c. Griqualand West.—Kolberg. Vaal River (*Mrs. Barber*).

216. (15.) Zeritis Argyraspis, Trimen.

Zeritis Malagrida (part), Trim., Rhop. Afr. Aust., ii. p. 344 (1866).
♂ ♀ *Zeritis Argyraspis*, Trim., Trans. Ent. Soc. Lond., 1873, p. 114, pl. i. ff. 7, 8.

Exp. al., (♂) 1 in. 4–6½ lin.; (♀) 1 in. 7–9 lin.

♂ *Orange-fulvous, with rather wide fuscous bordering; cilia wide, fuscous, conspicuously varied with pure white between extremities of ner-*

---
[1] *Mesembryanthemum crystallinum.*

vules. *Fore-wing*: fuscous border very broad in apical region, occupying outer half of costa, and emitting a short ray towards base along subcostal nervure; on third median nervule the border abruptly narrows, thence forming a rather wide and tolerably even band to anal angle; base slightly clouded with fuscous. *Hind-wing*: border similarly developed in apical region, but abruptly constricted a little above third median nervule, and thence to anal angle being very much narrower than the corresponding portion in fore-wing; base faintly tinged with fuscous-grey; inner margin clouded with an ill-defined fuscous bordering beyond middle. UNDER SIDE.—*Hind-wing and bordering of fore-wing fuscous-ochreous, with bright-silvery, narrowly black-edged spots. Fore-wing*: on costa, at base, a longitudinal silvery streak; two small, rounded, silvery-white spots in discoidal cell near base, and an elongated similar spot at extremity of cell; nine spots in apical region, viz., two minute ones close to costa; one small rounded one a little beyond end of cell; two larger ones placed so as to form an oblique elongate marking between small rounded spot and apex, and four in a row from apex along hind-margin as far as third median nervule, the lowest spot the largest; orange-fulvous replaced by a broad creamy band along inner margin; inner portion of hind-marginal border below apical region marked with a row of three good-sized blackish spots. *Hind-wing*: on basal lobe of costa a rather large subovate spot; beyond it, below costal nervure, two similar spots placed longitudinally; a small spot in discoidal cell close to base, followed by a dot; a very irregular, roughly V-shaped marking at extremity of cell; between this marking and inner margin are a similar but more elongate marking and a rather small subovate spot; beyond middle a very irregular transverse row of eight spots of various sizes and shapes, of which the second and fourth (of about equal size) are much larger than the rest; a hind-marginal row of nine spots, of which the three last form an oblique line to inner margin beyond middle; at anal angle a small, short, oblique, red mark, bordered on each side with white; quite at base, below origin of submedian nervure, a small round spot like the rest.

Fore-wing with hind-marginal outline strongly elbowed about extremity of radial nervures. *Hind-wing* with a slight anal-angular lobe, and two tails of moderate length, broad at the base and tapering gradually to a point at extremities of first median nervule and submedian nervure respectively.

♀ Similar to ♂, *but fuscous bordering relatively not so broad*, and scarcely any tinge of fuscous at bases. *Fore-wing*: bordering in apical region not so greatly wider than its other portion, and emitting no ray along subcostal nervure, but only two lines from costal edge on second and third subcostal nervules; near anal angle the bordering is wider than about middle of hind-margin. UNDER SIDE.—Rather paler in ground-colour, but otherwise as in ♂. *Fore-wing*: the lowest spot of hind-marginal row vertically divided into two well-separated spots.

In outline the wings of the ♀ are more even and blunted, the *fore-wings* being less elbowed below apex, but with a slight prominence at end of first median nervule.

A streak margining the front of each eye; another at lateral base of each palpus; and six spots on each side of the abdomen (the latter edged with black); silvery-white. General colouring of body fuscous-ochreous, with two longitudinal whitish stripes on each side of breast; legs ochreous, with whitish femora.

A ♀ example from Murraysburg differs from other specimens in the great breadth of the fuscous bordering, particularly in the fore-wings.

There is a series of variations which leads from *Z. Malagrida*, Wallgrn., in the direction of *Z. Argyraspis*, but the latter, which is the largest and finest form I have seen, is sufficiently well marked in both sexes to demand separation as a species. From the type *Z. Malagrida* (to which M. Wallengren, who has seen specimens which I have forwarded to him, informs me must be referred the "var. *Aglaspis*" of *Rhop. Afr. Aust.*, ii. p. 272) *Z. Argyraspis* seems to be constantly distinguished by *the very broad field of orange-fulvous, which extends to the costal edge of fore-wings* on the upper side, and by the much brighter, more clearly defined, and rounder silvery spots of the under side. As regards the latter, it is very noticeable that *the spots of the hind-marginal row in the hind-wings are not sagittate*, and that *those of the row beyond the middle are far more irregularly placed* than in *Malagrida*, and present no approach to the continuity which almost forms a stripe in that species. In addition to these distinctions should be mentioned the much larger size of *Argyraspis*, and the different outline of the wings, which latter consists in a marked prominence of the apical region of the fore-wings, and the production of the slight dentations in the hind-wings of *Malagrida* at the ends of the submedian nervure and first median nervule into distinct pointed tails.

Dr. Kannemeyer was the first to communicate to me this striking form in the shape of a ♀ taken near Burghersdorp, in the Albert Division of the Cape Colony. Mr. Muskett, in 1864 and in 1870, sent me examples of both sexes from Murraysburg. Mr. E. L. Layard showed me a specimen taken by him near Beaufort West, and early in 1871, Mr. J. H. Bowker sent me a very fine ♂, taken "on the way from Murraysburg to Somerset East."

In August 1873 I had the good fortune to meet with this butterfly in the Namaqualand District, capturing two of each sex. These specimens were all easily taken, sitting constantly on the bare ground. They differ from the Murraysburg specimens in their narrower fuscous border (especially in the fore-wings), and in the larger silvery spots of their under side.

Localities of *Zeritis Argyraspis*.

I. South Africa.
   B. Cape Colony.
      a. Western Districts.—Springbokfontein, Messkraal, and Buffel's River, Namaqualand District. Beaufort (*E. L. Layard*).
      b. Eastern Districts.—Murraysburg (*J. J. Muskett*). Between Murraysburg and Somerset East (*J. H. Bowker*). Burghersdorp (*D. R. Kannemeyer*).

217. (16.) **Zeritis Wallengrenii,** *sp. nov.*

♂ ♀ *Zeritis Malagrida* (part), Trim., Rhop. Afr. Aust., ii. p. 272, pl. 5, f. 3 [♂], 1886.

Exp. al., (♂) 1 in. 2–6 lin.; (♀) 1 in. $5\frac{1}{2}$–$8\frac{1}{2}$ lin.

♂ *Fuscous-black, with a large dull-orange basi-discal patch in each wing; bases rather widely tinged with fuscous-grey; cilia fuscous-black, with small but conspicuous inter-nervular white interruptions.* Fore-wing: costa edged narrowly with greyish yellow-ochreous, wider near base; orange patch commencing near base, occupying lower part of discoidal cell, covering median nervure and basal halves of its three nervules, and extending narrowly to inner margin about middle;· extremity of discoidal cell marked by a pale ochreous-yellow narrow spot indenting the black above. Hind-wing: orange patch occupying costa to about middle, and crossing to inner margin, near which, beyond middle, it is usually confluent with a partly oblique submarginal orange band commencing on third median nervule and ending at anal angle; projection at extremity of first median nervule short, rather blunt,—that at anal angle much longer, slightly curved, rather broad, rounded at tip. ochrey-yellow tinged with lake. UNDER SIDE.—*Hind-wing and apical area and hind-marginal border of fore-wing yellow- or vinous-ochreous, clouded with brownish between nervures, and with numerous slender, mostly arrow-shaped, bright-silvery, very thinly black-edged marks.* Fore-wing: costa narrowly bordered with ochreous, and near base edged with whitish; orange extends from base over whole of cell and almost to hind-margin, fading into whitish all along inner margin; two small round black-edged silvery spots in discoidal cell; a larger one externally angulated at extremity of cell; spots of apical area arranged as in *Argyraspis*, but the four spots along hind-marginal portion sagittiform; the lowest spot of these four interiorly touches a round one which is the first of an oblique row of three reaching to first median nervule (occasionally a fourth, nearer base, very faintly indicated beneath that nervule); outwardly bordering orange, between third median nervule and submedian nervure, a row of three large suffused fuscous sub-sagittiform spots, with silvery points externally. *Hind-wing:* neuration and a submarginal ray paler than rest of area; the following silvery markings, viz., a narrow spot close to costa and base; two opposedly sub-sagittiform marks between costal and subcostal nervures; two small round spots in discoidal cell; a large superiorly-acuminate spot at extremity of cell; three or four small spots below cell; a discal irregular row of eight or nine variously shaped marks, of which the two respectively between subcostal nervules and second and first median nervules are opposedly sagittiform; and a hind-marginal row, interrupted near costa and near anal angle, of sharply-sagittiform marks resembling those in the fore-wing. *Cilia*, mixed fuscous and ochreous, the white interruptions less conspicuous than on upper side.

♀ *Duller, paler. Fore-wing:* a submacular narrow transverse orange ray from subcostal nervure, not far before apex, joins discal orange on third median nervule; costal ochreous more developed. *Hind-wing:* submarginal orange ray beginning on radial nervule, separated from discal orange by an irregular fuscous stripe; anal-angular projection much shorter than in ♂, but that at extremity of first median nervule rather better developed,—a deep excavation between the two. UNDER SIDE.—As in ♂.

VAR. A. (♂ and ♀).

♂ Orange in fore-wing a little broader exteriorly; submarginal orange ray in hind-wing much broader on third median nervule, and usually completely confluent with discal orange. UNDER SIDE.— *Hind-wing and apical area of fore-wing pale-fuscous, the neuration yellow-ochreous in parts only; all the silvery spots (except disco-cellular ones of fore-wing) much thickened,* but decidedly less so than in *Argyraspis*, the arrow-head ones retaining their characteristic form. *Fore-wing:* edging of costa, from base to middle, silvery-white. *Hind-wing:* no marked projection on first median nervule, but anal-angular one very prominent.

♀ Orange very pale and dull; the ray in fore-wing indistinct or obsolete; that in hind-wing more even than in typical ♀. UNDER SIDE.—As in ♂.

(*Hab.*—Swellendam and Grahamstown, Cape Colony.)

A few examples (three ♂ s and a ♀) sent from near Grahamstown by Colonel Bowker and Mrs. Barber exhibit some divergence from the variety in their duller upper-side orange (especially dusky in basal half of hind-wing) and in the thicker, blunter silvery markings of the under side, some of which partly coalesce. The ♂ s, too, are unusually small (1 in. 1–3½ lin.), and the fore-wings less produced apically.

The markings of the head and body agree with those of *Argyraspis*, but are thinner and less conspicuous.

In outline of wing this species is in both sexes blunter and less elbowed hind-marginally than *Argyraspis*, and the ♂ has the anal-angular projection well developed in the hind-wing, and that at extremity of first median nervule small or obsolete—just the opposite of what occurs in *Argyraspis*. In most respects it holds a middle place between the species just mentioned and *Malagrida*, Wallengr., but the silvery markings are remarkably thin and acute, more so than in the latter. The variety, however, is unmistakably closer to *Argyraspis* as regards its under-side colouring and marking; but both forms of *Wallengrenii* differ greatly from that species in their very limited field of orange on the upper side, which is, on the contrary, much larger and brighter than in *Malagrida*.

I met with this butterfly rather numerously on hills near Stellenbosch in December 1862, and captured the paired sexes on the 20th of that month. It constantly settles on the ground, and can be taken with the fingers if cautiously approached.

It is with much pleasure that I dedicate this species to the well-known Swedish lepidopterist, Pastor H. D. J. Wallengren, the recorder of the rich South-African collections formed by the lamented Wahlberg.

Localities of *Zeritis Wallengrenii*.

I. South Africa.
B. Cape Colony.
 a. Western Districts.—Stellenbosch. Swellendam (*L. Taats*).—
  Var. A.
 b. Eastern Districts.—Grahamstown (*J. H. Bowker* and *M. E. Barber.*)—Var. A.

218. (17.) Zeritis Malagrida, Wallengren.

♂ *Cygaritis Malagrida*, Wallgrn., K. Sv. Vet.-Akad. Handl., 1857, Lep. Rhop. Caffr., p. 43.

♂ ♀ *Zeritis Aglaspis*, Trim., Trans. Ent. Soc. Lond., 3rd Ser., i. p. 286 (1862).

♂ ♀ *Zeritis Malagrida*, Var., Trim., Rhop. Afr. Aust., ii. p. 272 (1866).

Exp. al., (♂) 11 lin.—1 in. 2 lin.; (♀) 1 in. 2-7 lin.

♂ *Blackish-brown, with a dull-orange discal space in each wing; cilia long, blackish-brown, widely interrupted with white between nervules.* Fore-wing: basal area widely suffused with glossy yellowish-grey; orange longitudinal space very restricted, barely entering discoidal cell, covering three median nervules from their origins, and ending abruptly at some distance before hind-margin; a narrow pale-yellowish terminal disco-cellular spot touching orange space; usually a very faint, small, suffused, subapical orange spot near costa. Hind-wing: a rather wide submarginal orange band between discoidal nervule and submedian nervure, dentated both interiorly and exteriorly by three dark crossing nervules; a slight suffusion of orange about origins of median nervules; inner-margin bordered with whitish-grey from base to beyond middle. UNDER SIDE.—*Hind-wing and apical area and narrow hind-marginal border of fore-wing greyish or vinous-greyish, intersected by pale-yellowish neuration, and with rather dull silvery markings thinly edged with black.* Fore-wing: orange-yellow fading into white along inner margin; two small longitudinally elongate-ovate silvery-white black-ringed spots in discoidal cell, and a similar vertically elongate one at its extremity; two similar very small spots close to costa beyond middle; an oblique silvery streak from apex, commencing a bi-angulated row of small silvery spots extending to first median nervule; close to hind-margin a row of outwardly-produced sub-sagittiform silvery marks from apex to third median nervule, below which the row is continued to submedian nervure by three rather suffused blackish marks. Hind-wing: the following silvery markings, viz., a dot at base, close to costa; two in discoidal cell, and a large sub-trigonate spot at its extremity; two converging short curved streaks from costal to subcostal nervure, before middle; an irregular, discal, zigzag macular streak, rather suffused on its inner edge, continuous from second subcostal nervule to inner margin; a row, close to hind-margin, of sagittiform marks, becoming acuter inferiorly, edged interiorly with red and preceded by a whitish streak. *Cilia* greyish or vinous-greyish, with white inter-nervular interruptions.

♀ *Duller and paler. Fore-wing:* basal area more decidedly tinged with ochrey-yellow; a rather suffused macular orange ray from subcostal nervure near apex joining discal orange on third median nervule. *Hind-wing:* basal and inner-marginal area much suffused with ochrey-yellow; orange band less macular, more continuous, not so dentated by nervules, suffused on its inner edge. UNDER SIDE.—As in ♂. *Hind-wing:* discal macular silvery streak more completely interrupted on first and second subcostal nervules.

A ♀ sent to me from the Transvaal by Mr. David Arnot is considerably larger than usual, expanding 2 in. 7 lin.; the orange of the fore-wing, and to a rather less extent that of the hind-wing, is extended basewards and mingled imperceptibly with the ochrey-yellow suffusion, and the silvery markings of the under side are much enlarged and conspicuously whiter than usual.

The fore-wing in *Malagrida* has scarcely any trace of hind-marginal subapical prominence in the ♂, and none at all in the ♀. This, together with its smaller size, very restricted field of orange in both wings, and the continuous and comparatively regular course of the discal silvery streak on the under side of the hind-wings, readily distinguish it from its allies *Wallengrenii* and *Argyraspis*, Trim. It is also very noticeable that the anal-angular projection of the hind-wing is very short, and that on first median nervule scarcely perceptible.

I first met with this species in March 1860 on the Lion's Hill at Cape Town, mistaking the first example for a brightly-marked *Thyra*, Linn., as it pitched on the ground or on stones. It is extremely local, but numerous about its special haunts. Though tolerably swift on the wing, it is not easily roused, and never flies for more than a few yards at a time; it almost always settles on the ground, and I have taken several examples with my fingers. It seems to be strictly a late summer butterfly, as I have not met with it before the beginning of February or after the end of March.

Localities of *Zeritis Malagrida.*

1. South Africa.
    B. Cape Colony.
        a. Western Districts.—Cape Town.
    K. Transvaal.—Locality not noted (*D. Arnot*).

219. (18.) **Zeritis Thyra,** (Linnæus).

PLATE IX. fig. 5 (♀).

♂ *Papilio Thyra,* Linn., Mus. Lud. Ulr. Reg., p. 329, n. 147 (1764); and Syst. Nat., i. 2, p. 789, n. 227 (1767).
♂ *Papilio Nycteus,* Cram., Pap. Exot., iv. pl. ccclxxx, ff. F, G.
Papilio Eradrus, Fab., Mant. Ins., p. 89, n. 806 (1787); and (*Hesperia Eradrus*) Ent. Syst., iii. 1, p. 343, n. 306 (1793).
♂ *Alocides Thyra,* Hübn., Samml. Exot. Schmett., ii. pl. 88 (? 1806).

Polyommatus Ecadrus (Var. 2), Godt., Enc. Meth., ix. p. 672 (1819).
♀ *Zeritis Thyra*, Westw., Gen. Diurn. Lep., pl. 76, f. 9 (1852).
♂ ♀ *Zeritis Thyra*, Trim., Rhop. Afr. Aust., ii. p. 273, n. 170 (1866).

Exp. al., (♂) 1 in. 2–6 lin. ; (♀) 1 in. 2½–7½ lin.

♂ *Orange-red, with blackish borders. Fore-wing:* border rather narrow at base (where it is suffused with greyish-yellow-ochreous, which sometimes narrowly edges costa to apex), but widening greatly to apex, whence, abruptly narrowing on third median nervule, it is moderately broad along hind-margin to posterior angle, where it usually widens again ; base blackish mixed with ochreous. *Hind-wing:* border sometimes nearly obsolete on costa, but in many Cape Town examples rather broad ; always developed very broadly about apex, narrowed abruptly on third median nervule, and thence only a dentate edging to anal angle. UNDER SIDE.—*Hind-wing and border of fore-wing greyish or brownish ochreous, or often purplish-lake*, the former *with irregular whitish streaks and spots. Fore-wing:* orange usually paler, especially towards inner margin, which is bordered with whitish ; two black spots in cell, one closing it ; an inwardly-oblique, transverse, bi-angulated row of five or six beyond it, and a suffused submarginal row,—all spots above median nervure and its first branch silvery-white centred, those of cell very conspicuously ; below cell two blackish markings (one basal, one just beyond), usually confluent. *Hind-wing:* two small spots at base, one on costa, one in cell ; a transverse row of four before middle (one in cell) ; a long streak closing cell, usually confluent inferiorly with a most irregular, very sharply dentated, more or less suffused transverse stripe immediately succeeding it ; a small spot touching costal nervure about middle ; a submarginal dentated streak, often indistinct, always so towards costa. In both wings a rather indistinct hind-marginal row of spots, sometimes interiorly bordered by some whitish marks. *Cilia* blackish, interrupted with white between nervules.

♀ *Orange usually rather paler and duller, border not so dark ;* ochreous basal suffusion usually much broader. *Fore-wing:* at outer and upper edge of orange there is often an upward projection of that colour as far as first radial nervule ; border often very broad at posterior angle. *Hind-wing:* apical border usually broader than in ♂ ; usually a suffused blackish spot between first median nervule and submedian nervure near anal angle. UNDER SIDE.—As in ♂.

This species is very variable, chiefly in the colour of under side of hind-wing and the distinctness of its markings, but also in the extent and tint of the orange of upper side, which in some specimens is very limited in extent, in others very broad. I have no doubt of this being the true *Thyra* of Linné, having taken one specimen near Cape Town which answers in every particular to the details given in "Museum Ludovicæ, &c." In the majority of examples the under-side markings are not so bright as Linné notes them.

The Cape Town specimens, in both sexes, have the upper side red deeper in colour, and more restricted in area (especially as regards the hind-wing) than those from more eastern localities.[1] In the latter, the red in both wings commences close to base, and in the hind-wing always occupies basal half of costa, while in the fore-wing it is usually confluent with the basal ochrey-yellow border of costa. In a very large pair (♂ 1 in. 6 lin., ♀ 1 in. 7½ lin.), which I took in Bain's Kloof, on the Worcester side, the ♂ presents a very broad area of bright orange-red, while the ♀, though the red is well developed basally, has the dark border so enlarged as to cover the outer half of each wing. Specimens from the neighbourhood of Port Elizabeth, and in a less degree those from near Grahamstown, are not only smaller than usual, but exhibit on the under side of the hind-wing rather less irregularity in the discal whitish streak, and about the middle of the submarginal streak a small but distinct whitish cloud or suffusion.[2] The paired sexes of this variation, captured in October 1879, at Zwartkops railway cutting, were sent to me by Colonel Bowker. Examples from Kaffraria Proper and Natal are also small, and in them the very wide red field is decidedly paler and yellower, and on the under side a more or less decided lake hue seems constantly present, together with more glistening whitish markings in the hind-wing. Cramer's figures (*op. cit.*) roughly represent a ♂ in which the upperside orange is rather pale and of restricted area, and indented in both wings by a disco-cellular dark spot; the under side is of the purplish-lake coloration. Hübner's excellent illustrations are evidently from Cape Town specimens; they exhibit ♂s of both colorations, but Hübner has given the larger (lake beneath) as the ♀.

PUPA.—Dull yellowish-green; back of thorax greener than the other parts. Abdomen with a median dorsal dark-ashy streak. Spiracles raised, brownish. About half an inch in length; thick, rounded, smooth; head blunt; dorso-thoracic prominence elevated but blunt. (Lying, quite free of any attachment or cocoon, under a stone, near Cape Town, 27th September 1874.)

The larva's skin, which still surrounded the tail of this pupa, was dull-brown, rather closely set with short black (and some white) spines; the head rather large, shining dark-brown.

A ♂ imago emerged on 20th October.

---

[1] A remarkable exception occurs in a ♂ taken by M. L. Péringuey at Hex River Mountain in the Worcester District, where the red is much more restricted than in any Cape Town example that I have seen, forming in the fore-wing only a narrow discal ray, and in the hind-wing only a discal patch wholly beyond middle.

[2] Very near these, but paler and duller generally, and with the whitish cloud much less apparent, are a ♀ from Basutoland, and another (with border of fore-wing broader than usual) from Griqualand West, both taken by Colonel Bowker. Two larger ♀s from Little Namaqualand are similar, but have the lake coloration on the under side. A small ♂ from the Eastern Transvaal has the fore-wing border remarkably narrow, and the strix of the under side (lake) of the hind-wing glistening and unusually regular.

This widely-distributed species is common in many localities, preferring pathways and other spots of bare ground on hillsides on which it constantly settles, taking short rapid flights, and returning to the same or an immediately adjacent spot. It seldom visits flowers, but I have occasionally noticed it on *Brunia* and some others. Though most prevalent from October to March, I have taken it as early as August and as late as the end of April.

Localities of *Zeritis Thyra*.

I. South Africa.
  B. Cape Colony.
   a. Western Districts.—Springbokfontein, Namaqualand. Cape Town. Stellenbosch. Bain's Kloof (Worcester side). Hex River Mountain (*L. Péringuey*). Genadendal, Caledon District (*G. Hettarsch*). Knysna.
   b. Eastern Districts.—Port Elizabeth. Between Zwartkops and Coega Rivers (*J. H. Bowker*). Grahamstown.
   c. Griqualand West.—Vaal River (*J. H. Bowker*).
   d. Basutoland.—Koesberg (*J. H. Bowker*).
  D. Kaffraria Proper.—Bashee River (*J. H. Bowker*).
  E. Natal.
   a. Coast Districts.—D'Urban and Pinetown (*J. H. Bowker*). Great Noodsberg.
  K. Transvaal.—Lydenburg District (*T. Ayres*).

220. (19.) Zeritis Aranda, (Wallengren).

♂ *Cygaritis Aranda*, Wallgrn., K. Sv. Vet.-Akad. Handl., 1857,—Lep. Rhop. Caffr., p. 43.
♂ ♀ *Zeritis Mars*, Trim., Trans. Ent. Soc. Lond., 3rd Ser., i. p. 285 (1862).
♂ ♀ *Zeritis Pierus*, Cram., Var. A., Trim., Rhop. Afr. Aust., ii. p. 275.

Exp. al., (♂) 11 lin.—1 in. 1 lin.; (♀) 1 in.—1 in. 3½ lin.

♂ *Orange-yellow, with blackish-brown borders. Fore-wing*: base below median nervure very narrowly tinged with dark-grey; orange-yellow covers costa to a point a little before extremity of discoidal cell, but is there slightly paler and duller; a small, more or less distinct, fuscous, terminal disco-cellular spot; border commencing abruptly and rather broadly on costa, widening considerably to apex, and continued below third median nervule, evenly and rather broadly, along hind-margin to posterior angle. *Hind-wing*: sometimes an indistinct terminal disco-cellular fuscous lunule; border commencing with a large, broad, inferiorly-rounded, costal-apical patch, very abruptly narrowing on discoidal nervule into a streak (interiorly dentated between nervules) along hind-margin to anal angle; an ill-defined fuscous spot near anal angle, between first median nervule and submedian nervure (sometimes obsolete); a short, acute, anal-angular projection at extremity of submedian nervure. *Cilia* glistening fuscous-grey mixed with purplish-lake, *not* interrupted with white. UNDER SIDE.—*Hind-wing and costal, apical, and hind-marginal border of fore-wing variable in tint, ochrey-brownish, ferruginous-brownish, or purplish-lake, marked with small,* mostly rather ill-defined, glistening sub-metallic spots with dark edges. *Fore-wing:* three disco-cellular spots, one below cell; an irregular, bi-angulated, oblique discal row of six, all black with brilliant silvery-white centres; a submarginal regular row of six spots, of which the three superior are small and indistinct, but the three inferior large and black, occasionally with minute silvery dots on their inner side. *Hind-wing:* three rather ill-defined transverse rows of small sub-metallic spots,—the first of four or five before middle including a terminal disco-cellular lunule,—the second of eight or nine discal, sinuate, irregular,—the third submarginal, regular, but becoming obsolete superiorly; near base three glistening small spots, viz., one close to costa and two in discoidal cell.

♀ *Similar; dark borders broader.* *Fore-wing:* border more deeply excavating orange-yellow immediately beyond extremity of discoidal cell, and sometimes emitting an ill-defined suffused ray across median nervules. *Hind-wing:* apical dark patch larger, usually a little suffused; anal-angular projection blunter. UNDER SIDE.—As in ♂, but all the spots usually better defined.

In a ♀ which I captured at Knysna, in the Cape Colony, the dark border of the fore-wing is so enlarged as to occupy the outer two-thirds of the whole area, with the exception of a small indistinct discal orange-yellow spot; and the costal-apical patch of the hind-wing is much larger than usual, as well as the spot near anal angle.[1]

Specimens from the Basuto Territory and the Transvaal exhibit in both sexes a narrower dark border, especially near the apex of the fore-wing, where it is, moreover, usually penetrated (in some cases almost divided) by an upward projection or short ray from the discal part of the orange-yellow.[2] Examples from the Coast of Natal resemble these, and some of the ♂s have the costal part of the border of the fore-wing still narrower, while in both sexes the anal-angular projection of the hind-wing is longer.

In colouring and in the extent of the orange on the upper side, *Aranda* much resembles the Eastern specimens of *Thyra*, Linn., but the under-side markings, particularly those of the hind-wings, and the greater proportion of silvery-centred spots on the fore-wings, show that it is more nearly allied to *Pierus*, Cram., *Taïkosama*, Wallengr., and *Almeida*, Feld., though differing from all of these in the entire absence of the row of fuscous spots close to the hind-margin. The anal-angular projection of the hind-wing is much acuter and longer in *Aranda* than in its near allies; the butterfly is also the smallest of the group.

I met with this species pretty commonly at Knysna in 1858–59 from the middle of October to the middle of January, and again from the beginning of March to that of April. It was almost confined to the hills, frequenting dry sandy spots, and in habits did not differ from its near congeners. In Feb-

---

[1] There are two similar but larger ♀s (unfortunately without locality) in the South African Museum; they have the bases, however, much suffused with fuscous grey, so that the orange is greatly obscured or obsolete on costa, and in the hind-wing as well as in the fore-wing a more or less marked fuscous discal ray divides the orange. Two ♂s accompanying them have the orange redder, and reduced by wider dusky bases and dark borders; in both there is an indication of the dark discal ray across the orange of each wing.

[2] Of two ♀s from Kaffraria Proper, one possesses and the other (the largest specimen I have seen) does not possess this short orange ray. Both have the border of fore-wing broadly developed, especially on costa beyond middle.

ruary 1867 I took a ♂ specimen of the Natal variation on a hillside near Verulam, and Colonel Bowker has lately (1884–85) sent several examples of both sexes found near Pinetown. The only Cape Town specimen I have seen was taken by myself on 30th November 1862.

Localities of *Zeritis Aranda*.

I. South Africa.
   B. Cape Colony.
      a. Western Districts.—Cape Town. Knysna.
      b. Eastern Districts.—Fort Murray and Fort Hare, British Kaffraria (*W. S. M. D'Urban*).
      d. Basutoland.—Maseru (*J. H. Bowker*).
   D. Kaffraria Proper.—Bashee River (*J. H. Bowker*).
   E. Natal.
      a. Coast Districts.—Pinetown (*J. H. Bowker*). Verulam.
      b. Upper Districts.—Riet Spruit, near Upper Umgeni (*W. Morant*).
   K. Transvaal.—Potchefstroom and Lydenburg Districts (*T. Ayres*).

221. (20.) Zeritis Almeida, (Felder).

♂ *Nais Almeida*, Feld., "Verh.' Zool.-Bot. Gesell. Wien., xii. 1, p. 478 (1862);" and Reise d. Novara, Lep., ii. p. 264, pl. xxxii. ff. 25, 26 (1865).

♂ ♀ *Zeritis Pierus*, Trim. [part], Rhop. Afr. Aust., ii. p. 274, n. 171 (1866).

Exp. al., (♂) $11\frac{1}{2}$ lin.—1 in. $2\frac{1}{2}$ lin.; (♀) 1 in. 2–4 lin.

♂ *Dark orange-red, with very broad blackish-brown borders; bases widely suffused with mixed ochrey-yellow and fuscous-grey; cilia* greyish mixed with fuscous, at posterior angle of fore-wing slightly whitish. *Fore-wing*: orange-red very restricted, commencing at about origin of first median nervule, scarcely extending below submedian nervure or above median nervure and its third nervule (except, rarely, at its outer superior extremity, where a narrow projection sometimes reaches subcostal nervure), and much narrowed inferiorly by broadening inward of hind-marginal border from near posterior angle. *Hind-wing*: basal suffusion extending more or less along inner margin beyond middle; costal-apical patch very large, extending to base of third (and sometimes of second) median nervule; hind-marginal border very narrow, sharply and regularly dentating orange between nervules; orange-red just enters extremity of discoidal cell. UNDER SIDE.—*Hind-wing and border of fore-wing greyish or ferruginous-brown (rarely purplish-lake), the former with transverse rows of indistinct dark-edged shining-grey or sub-metallic spots, and usually with more or less whitish suffusion on disc. Fore-wing*: costa near base thinly edged with white; ordinary black spots arranged as in *Aranda*, Wallengr., but very rarely any silvery white-centred ones below median nervure; last blackish spot of submarginal row usually much suffused interiorly; posterior angle whitish; an additional hind-marginal row of rather indistinct elongate dark spots. *Hind-wing*: spots arranged as in *Aranda*, but (as in fore-wing) an indistinct hind-marginal row of dark spots; often a faint whitish or hoary suffusion immediately preceding indistinct spots of submarginal row.

♀ *Rather paler and duller;* cilia paler generally, but with darker interruptions on nervules, whiter at posterior angle of fore-wing. *Fore-wing*: orange usually rather more extended at extremity of discoidal cell and on inner margin about middle, and almost always emitting a more or less developed upward discal ray to or towards subcostal nervure. UNDER SIDE.—As in ♂. *Hind-wing*: discal whitish rather more diffused.

A ♂ from between the Zwartkops and Coega Rivers, and a ♂ and two ♀ s from Murraysburg, are considerably larger (♂ s 1 in. 4 lin., ♀ s 1 in. 5–6 lin.) than the more Western examples. On the under side these ♂ s have the whitish at posterior angle of fore-wing more pronounced, and extending along inner margin, and the discal whitish suffusion of hind-wing immensely developed; while in the ♀ s the former character is equally, the latter less markedly, exhibited. In the ♀ s, moreover, the former character is unusually well manifested in the cilia on the upper side, and on the same surface the orange-red is much developed towards base, the customary basal suffusion being in one specimen almost obsolete.

A ♂ I took at Knysna much resembles the darkest of the dusky ♀ s of *Aranda* above mentioned, the orange-red being on the upper side limited in the fore-wing to a small patch before middle and a small discal spot, and in the hind-wing to a faint central-inferior ray and a patch along lower half of hind-margin. Another ♂ from the same locality is, however, remarkable for the increased size and brighter tint of the orange-red in both wings.

In general aspect and colouring of the upper side *Almeida* bears considerable resemblance to *Thyra*, Linn., especially in the larger ♀ examples. Besides the much more rounded outline of the wings in the ♂, and the absence of white in the cilia, the markings of the under side of the hind-wing are altogether different, the central irregular whitish streaks and the submarginal similar streak of *Thyra* being replaced by rows of small shining dark-edged spots arranged as in *Aranda*. To the latter *Almeida* is very closely allied, though its much darker colouring and more limited orange-red field (especially in the fore-wings) give it a very different aspect. The additional row of spots close to hind-margin seems the most constant distinction of *Almeida*, but this character recurs in all the nearly allied forms except *Aranda*.

The habits and haunts of *Almeida* do not differ from those of its near allies; it constantly settles on the bare ground in hilly places, and is very easily captured. I found it rather numerously at Knysna from the middle of October to the middle of March. Near Cape Town it is certainly rare, and I have only met with a very few individuals in the neighbourhood of Wynberg. A single example only, taken by Mr. T. Ayres, has reached me from the Transvaal.

Localities of *Zeritis Almeida*.

I. South Africa.
B. Cape Colony.
   a. Western Districts.—Cape Town. Caledon (*T. D. Butler*). Knysna. Plettenberg Bay.
   b. Eastern Districts.—Between Zwartkops and Coega Rivers, Uitenhage District (*J. H. Bowker*). Murraysburg (*J. J. Muskett*).
K Transvaal.—Potchefstroom (*T. Ayres*).

222. (21.) Zeritis Pierus, (Cramer).

PLATE IX. fig. 6 (♂).

♀ *Papilio Pierus*, Cram., Pap. Exot., iii. p. 84, pl. ccxliii. ff. E, F (1782).
Hesperia Suetonius, Fab., Ent. Syst., iii. 1, p. 320, n. 213 (1793).
♂ ♀ *Zeritis Pierus*, Var. B., Trim., Rhop. Afr. Aust., ii. p. 275 (1866).

Exp. al., (♂) 1 in. 2–4½ lin. ; (♀) 1 in. 2–5 lin.

♂ *Greyish-brown, with a wide basal suffusion of rather glistening hoary-grey; a small orange marking in each wing, sometimes obsolete, or nearly so, in fore-wing.* Fore-wing: orange limited to two or three short central rays about the median nervules, and often only represented by two small indistinct spots or a few scales,—occasionally altogether wanting. Hind-wing: hoary-grey suffusion extends along inner-marginal area; orange forming a small patch of two or three rays near hind-margin between third median nervule and submedian nervure, sometimes much obscured and indistinct. Cilia greyish-white (becoming white about posterior angle of fore-wing) interrupted with greyish-brown on nervules. UNDER SIDE.—*Varying from whitish to greyish and ferruginous brown, with paler glistening dark-edged spots in hind-wing.* Fore-wing: orange; silvery-white-centred, black discocellular, and other spots, quite as in *Aranda*, Wallengr.; additional hind-marginal row of spots as in *Almeida*, Feld., but more distinct. Hind-wing: spots arranged for the most part as in *Almeida*, but considerably larger, paler, and with their dark edges much more strongly marked, especially exteriorly,—those of discal row elongated and contiguous, forming a macular streak interrupted only on third median nervule and submedian nervure; row of submarginal spots obsolete; hind-marginal row almost so, except the two spots next anal angle, which are blackish, immediately preceded by a paler mark.

♀ *Basal grey suffusion wider, mixed with dull-yellowish; orange paler, more developed in both wings.* Fore-wing: orange interiorly blending insensibly with basal suffusion, reaching narrowly as far as submedian nervure, and from its upper outer extremity on third median nervule emitting (as in *Almeida*, &c.) a ray as far as subcostal nervure. Hind-wing: orange largely developed over hind-marginal area (as in *Almeida*, &c.); dentated hind-marginal edging sometimes reduced to inter-nervular fuscous spots. UNDER SIDE.—*Usually rather paler than in ♂, and with the spots of the hind-wing and the marginal ones of both wings more distinct.*

This is a very close ally of *Almeida*, Feld., but distinguishable by the much paler brown of the upper side in both sexes, the remarkable limitation of the orange on the upper side of the ♂, and the disposition and development of the spots on the under side of the hind-wings, which are so strongly bordered with brown, and in the discal row form an almost continuous streak. It is the exaggeration of this last feature in Cramer's figure above cited which convinced me that this was the actual butterfly intended to be delineated by that very rough representation, which in its darker colouring rather more resembles the ♀ *Almeida*.

This is by far the commonest *Zeritis* about Cape Town, usually making its appearance early in October, and remaining out till the end of April. In 1870 I saw a specimen on the wing as early as the 9th September. It is to be noticed everywhere on dry open spots, especially frequenting pathways, and sitting so close as almost to be trodden on by the passing foot. Both sexes present rather a dull appearance when taking their very short flight, but the male looks particularly dingy. The under-side colouring so resembles the soil as to be an excellent protective disguise when the butterfly is at rest. The only specimen I took at Knysna is a ♂ with a larger orange patch than usual on the upper side of both wings, and with the brown borders of the spots on the under side of the hind-wing unusually broad.[1]

Localities of *Zeritis Pierus*.

I. South Africa.
   B. Cape Colony.
      a. Western Districts.—Cape Town. Wellington. Stellenbosch. Vogel Vley, Tulbagh District. Robertson. Swellingdam (*A. C. Harrison*).
      b. Eastern Districts.—Port Elizabeth. Uitenhage (*J. H. Bowker* and *S. D. Bairstow*). Between Zwartkops and Coega Rivers, Uitenhage District (*J. H. Bowker*). King William's Town (*W. S. M. D'Urban*). Between Somerset East and Murraysburg (*J. H. Bowker*).
   D. Kaffraria Proper.—Bashee River (*J. H. Bowker*).

223. (22.) Zeritis Taikosama, (Wallengren).

♀ *Cygaritis Taïkosama*, Wallengr., K. Sv. Vet.-Akad. Handl., 1857,—Lep. Rhop. Caffr., p. 43.
♂ ♀ *Zeritis Pierus*, Var. B. [part], Trim., Rhop. Afr. Aust., ii. p. 275 (1866); and Trans. Ent. Soc. Lond., 1870, p. 372.

Exp. al., (♂) 11 lin.—1 in. 1½ lin.; (♀) 1 in.—1 in. 4 lin.
Very closely allied to *Pierus*, Cram.

♂ *Brownish-grey, paler, and with a faint glistening-yellow suffusion over basal area; a narrow discal, macular, pale, dull orange-yellow streak in fore-wing, and a similar larger hind-marginal marking in hind-wing; cilia greyish, indistinctly interrupted with fuscous on nervules. Fore-wing:* pale, dull, orange-yellow streak narrow, submacular, lying between subcostal nervure and first median nervule, paler and less distinct in its superior portion, sometimes wholly very indistinct, and rarely quite obsolete. *Hind-wing:* hind-marginal pale, dull, orange-yellow markings, almost as variable in extent of development as the streak of fore-wing, sometimes all but obsolete, but when fullest developed forming a moderately broad band between radial nervule and anal angle, rather sharply indented with fuscous exteriorly between nervules. UNDER SIDE.—*Hind-wing and border of fore-wing greyish or reddish-brown, the former with*

[1] In a series of examples taken by Colonel Bowker between Somerset East and Murraysburg, there are gradations of the ♂ in which the orange of the fore-wing on the upper side varies from a few scales only to a patch almost as large as in *Almeida* ♂.

three transverse rows of sub-metallic or glistening dark-edged spots. Fore-wing: orange-yellow field very pale, whitish towards inner margin; black spots as in *Pierus*, Cram., but their silvery-white centres more metallic, and *those of the submarginal row also bearing a silvery-white dot on their inward side*. *Hind-wing*: spots smaller than in *Pierus*, those near base and in row before middle similarly disposed; but those of discal row all separate, not forming an almost continuous streak, and with the third spot from costa (between second subcostal and radial nervules) out of line with and nearer base than the rest, and the three lower spots not inclining towards anal angle; *an additional row of similar spots midway between discal row and hind-margin, sinuated similarly, but not quite parallel to the former*.

♀ Basal suffusion more tinged with yellow; orange-yellow varying in extent from being little more developed than in ♂ to presenting a broad discal field much as in *Pierus*, though of a paler and duller tint. Fore-wing: orange-yellow either entire from outer part of basal suffusion, or narrowly, broadly, or very broadly divided into two by the ground-colour on median nervules. *Hind-wing*: orange-yellow always better developed than in ♂, sometimes merging into basal suffusion, its outer edge strongly indented by fuscous, which sometimes forms separate inter-nervular spots. UNDER SIDE.—As in ♂, but the spots of hind-wing less metallic; the hind-marginal fuscous spots better defined, especially in hind-wing.

The characters italicised above seem constantly to distinguish this form from *Pierus*,—the principal ones being, on the under side, the smallness and separateness of the spots of the discal row and the presence of an outer row of spots besides the latter in the hind-wings, and the more or less conspicuous silvery-white dots marking inwardly the black spots of the submarginal row in the fore-wings. Usually *Taïkosama* is smaller than *Pierus*, but two or three females from the upper districts of Natal, in which the upper-side orange, though dull in tint, occupies a wider space than usual, are quite as large. Much resembling these ♀s is a ♂ from Zululand of unusually large size, with the upper side orange, though paler, developed quite as largely as in *Almeida*, Feld.

This form seems peculiar to the interior of South Africa, where it has evidently a considerable range. It was first brought prominently to my notice as probably separable from *Pierus* by a long series of specimens taken in Basutoland by Colonel Bowker; and Mr. W. Morant subsequently sent me several examples from Natal, the Orange Free State, and the Transvaal, all taken in September and October 1870.

Localities of *Zeritis Taïkosama*.

I. South Africa.
  B. Cape Colony.
    b. Eastern Districts.—Burghersdorp, Albert District (D. R. Kannemeyer).
    d. Basutoland.—Koesberg, Maseru, and Maluti Mountains (J. H. Bowker).
  C. Orange Free State.—Hebron, between Caledon and Orange Rivers (W. Morant).

D. Kaffraria Proper.—Tsomo River (*J. H. Bowker*).
E. Natal.
 b. Upper Districts.—Blue Krantz River and Colenso (*W. Morant*).
F. Zululand.—Napoleon Valley (*J. H. Bowker*).
K. Transvaal.—Special localities not noted (*W. Morant, H. Barber, and T. Ayres*).
II. Other African Regions.
A. South Tropical.
 b1. Interior.—Tauwani River, near Bamangwato (*F. C. Selous*).

224 (23.) Zeritis Molomo, Trimen.

♀ *Zeritis Molomo*, Trim., Trans. Ent. Soc. Lond., 1870, p. 373, pl. vi. f. 9.

Exp. al., (♂) 1 in. 2½ lin.; (♀) 1 in. 4–5½ lin.

♂ *Orange-yellow, with a blackish-brown costal patch and hind-marginal border in fore-wing, and costal-apical patch and hind-marginal edging in hind-wing.* Fore-wing: costa from base to a little before middle rather broadly bordered by ochre-yellow; costal patch united narrowly on costa to apical commencement of hind-marginal border, acuminate inferiorly, piercing the orange field as far as third median nervule, so as to leave a narrow orange ray beyond its outer edge; hind-marginal border broad, of even width from apex to posterior angle, slightly denticulate on its inner edge; an indistinct paler lunule at extremity of discoidal cell. *Hind-wing*: costal-apical patch like that in fore-wing and also piercing the orange as far as third median nervule, but narrower and much closer to hind-margin and more intimately united to the commencement of hind-marginal edging; the latter almost linear below second subcostal nervule, but distinct, well-defined, and rather sharply dentating the orange between nervules. *Cilia* broad, whitish, with wide fuscous interruptions on nervules. UNDER SIDE.—*Hind-wing and costal, apical, and hind-marginal border of fore-wing very pale creamy-ochreous; hind-wing with large dark-edged dull-silvery spots, and with a dark-brownish suffusion over basal and inner marginal area, and also between central and discal rows of spots.* Fore-wing: silvery-white centred black spots quite as in *Taikosama*, Wallengr., and *Orthrus*, Trim., but more brilliant; spots of hind-marginal row minute but distinct, very slightly edged with orange. *Hind-wing*: spots arranged as in the species just named, except that the second and third rows are considerably closer together, and that the spots of the hind-marginal row are very minute; a whitish suffusion between discal and hind-marginal rows of spots; an additional metallic spot near base, just below and beyond first spot of first transverse row.

♀ *Similar to ♂, orange-yellow rather deeper in tint, dark marginal markings in fore-wing rather broader.* Fore-wing: between hind-marginal border and cilia an orange-yellow line, interrupted by small nervular black spots corresponding with the dark interruptions of the cilia. *Hind-wing:* costal-apical patch blunter inferiorly, with (in two examples) faint indications of a linear continuation over lower part of disc; hind-marginal edging broken into a series of quite free and separate inter-nervular spots and an outer series of nervular ones. UNDER SIDE.—Rather deeper in tint; spots generally better defined; the brown clouding in hind-wing inclining to ferruginous.

Variety A. (♂ and ♀).

♂ *Fore-wing:* costal patch much longer, beginning at a point below costal nervure about middle of superior margin of discoidal cell; hind-marginal border rather narrowed about the middle, more deeply dentated with orange on nervules along its inner edge. *Hind-wing:* costal-apical patch considerably broader, its inferior extremity much farther from hind-margin, not far from origin of third median nervule.

♀ Rather paler and duller. *Fore-wing:* costal border more or less tinged with greyish; costal patch smaller, sometimes suffused and ill-defined; *hind-marginal border more or less deeply pierced by orange rays on median nervules and second radial nervule (in one specimen as far as hind-margin itself*). *Hind-wing:* costal-apical patch occupying same position as in ♂, but usually suffused and much reduced in size.

UNDER SIDE in both sexes with the spots smaller and duller, and with the tint of the hind-wing rather more reddish and with less distinct brownish clouding.

(*Hab.*—Griqualand West, Eastern Interior of Cape Colony, Delagoa Bay.)

On the upper side this species bears much resemblance to *Aranda*, Wallengr., but its under-side markings plainly indicate its nearer alliance to *Orthrus*, Trim., and *Taïkosama*, Wallengr. The larger size and greater lustre of the metallic spots on the under side of the hind wings perhaps best distinguish it from the two latter, though the very great difference in the upper-side colouring constitutes the most immediately obvious distinction. I do not think it advisable to separate the variety just described from *Molomo*, though the invasion of the hind-marginal border of the fore-wing by the orange ground-colour gives the ♀ a very peculiar aspect.

The ♂ of this butterfly was taken by Mr. W. Morant in the Orange Free State in November 1868, and in the Transvaal on 25th October 1870. Colonel Bowker had previously (October 1864) sent ♀s from Kaffraria Proper, and in December 1868 forwarded another from Basutoland. The latter naturalist also sent me all the examples of the variety I have seen, except a ♂ which I took at Klipdrift, Griqualand West, on 19th September 1872, a ♀ from Kimberley (in which the dark markings are almost obsolete) belonging to Mr. H. Grose Smith, and a ♀ from Delagoa Bay given me by Mrs. Monteiro.

Localities of *Zeritis Molomo*.

I. South Africa.
  B. Cape Colony.
    b. Eastern Districts.—Between Somerset East and Murraysburg (*J. H. Bowker*).
    c. Griqualand West.—(Var. A.) Klipdrift, Vaal River. Kimberley (*H. G. Smith*). Vaal River (*J. H. Bowker*).
    d. Basutoland.—Koro-Koro (*J. H. Bowker*).

C. Orange Free State.—Special locality not noted (*W. Morant*).
H. Delagoa Bay.—Lourenço Marques (*Mrs. Monteiro.*—Var. A.)
K. Transvaal.—Special locality not noted (*W. Morant*).

225. (24.) **Zeritis Orthrus,** Trimen.

♂ *Zeritis Orthrus*, Trim., Trans. Ent. Soc. Lond., 1874, p. 340, pl. ii. f. 10.

Exp. al., (♂) 1 1½ lin.—1 in. 2½ lin.; (♀) 1 in. 1–5 lin.

♂ *Fuscous-grey, with pale-grey submarginal markings.* Fore-wing: costa narrowly clouded with pale-grey from base to near apex; a transverse submarginal stripe, ill-defined and somewhat dentated on its edges, extending from costa to first median nervule; usually a thin terminal pale disco-cellular lunule. *Hind-wing:* between radial nervule and anal angle *a narrow, sharply-dentated streak*, of which the outward dentations reach the hind-marginal edge upon the extremities of the nervules. Cilia of the ground-colour irregularly varied with pale-grey, chiefly between nervules. UNDER SIDE.—*Hind-wing and costal, apical, and hind-marginal border of fore-wing very pale creamy-grey, with sub-metallic spots; the markings generally almost identical with those of Z.* Taïkosama, *Wallgrn.* Fore-wing: the two spots before middle, below median nervure and its first nervule, smaller and much more indistinct than in *Taïkosama*, the white centre often wanting in one or the other of these spots; the outer of the two submarginal rows of spots better marked than in *Taïkosama*. *Hind-wing: the third row of spots (beyond middle) obsolete*, only represented by some scarcely traceable darker clouding, mixed with a few sub-metallic scales; a rather well-marked hind-marginal row of pale fuscous spots ending with a black spot at anal angle; a little before the last-mentioned spot a black dot.

♀ *Like the* ♂, *but with the pale-grey stripe in fore-wing suffused and indistinctly radiating along nervules to hind-margin; or considerably darker and browner, with the stripe in both wings ochre-yellow, usually more or less radiating in fore-wing.* Fore-wing: stripe varying much in length, width, and distinctness,—in one example grey superiorly and ochre-yellow inferiorly, usually emitting more or less prolonged projections on nervules towards hind-margin; terminal disco-cellular lunule usually indistinct or obsolete. *Hind-wing:* hind-marginal stripe always narrow, usually commencing on second, but sometimes on first, subcostal nervule; the deep dark inter-nervular indentations along its outer edge sometimes (as in *Taïkosama*) forming distinct spots. UNDER SIDE.—*Always rather darker than in* ♂, *and often more or less tinged with ferruginous in hind-wing and along border of fore-wing.* *Hind-wing:* spots generally more distinct and shining, especially those of third transverse row, which are more or less arrowhead-shaped and form a pretty even series quite across wing; preceding (median) row more constantly complete than in ♂, where some of the middle spots are often wanting.

There is a slight approach in several examples (of both sexes) to caudation of the hind-wing on first median nervule and submedian nervure, and in specimens from Delagoa Bay (two ♂ s, three ♀ s) this tendency is more pronounced.

This near ally of *Taïkosama* is distinguished by its darker, *more ashy ground-colour* of the upper side, with (in the ♂ constantly and in the ♀ occasionally) *pale-grey* submarginal marking, which is always restricted (whether grey or yellow) in the hind-wing to a narrow hind-marginal stripe; by the absence or slight development of any basal suffusion; and on the under side by the larger and brighter spots of the hind-wing, and the larger silvery-white centres of the spots of the fore-wing. The under-side colouring is also (especially in the ♂) much paler, and the spots of the third transverse row (all but obsolete in the ♂) of the hind-wing are distinctly sagittiform instead of rounded or sublunulate. The females in which the stripes are grey come from Delagoa Bay, as well as one in which that of the fore-wing is half grey and half yellow.

Mr. Walter Morant was the first to bring this form to my notice, in the year 1872; the two males he sent having been taken in the upper districts of Natal in September 1870. Mr. J. M. Hutchinson and Colonel Bowker have subsequently forwarded examples of both sexes from the same region; and I have received others from the Transvaal, collected by Mr. T. Ayres, and from Delagoa Bay, taken by Mrs. Monteiro.

Localities of *Zeritis Orthrus*.

I. South Africa.
 E. Natal.
  b. Upper Districts.—Estcourt (*J. M. Hutchinson*). Bushman's River and Colenso (*W. Morant*). Between Tugela and Mooi Rivers (*J. H. Bowker*).
 F. Zululand.—Napoleon Valley (*J. H. Bowker*).
 II. Delagoa Bay.—Lourenço Marques (*Mrs. Monteiro*).
 K. Transvaal.—Lydenburg District (*T. Ayres*).

II. Other African Regions.
 A. South Tropical.
  b1. Interior.—Tauwani River, near Bamangwato (*F. C. Selous*).

226. (25.) **Zeritis Barklyi**, Trimen.

♂ ♀ *Zeritis Barklyi*, Trim., Trans. Ent. Soc. Lond., 1874, p. 338, pl. ii. ff. 8, 9.

Exp. al., (♂) 1 in. 4–4⅜ lin.; (♀) 1 in. 4–5½ lin.

♂ *Pale silvery-grey, spotless. Fore-wing:* an ill-defined, dull-fuscous, macular or submacular, hind-marginal border, forming rather a broad apical mark, but narrowing lower down, and usually more or less interrupted by the ground-colour on nervules. *Hind-wing:* an elongate, ill-defined, apical, fuscous marking lying between first subcostal and radial nervules. UNDER SIDE.—*Hind-wing and* narrow border of fore-wing ashy brownish grey; rest of fore-wing orange-red. *Fore-wing*: grey border widening slightly just before and at apex; a dark fuscous dash at base inferiorly edging median nervure; black spots in number and arrangement like those of *Thyra*, Linn., and allies, viz., three white-centred cellular spots, a transverse, discal, irregular row of six, of which the four upper are white-dotted interiorly, and two parallel submarginal rows of seven each, of which the outer row is composed of very small spots indistinctly marking the inner edge of the hind-marginal grey border. *Hind-wing*: the following indistinct, hardly sub-metallic, fuscous-edged, greyish spots, viz., three cellular (basal one minute, middle one near third), two supra-cellular (wide apart), two infra-cellular, and six sub-confluent, forming an irregular discal row; a submarginal row of seven thin fuscous lunules, and a marginal spot of seven fuscous sublunulate spots (of which the three lower are well marked); between the two rows *one of pale orange lunular marks* (the three lower of which are usually well marked); on each side of the irregular discal row of spots *several more or less indistinct pale orange marks*. Cilia fuscous, varied between nervules with greyish-white.

♀ *Not so glossy; fore-wing with a subapical pale orange-yellow rounded patch*. *Fore-wing*: patch lies between third subcostal and first median nervure, extending from extremity of discoidal cell, and bounded externally by the hind-marginal fuscous band, which is darker than in ♂, broad, even, and not macular. *Hind-wing*: apical fuscous marking smaller, less distinct. UNDER SIDE.—As in ♂, but paler and duller throughout.

(Described from ten ♂ and three ♀ specimens.)

In structure and in the colouring and marking of the under side of the wings this species is plainly referable to the group of which *Z. Pierus*, Cram., may be considered as the type, but the *silvery-grey of the upper side* is a most striking distinctive character, quite unique in the genus. This peculiar colour is so pale that at first sight the expanded ♀, with its orange apical patch in the fore-wings, might almost be taken for a small dull ♂ *Anthocharis* or *Teracolus*. The under-side markings combine to some extent the characters of *Z. Thyra*, Linn., and *Z. Pierus*, Cram., but the brownish-grey inclines much more to ashy than in either of the species named, and *the lunular and scattered marks of pale orange in the hind-wings* are only found in *Z. Barklyi*.

I have named this butterfly after his Excellency Sir Henry Barkly, the Governor of the Cape Colony, to whose kindness I owe the opportunity of visiting Namaqualand, and who first called my attention to the species as something unusual. It was on the 17th August 1873, between Koekfontein and the Komaggas Mission Station, that the insect was first observed, settling on the small pink flowers of a species of *Mesembryanthemum* which carpeted the sides of the waggon-road. Other localities where it was subsequently seen were on the road from Komaggas to Spectakel; near Steinbokfontein (between Spectakel and Abbevlakte); and at Oograbies (about fifteen miles inland from Port Nolloth). It is very conspicuous on the wing, the pale upper side (of the ♂ especially) flashing like silver in the sunshine. Though settling frequently on flowers and on the ground, it is by no means so easy of capture as most of its allies, being unusually wary of approach and swift in flight. While in motion it has much the appearance of a large pale *Lycæna*, such as *L. Corydon*, Scop., or *L. Asteris*, Godt.; but when it has settled, its under-side colouring renders it as inconspicuous as its near congeners in repose usually are. In the elevated hilly country *Z. Barklyi* seemed to be rather widely dispersed, but was local in its haunts, being numerous in a few spots only, on the sunny slopes of hills. In the flat sandy country near the coast no examples were to be found.

Localities of *Zeritis Barklyi*.

1. South Africa.
   B. Cape Colony.
   a. Western Districts.—Oograbies, Steinbokfontein, between Komaggas and Spectakel, and between Komaggas and Koekfontein, Namaqualand District.

Genus PENTILA.

Pentila, Westwood [part], Gen. Diurn. Lep., ii. p. 503 (1852); Hewitson, Exot. Butt., iii. p. 119 (1866); Trimen, Rhop. Afr. Aust., ii. p. 284 (1866).

IMAGO.—*Head* small, densely scaled in front, vertex with closely-appressed hair; *eyes* large, globose, naked; *palpi* inferior, exceedingly small and short,—basal and second joints clothed beneath with long scales of unequal length, — terminal joint minute, short, obtuse; *antennæ* short, rather thick, with abruptly-formed elongate-ovate club, obtuse at tip.

Thorax very short and slender, thinly scaled. *Fore-wings* large, long; hind-margin convex, entire; costa moderately arched; costal nervure ending about middle of costa; subcostal nervure five-branched, —first and second nervules given off, not far from each other, some distance before extremity of discoidal cell,—third about midway between cell and apex,—fourth about midway between third and apex, and terminating at apex; discoidal cell unusually long; first disco-cellular nervule long, lying longitudinally,—second much shorter, oblique,—third the longest, curved, meeting first median nervule at a decided angle a good way beyond latter's origin. *Hind-wings* elongate; hind-margin very convex, entire; precostal nervure distinct; costal nervure short, ending not far beyond middle; first subcostal nervule emitted considerably before extremity of cell and ending at apex; upper disco-cellular nervule short, oblique,—lower twice as long as upper, much curved, meeting first median nervule as in fore-wing; submedian nervure extending to anal angle; internal nervure very long, ending not far before anal angle; discoidal cell rather long and narrow. Legs short, very stout, thinly scaled, without hairs; tibiæ without terminal spurs; tarsi spinose beneath: *fore-legs of ♂* with tibia very finely spinose beneath,—tarsus very short, not articulated, blunt at tip, spinose beneath.

Abdomen long, much thickened posteriorly, where it is also clothed with short appressed hair.

This curious genus consists of nine or ten species of yellow or reddish-ochreous or yellowish-white butterflies, more or less spotted and bordered with blackish, whose weak structure and sub-diaphanous, thin, lustreless wings give them the aspect of moths of the *Geometræ* group. The very small palpi, short and somewhat thick antennæ, very small thorax, long discoidal cell in both fore and hind wings, ample wings, and long incrassate abdomen, render *Pentila* easily recognised, and very unlike the mass of Lycænideous genera. In the minute palpi it resembles *Alæna*, and in the development of the superior (first) discocellular nervule of the fore-wings, and short, thick, hairless legs, it is like *Deloneura*; but in most respects it is a very isolated form, its general habit somewhat recalling that of *Pontia* (= *Nychitona*, Butl.) among the *Pierinæ*.

All the known species inhabit Tropical Africa, and, with the exception of *P. Peucetia*, Hewits., and *P. tropicalis* (Boisd.), appear to be confined to the West Coast. The two species named inhabit Eastern Africa (the Zambesi and Mozambique respectively), and *Tropicalis* is abundant on the coast of Natal. It is very doubtful whether the latter also occurs in West Africa, Hewitson having associated with it the *P. Abraxas* of Westwood, an Ashanti form, which appears to be quite distinct.

As might be inferred from its structure, *Pentila* is an exceedingly slow flyer. *P. tropicalis*, which I had many opportunities of observing in Natal, is of somewhat gregarious habits, and many examples may be met with about a particular spot, flitting feebly among shrubs or long grass.

227. (1.) Pentila tropicalis, (Boisduval).

♂ *Tingra tropicalis*, Boisd., App. Voy. Deleg. l'Afr. Aust., p. 589, n. 46 (1847).
♀ „ „ Var., Wallengr., K. Sv. Vet.-Akad. Handl., 1857,— Lep. Rhop. Caffr., p. 46.
♂ *Pentila tropicalis*, Hopff., Peters' Reise Mossamb., p. 413 (1862).
♂ „ „ Hewits., Exot. Butt., iii. pl. 60, f. 2 (1866).
♂♀ „ „ Trim., Rhop. Afr. Aust., ii. p. 284, n. 176 (1866).

Exp. al., (♂) 1 in. 3½–5½ lin.; (♀) 1 in. 3–7 lin.

♂ *Ochre-yellow, with a few small rounded blackish spots; in both wings a terminal disco-cellular spot; fore-wing with a rather wide apical hind-marginal blackish border; cilia ochre-yellow, in fore-wing with black nervular interruptions. Fore-wing:* costal border more or less thickly sprinkled with blackish atoms, especially near base; border rather broad at apex, usually marked exteriorly by linear inter-nervular ochre-yellow marks (which become continuous below third or second median nervule), but usually broken into separate spots below third or second median nervule; above cell two or three small spots, in it generally two or three others centrally situated, below it two (of which the inner is sometimes obsolete), one on each side of third median nervule. *Hind-wing:* a spot on costa before middle; on hind-marginal edge a series of minute usually sublinear nervular spots. UNDER SIDE.—*Same colour as on upper side, but everywhere, except on inner-marginal area of fore-wing, thinly sprinkled with blackish atoms;* common to both wings, an irregular discal row, and a regular sub-marginal row of small blackish spots, and a series of minute nervular spots on hind-marginal edge; terminal disco-cellular spots as on upper side. *Fore-wing:* blackish border wanting, except as slightly represented by the spots of submarginal row, of which the lowest spot is much larger than the rest. *Hind-wing:* four very small spots near base, viz., two above, one in, and one below cell; one or two similar spots in cell, farther from base; two larger spots below cell, one on each side of first median nervule.

♀ *Like ♂, except that the spots generally are smaller and fainter, and that in fore-wing the costal irroration is thinner, and the apical hind-marginal border almost or entirely obsolete.* UNDER SIDE.—As in ♂, but irroration fainter and sparser.

The nearest allies of this species are *P. Amenaida* and *P. Rotha*, Hewits., from Angola and Gaboon respectively, but both possess larger, differently situated spots, and a broad blackish border to both wings. *P. Abraxas*, Westw., associated by Hewitson (*op. cit.*, p. 119) with *Tropicalis* as a variety of the ♀, has every appearance of being quite distinct; it is yellowish-white, with both discal and marginal spots large and rounded, and is recorded from several parts of Western Africa.

I met with this curious butterfly not uncommonly on the coast of Natal in 1867, from the end of February to the end of March. It flies very slowly, always frequenting shady spots on the edge of woods. It is usual to find a good many specimens about some particular spot; at the Itongati River I met with quite a company of them settled on stems of grass, and flitting deliberately about in an avenue just at sunset. On almost all other occasions of noticing this species, I took them while flying slowly in bushy places at some height (10 to 15 feet) from the ground. Among examples received from Colonel Bowker were several noted as captured near D'Urban in the months of November and December.

Hopffer (*op. cit.*) remarks that the three specimens (♂) which he describes from Querimba were considerably smaller than Boisduval's from Natal, measuring only 1 in. 1 lin. across the fore-wings.

Localities of *Pentila Tropicalis*.

I. South Africa.
 E. Natal.
  a. Coast Districts.—D'Urban, Verulam, and Itongati River.
  II. Delagoa Bay.—Lourenço Marques (*Mrs. Monteiro*).
II. Other African Regions.
 A. South Tropical.
  b. Eastern Coast.—" Querimba " (*Hopffer*).
 B. North Tropical.
  a. Western Coast.—" Old Calabar " (*Hewitson*).

Genus D'URBANIA.

D'Urbania, Trim., Trans. Ent. Soc. Lond., 3rd Ser., vol. i. p. 400 (1862); Rhop. Afr. Aust., ii. p. 282 (1866).
Pentila, Westw. [part], Gen. Diurn. Lep., ii. p. 503 (1852).
Liptena [part], Hewits., Exot. Butt., iii. p. 119 (1866).

IMAGO.—*Head* small, clothed on front and on vertex with very short down; *eyes* smooth; *palpi* rather long, scaly, divergent, not or barely rising above top of head,—second joint long and stout, third short, slender, acuminate; *antennæ* short, rather stout, with club rather abruptly formed, sub-cylindrical, blunt at tip.

Thorax short, rather slender, laterally and posteriorly with a few short hairs above; prothorax superiorly coated densely with short hair; breast scaly and downy. *Fore-wings* more or less elongate; costa usually but slightly or moderately arched,—sometimes more decidedly so; hind-margin convex, entire; posterior angle rounded; costal nervure ending about middle of costa; subcostal nervure five-branched, first nervule emitted at half the length of discoidal cell, second midway between first and extremity of cell, third and fourth nearer apex than cell, fourth ending on costa just before apex, fifth ending a little below apex; discoidal cell short or very short; upper radial nervule joined to subcostal nervure at or a little beyond extremity of cell; inferior disco-cellular nervule considerably longer than middle one, curved, joining third median nervule at an angle some distance from latter's origin; first median nervule given off before or considerably before middle. *Hind-wings* elongate; costa prominent or very prominent at base, thence little arched or nearly straight; hind-margin entire, very convex; apex and anal angle much rounded; inner margins but slightly convex, only covering sides of abdomen; costal nervure ending about or a little beyond middle; first subcostal nervule emitted at or a little before extremity of discoidal cell; cell short; upper disco-cellular nervule rather short, transverse; lower one oblique, long, curved, joining first median nervule at an angle a good way from latter's origin; submedian nervure short. *Legs* stout, scaly, without hairs; under side of tibiæ with two or three minute spines, but no terminal spurs; tarsi long, spinose beneath; *fore-legs* of ♂ more slender, with tarsus not articulated, spinose beneath, and ending in a fringe of small spines.

Abdomen of moderate length, stout in ♀.

LARVA.—Elongate, not onisciform, the segments well marked, clothed with down and with long hair. (Described from drawings by Mrs. Barber.)

PUPA.—Broad, rather thick, convex above, flattened below; back of thorax, and especially back and sides of abdomen, with fascicles of long hair.

The rather long, divergent, almost horizontally projecting palpi, short, abruptly-clavate antennæ, short discoidal cell, and very long first subcostal and first median nervules in the fore-wings, short submedian nervure of hind-wings, and stout hairless legs are characteristic features of this genus, which I founded for the reception of a species discovered in British Kaffraria by Mr. W. S. M. D'Urban, F.L.S. Since the publication of *Rhopalocera Africæ Australis*, two closely allied species (*D. limbata* and *D. saga*, Trim.) have been found in South Africa, and to them I consider must be added the little species *Aslauga*, Trim., which in 1873 I referred to the genus *Liptena* of Hewitson. On investigating, also, some of the species referred by Hewitson to *Liptena*,[1] I have come to the conclusion that *Isca*, Hewits., *Acræa*, Westw., and probably also *Ilma*, *Libyssa*, *Lagyra*, *undularis*, and *Lircæa*, Hewits., should be included in *D'Urbania*. The species named *L. Eleaza* by Hewitson is (to judge by the figures) rightly referred by Mr. Kirby to Butler's genus *Pseuderesia* (*Trans. Ent. Soc. Lond.*, 1874, p. 532), founded on a Gold Coast butterfly which I have not seen.

The species mentioned are all purely African; out of the eleven held to belong to *D'Urbania*, three (*Amakosa*, *limbata*, and *saga*) are peculiar to the Southern Sub-Region, one (*Aslauga*) is common to South Africa and Zanzibar, and the rest are only known from Western Tropical Africa. The more typical species (*Amakosa* and near allies) are dark-brown, with spots or patches of some shade of red, and have the under side much mottled with fuscous and grey; but the longer-winged *Acræa*, Westw., has an ochrey-yellow under surface barred marginally with black and white; and the *Lagyra* group consists of white almost unspotted species with blackish margins.

The only species that I have seen in life is *D. Amakosa*, and that only on one occasion; the ♀ which I captured flew very slowly and looked like a small species of *Acræa*. This species, however, as Mrs. Barber and Colonel Bowker inform me, habitually settles on stones;

---

[1] Mr. Hewitson was aware that he had brought together under this genus butterflies of considerably differing characters, for he wrote (*Exot. Butt.*, v. p. 84), "I have preferred to place several heterogeneous species in the genus *Liptena*, rather than to make new genera to receive them."

and so does *D. saga*, as I learn from M. Péringuey, who discovered the latter in the Hex River Mountains of the Cape Colony. *D. limbata* is stated by Mr. J. M. Hutchinson to be similarly attached to stones on the hills of the interior of Natal; but *Aslauga* was found by Colonel Bowker to frequent dry branches and twigs.

D. Amakosa has a wide range over Eastern South Africa, but is not known to have occurred west of Grahamstown. The Hex River Mountains, in the west of the Cape Colony, are the only recorded locality for *D. saga*; and *D. limbata* appears to be peculiar to Upper Natal, while *Aslauga* is limited to the coast of that Colony.

It is worthy of note that the larva departs widely from the ordinary onisciform Lycaenide type, and that both it and the pupa are very hairy.

228. (1.) D'Urbania Amakosa, Trimen.

♂ ♀ *D'Urbania Amakosa*, Trim., Trans. Ent. Soc. Lond., 3rd Ser., i. p. 401 (1862); and Rhop. Afr. Aust., ii. p. 283, pl. 5, ff. 4, 5 (1866).

Exp. al., (♂) 1 in. 1–4 lin.; (♀) 1 in. 5–9 lin.

Dark-brown, inclining to fuscous; each wing with a submarginal, curved, transverse row of orange-yellow spots.

♂ *Fore-wing*: row of six spots, forming almost a semicircle, extending from subcostal to submedian nervure,—sometimes indistinctly marked. *Hind-wing*: somewhat paler than fore-wing; only four spots in transverse row, which is not markedly curved, extending from second subcostal to third median nervule. *Cilia* of both wings conspicuously chequered brown and white. UNDER SIDE.—*Fore-wing: before transverse row of spots rather thinly, beyond it thickly, irrorated with whitish;* row of orange spots increased to a broad macular stripe, narrowly edged with black on each side; the irrorations immediately succeeding stripe forming a series of acute, rather indistinct lunules. *Hind-wing: universally and densely irrorated with whitish;* the position of the transverse row of upper side faintly indicated by some scarcely distinguishable whitish spots, followed by some indistinct fuscous lunules. *Cilia* not so conspicuously chequered as above.

♀ *Orange spots in both wings enlarged and confluent, forming a broad band,* which in fore-wing widens at its lower extremity. *Cilia* as in ♂. UNDER SIDE.—Quite similar to that of ♂; orange band of fore-wing paler than on upper side.

Both sexes are exceedingly variable as regards the orange markings. The typical examples from near King William's Town and others from Kaffraria Proper have these markings in their least developed condition, and in the ♂ especially the spots are always very small, and some of them occasionally obsolete. Examples from the Natal Coast and the

Transvaal exhibit an intermediate development,[1] the spots in the forewing of the ♂ being enlarged and elongated, while those of the hindwing form a confluent band as broad as in the typical ♀; and the same markings in the ♀ are much widened, and in the fore-wing connected with the base by a partly interrupted broad inner-marginal orange ray. The examples in which these markings attain their utmost expansion are from near Grahamstown and other localities in the Albany District, where in the ♂ the hind-wing band is suffusedly extended towards base, and in the ♀ really occupies all the area except a more or less suffused central costal patch and rather narrow hind-marginal border; while in the fore-wing the spots of the ♂ are not much enlarged, but in the ♀ the inner-marginal orange is broad and uninterrupted almost to base.

In some examples of both sexes the whitish spots of the discal row on the under side of the hind-wings are outwardly marked with red.

LARVA.—Ashy-grey; a dorsal central darker line bordered on each side by a row of rather ill-defined whitish spots; on each side inferiorly a row of round red spots—those along anterior half of body with white marks between them. Head red. Rather sparsely set with moderately-long ashy-grey hairs, and (apparently) with a shorter closer clothing of ochrey-yellow hairs along the back and on each side; the long hairs apparently springing from series of tubercles. Length, 6 lin. "On a common species of grass, *Anthistiria ciliata*."—M. E. Barber. PLATE II. ff. 2 (from drawings by Mrs. Barber).

PUPA.—Pale-brown; some darker lines indicating outline of limbs. Under side smooth; a flat silken coating covering most of abdominal surface (only basal segment bare). Back and sides very hairy; the thorax and basal half of abdomen with tufts of short sandy and longer whitish hairs; the terminal half with four rows (two dorsal and two lateral) of tubercles supporting fascicles of very long bristly brown and whitish hairs, the lateral tufts being the longest. Length, 5-6 lin. —PLATE II. ff. 2a (from my own drawings).

The above descriptions are made from drawings (of the larva) and specimens (of the pupa) received from Mrs. Barber, who wrote that these earlier stages of *Amakosa* were discovered by Miss Fanny Bowker at Pembroke, near King William's Town. The caterpillars were feeding on the grass above named, and were gregarious; on a flat rock beneath the grass numerous chrysalides were attached in a group, five or six within a square inch, by a slight silken web; and among several of the butterflies close at hand there were some quite fresh from the chrysalis and unable to fly. Mrs. Barber notes that the chrysalis state

---

[1] Two ♂s and a ♀ from the high country near the source of the Kraai River, in the extreme north-east of Cape Colony, are smaller than usual; in the ♂ the hind-wing spots are confluent, but form a very narrow stripe, and in the ♀ the orange is of considerable width, but has scarcely any baseward extension in fore-wing, and none in hind-wing.

is of very short duration. A good many butterflies emerged on the journey to Grahamstown, and I extracted a fully-developed dead ♀ from one of the pupæ sent to me.

This species was first brought to my notice by Mr. W. S. M. D'Urban, F.L.S., who found it commonly near King William's Town in November 1860 and January 1861, sitting on rocks and stones. The same habits have been recorded for it by Mrs. Barber and Colonel Bowker. During my stay at Grahamstown in 1870, the season appeared to be an unfavourable one for the species, and I only met with a single specimen—a ♀ with very largely developed orange markings, which was flying slowly on the Beacon Hill at Highlands, and looked on the wing like *Acræa Rahira*. With a wide range over Eastern South Africa, *Amakosa* does not seem to be recorded from any station westward of the Albany district of Cape Colony.

Localities of *D'Urbania Amakosa*.

I. South Africa.
  B. Cape Colony.
    b. Eastern Districts.—Grahamstown. Fort Brown, Fish River, Albany District (*M. E. Barber*). King William's Town (*W. S. M. D'Urban*). "Windvogelberg, Queenstown District."—W. S. M. D'Urban. Pembroke, near King William's Town (*Miss F. Barker*). Heads of Kraai River, Barkly District (*J. H. Bowker*). Fort Warden, Kei River (*J. H. Bowker*).
  D. Kaffraria Proper.—Tsomo and Bashee Rivers (*J. H. Bowker*).
  E. Natal.
    a. Coast Districts.—Pinetown (*J. H. Bowker*).
  K. Transvaal.—Lydenburg District (*T. Ayres*).

229. (2.) D'Urbania limbata, *sp. nov.*

Exp. al., (♂) 1 in. 0½–3 lin.; (♀) 1 in. 4½ lin.
Closely allied to *D. Amakosa*, Trim.

♂ *Blackish-brown, with a hind-marginal orange-red band, macular in fore-wing, very broad and unbroken in hind-wing*. *Fore-wing*: band lying very close to hind-margin, composed of five elongate-ovate spots, between subcostal nervure and first median nervule, gradually diminishing downward; a small or minute orange spot immediately above first spot of band, and the trace of a still smaller spot just before it. *Hind-wing*: band more than twice as broad as in fore-wing, except the first spot (between subcostal nervules), which is small, and in two out of three examples partly separate from band, and extending to submedian nervure, its outer edge sinuated by the very narrow dark hind-marginal streak. UNDER SIDE.—*Hind-wing and narrow apical and hind-marginal border of fore-wing very pale greyish-brown irregularly speckled with white*. *Fore-wing*: orange band situated as on upper side, but not macular, the five spots being enlarged and confluent,—its outer edge with a series of blackish nervular indentations, its inner edge bounded by a conspicuous black streak; costal area from base to band rather widely speckled with white; inner discal area smooth fuscous, not speckled; in discoidal cell two blackish white-ringed spots, one near base rounded, the other 8-shaped about middle; a similar larger, curved, sublunulate mark at extremity of cell. *Hind-wing*: a double parallel series of white annulets near base, —two annulets being above discoidal cell, two 8-shaped ones in cell, and two below cell; an imperfect elongate suffused white lunule at extremity of cell; *a series of nine elongate, orange, outwardly black and brown dotted, white-ringed spots, submarginal for the greater part of its course, round the wing from costa beyond middle to inner margin close to base*. *Cilia* brownish, with very small white inter-nervular interruptions.

♀ *Similar, slightly paler*. *Fore-wing*: band broader, its upper three spots confluent; a very small additional (sixth) spot between first median nervule and submedian nervure. *Hind-wing*: band broader, except as regards the first spot. UNDER SIDE.—As in ♂, but subocellate orange and blackish white-ringed spots in the hind-wing enlarged and slightly suffused.

The *position of the orange band on the upper side of both wings* at once distinguishes this very handsome form from *D. Amakosa*, as in none of those of the latter in which the orange is most developed does its external edge come nearly so close to the hind-margin; *the brighter, redder tint of the orange* is also very noticeable in *Limbata*, as well as the more important distinction that, in both sexes, *the band of the fore-wings is broader in its upper than in its lower portion* (the reverse being the case in *Amakosa*); in the ♂, also the sixth (lowest) spot in the fore-wing band is absent, and in the ♀ it is very small. On the under side, the browner less ashy tint of the hind-wing and border of fore-wing, with the distinct black cellular markings and inner edging of orange band, and in the hind-wing the very distinct basal annulets and the long and conspicuous series (following three-fourths of the circuit of the wing) of white-ringed red-and-blackish spots—only a portion of which are indistinctly represented in some specimens of *Amakosa*, are also noticeable distinctions. There is much less white in the *cilia* of *D. limbata*.

The discovery of this butterfly is due to Mr. J. M. Hutchinson of Kimbolton, near Estcourt, Natal, who sent me a ♂ in June 1882. I thought it possible that this might be merely a sport of the variable *D. Amakosa*, but recommended Mr. Hutchinson to look out for other specimens; and in July 1884 I was delighted to receive from him two more ♂ s and a ♀, with the information that seven had been captured, including the paired sexes, about three miles from Estcourt, and that these examples did not present any variation, particularly in the important feature of the position of the orange bands.

Mr. Hutchinson writes that the insect occurs in several spots along the Bushman's River, but always in the same kind of station, viz., the sides or summits of rocky hills. The specimens noticed were all—with one exception, which was on the wing—sitting on stones, and were easily boxed while settled.

Localities of *D'Urbania limbata*.

I. South Africa.
   E. Natal.
   b. Upper Districts.—Estcourt and Bushman's River (*J. M. Hutchinson*).

230. (3.) **D'Urbania saga,** Trimen.

♂ *D'Urbania saga*, Trim., Trans. Ent. Soc. Lond., 1883, p. 354.

Exp. al., (♂) 1 in. 1–5 lin. ; (♀) 1 in. 6 lin.

♂ *Dark-brown ; a discal row of dull pale ochreous-yellow elongate spots in each wing*. *Fore-wing :* discal row of five spots strongly incurved, so that the last spot (between first and second median nervules) is rather nearer base than the first spot (immediately above first radial); a faint indication of a sixth spot just below first median nervule ; a row of three small and very indistinct spots of the same colour a little beyond and parallel with upper part of discal row ; at extremity of discoidal cell a scarcely perceptible ochreous-yellow spot, preceded by a similar not quite so indistinct marking in the cell not far from its extremity. *Hind-wing :* discal row of five spots usually not so distinct as in fore-wing, not curved, but rather irregular,—the first and second spots sublinear and rarely confluent,—the second being much longer than any of the others,—the last (below first median nervule) minute. *Cilia* of both wings dark-brown, conspicuously interrupted with white between the nervules. UNDER SIDE.—*Dark-brown, variegated with whitish*. *Fore-wing :* first spot of discal row small, white,— the remainder larger than on upper side, acuminate exteriorly, and of a paler yellow ; in discoidal cell a longitudinal whitish streak from base and a terminal whitish spot ; between end of cell and discal spots a transverse row of three short whitish rays ; spots beyond discal row distinct, white ; three or four small white marks between nervules on costal edge beyond middle. *Hind-wing : a conspicuous irregular discal white stripe,* well defined internally, but not externally, and *very sharply angulated on radial nervule ;* the following whitish marks before discal stripe, viz., one on costa at base, one in discoidal cell at base, and another just before extremity of cell, confluent with one immediately above it on costa ; between white stripe and hind-margin a row of thin internervular lunules, inwardly bordered by dark sagittiform marks.

♀ *Similar, larger ; spots on upper side larger and deeper in colour*. UNDER SIDE.—Duller ; all the whitish markings much reduced, especially the stripe in hind-wing, which is not nearly so widely suffused exteriorly.

Though a close ally of the very variable *D. Amakosa,* Trim., this species may at once be distinguished by the conspicuous sharply-angulated white stripe on the under side of its hind-wings. The under side of the wings generally also wants the whitish irroration always prevalent in *Amakosa*, so that the whitish markings show plainly out on the dark ground-colour. On the upper surface the discal row of spots is in the fore-wings more sharply incurved, and in the hind-wings not incurved and more irregular. The shape of the wings is also characteristic, as they (especially the fore-wings) have a much less curved costa and more produced apical region.

This very interesting addition to the genus *D'Urbania* was discovered by M. L. Péringuey, of Cape Town, on the Ilex River Mountain in the district of Worcester, Cape Colony, on the 2d January 1882. On that occasion he secured only one of the two specimens observed; but in December 1884 he was fortunate enough to capture ten more examples in the same locality, one of them being the ♀ above described. The butterfly frequented a steep ravine, and all the examples observed settled on the sides or face of the bare rocks in the full sunshine. When at rest, they could with difficulty be distinguished, so assimilated is their under surface to the colour of the rock; and on the wing they were very inconspicuous, flying in a slow, wavering manner, and suddenly dropping to settle.

Locality of *D'Urbania saga*.

I. South Africa.
  B. Cape Colony.
    a. Western Districts.—Ilex River Mountain, Worcester District (*L. Péringuey*).

231. (4.) D'Urbania Aslauga, (Trimen).

PLATE IX. figs. 9 (♂), 9a (♀).

♂ ? *Liptena Aslauga*, Trim., Trans. Ent. Soc. Lond., 1873, p. 117.

Exp. al., (♂) 1 1/2 lin.—1 in. 1 1/2 lin.; (♀) 1 in.—1 in. 2 lin.

♂ *Pale orange-ochreous; fore-wings with blackish bordering.* Fore-wing: border rather wide from base along costa, abruptly interrupted on costal edge beyond middle, but thence forming a broad apical border, which rapidly narrows along hind-margin to a point at anal angle; border marked on costal edge by four sub-quadrate spots of the ground-colour, and emitting at the point of abrupt interruption a broad ray downward as far as second median nervule; before this ray, and united to the border, a blackish marking defines extremity of discoidal cell; in some examples a similar blackish spot about middle of cell, and a smaller more indistinct one in cell nearer base. *Hind-wing:* spotless; a narrow, ill-defined, reddish-fuscous, hind-marginal edging. *Cilia* fuscous, with paler dull-whitish interruptions (more visible in *fore-wing*). UNDER SIDE.—*Hind-wing and markings of fore-wing shining leaden-grey, varied with reddish-ochreous spots.* Fore-wing: ground-colour towards inner margin paler than on upper side, but darker near costa and hind-margin; markings similar in position to those of upper side; mark at extremity of cell broader, and preceded by two similar cellular marks; ray from costal border before middle prolonged, with an inward curve, to first median nervule; cellular grey marks indistinctly prolonged below median nervure; costal spots of the ground-colour distinct; apical grey border intersected by two macular streaks of the ground-colour from costal edge, joining the ground-colour at their lower extremity; the outer of these rows is the longer, composed of more lunulate, often united spots, and situated immediately before the hind-marginal narrow grey edging. *Hind-wing*: reddish-ochreous spots arranged in five transverse rows at about equal distances apart, viz., the first, near base, of five minute elongated spots; the second, before middle, of five or six linear more or less united spots; the third, about middle, conspicuous, very irregular, of eight elongate spots touching each other and edged with blackish (which is suffused exteriorly); the fourth and fifth each of seven separated sagittiform spots, the outermost row corresponding with that on hind-margin of fore-wing; a spot at extremity of cell; three spots at base. *Antennæ* ringed alternately with black and white.

♀ *Ground-colour the same, but costal blackish bordering of fore-wing to beyond middle reduced to three or four small dusky spots on edge of costa, the outermost and darkest of which represents the abrupt termination of the border;* apical portion of border narrower and duller than in ♂, and usually terminating hind-marginally on second median nervule. UNDER SIDE.—As in ♂.

There is a little variation as regards the reddish-fuscous hind-marginal edging on the upper side of the hind-wing, which in both sexes is sometimes slightly widened and with more or less marked inward projection on nervules. Two specimens from Zanzibar in the Hewitson Collection are rather paler than the typical examples.

This very distinct little species appears to have no near congener. The pattern of the upper side and its colouring are not dissimilar from those of the ♀s of *D. Amakosa*, Trim., in which the orange is more than usually developed, but the under side is very different from that of any other *D'Urbania*.

I originally described *Aslauga* from a ♂ lent to me in 1869 by Mr. W. Morant, who took it in his garden at Pinetown, Natal, on the 19th May in that year. Mr. W. D. Gooch informed me that he had noticed it near D'Urban very rarely, but I did not receive any more specimens until the end of 1878, when Colonel Bowker sent seven ♂s and three ♀s, captured near that town on the 30th November and 1st December. The latter observer wrote that the butterfly was local, but of sociable habits, frequenting dry vine-stems and dead branches and twigs at about ten feet from the ground, and settling in little groups, repeatedly opening and closing the wings towards the sunshine.

Localities of *D'Urbania Aslauga*.

I. South Africa.
 E. Natal.
  a. Coast Districts. D'Urban (*J. H. Bowker*). Pinetown (*W. Morant*).
II. Other African Regions.
 A. South Tropical.
  b. Eastern Coast.—Zanzibar.—Hewitson Coll.

GENUS ALÆNA.

Alæna, Boisd., Voy. Deleg. dans l'Afr. Aust., App., p. 591 (1847).
Aeræa, Trim. [part], Rhop. Afr. Aust., i. p. 111 (1862).

IMAGO.—*Head* small, hairy (especially in front); *eyes* smooth; *palpi* extremely short, inferior hairy,—terminal joint minute, not visible; *antennæ* short, with large, broad, abruptly-formed, flattened, spoon-shaped club.

Thorax short, rather broad, slightly hairy on sides of back, pubescent on prothorax. *Fore-wings* elongate; costa almost straight; apex slightly rounded; hind-margin entire; inner margin slightly hollowed, thinly ciliated; costal nervure short, ending a little beyond middle; subcostal nervure five-branched,—first and second nervules given off before extremity of discoidal cell,—third and fourth not far from each other, nearer to apex than to extremity of cell (fourth terminating at apex); upper radial united to subcostal nervure a little beyond end of cell; discoidal cell of moderate length, rather narrow; disco-cellular nervules transverse, the inferior one slightly inclined outward and joining third median nervule at some little distance from latter's origin. *Hind-wings* elongate; costa very prominent at base, thence straight, ciliated throughout; costal nervure rather short, ending some distance before apex; subcostal nervure branched at extremity of discoidal cell; discoidal cell rather short and truncate; disco-cellular nervules almost equal in length, slightly oblique,—the lower one joining third median nervule (which is angulated at point of junction) at some little distance from latter's origin; internal nervure extending to about middle of inner margin, which is fringed throughout with long hairs. *Fore-legs* of ♂ stout, very hairy (especially the tibia),—tarsus short, downy, without distinct articulations, finely spinose beneath, without terminal claws;—of ♀ fully developed, less hairy than in ♂. *Middle* and *hind legs* stout, rather short, very hairy, with femora, tibiæ, and tarsi about equal in length,—tarsi strongly spinose beneath, hairy above and laterally, with terminal claws robust, strongly hooked.

Abdomen of moderate length, thick (especially in ♀), arched dorsally, blunt at tip, hairy laterally and very hairy beneath (especially in ♂) towards extremity.

This genus has undoubtedly been misplaced (from its foundation) among the *Acræinæ*, to which group it has merely a superficial likeness, and that not a strong one. Mr. H. W. Bates many years ago expressed to me his doubt whether the then only known species, *A. Amazoula*, Boisd., was really allied to *Acræa*; and the receipt since of ample material for dissection has enabled me to ascertain that *Alæna* is an aberrant form of Lycænidæ related to *D'Urbania* and *Pentila*. In neuration it more nearly agrees with the former genus, but differs in the longer discoidal cells and longer costal nervure and upper disco-cellular nervule of the hind-wings. Its very short palpi link it to *Pentila*, but are hairy instead of smooth; while its antennæ, with their large broad club, differ much from those of both genera. The very hairy legs to some extent resemble those of *Lachnocnema*, but the foretarsi of the male are of the ordinary non-articulate form, instead of being completely developed like those of the female, as in the latter genus.

Three species are now recorded, all African. The type, *A. Amazoula*, is ochre-yellow above and cream-colour below, without spots, but with the margins and neuration generally on both surfaces defined conspicuously with blackish. *A. Nyassa*, Hewits. (described in 1877), presents a very different appearance above, being blackish with a common central white band, but beneath, though whiter, is not unlike *Amazoula*. This species has been taken by Mr. Selous on the Shashani River, and may perhaps extend south of the Tropic. A third species, *A. interposita*, Butl. (described in 1883), appears intermediate in character between the two, but nearer to *A. Nyassa*.

A. Amazoula has a wide range in Eastern South Africa, from King William's Town to Zululand, but is not known from beyond those limits. It is by no means of general distribution, but extremely local, keeping to certain spots of very limited extent. Its flight is exceedingly slow, weak, and near the ground, and it settles at very short intervals.

232. (1.) Alæna Amazoula, Boisduval.

Acræa (Alæna) Amazoula, Boisd., App. Voy. de Deleg. dans l'Afr. Aust., p. 591, n. 60 (1847).
Acræa Amazoula, Trim., Rhop. Afr. Aust., i. p. 111, n. 71 (1862), and pl. 3, f. 3 [♂], (1866).

Exp. al., 11 lin.—1 in. 3 lin.

♂ *Blackish-brown, rayed between nervures with yellow-ochreous;* in both wings a longitudinal disco-cellular ray, indistinct or obsolete near base, and a curved discal transverse row of 7-8 more or less acuminate rays, of which the lowest (surmounting submedian nervure) is very much the longest, extending from just before posterior angle almost to base; *cilia* whitish, with faint brownish interruptions at extremities of nervules. UNDER SIDE.—*Hind-wing, and narrow costal, wide apical, and moderate hind-marginal border of fore-wing white, with all the crossing nervures strongly defined with black;* hind-margin edged with a black line. *Fore-wing:* field of wing pale yellow-ochreous; subcostal and median nervures clouded with black from base; extremity of discoidal cell closed by a black lunule.

♀ *All the yellow-ochreous markings enlarged and confluent, occupying all the field except a narrow blackish border.* *Fore-wing:* base very narrowly blackish; costa rather broadly bordered as far as extremity of discoidal cell, which is itself marked by a black closing lunule; a rather broad apical and hind-marginal border, more or less radiating on nervules; subcostal nervure clouded with black from base to about middle. *Hind-wing*: all the nervures and nervules defined with black except radial; hind-marginal border narrower than in fore-wing, but radiating strongly on nervules; base more or less clouded with blackish, sometimes extending along costa to about middle. UNDER SIDE.— As in ♂.

In a few ♂ individuals I have found the disco-cellular yellow-ochreous ray all but obsolete.

I met with this remarkable little butterfly in Natal in March and April 1867, but only in two localities, viz., on the Intzutze in the Great Noodsberg, and at the Umgeni Falls near Maritzburg. On each occasion there were a few specimens only, flitting slowly about the grass and herbage on the side of a ravine, and constantly settling on the stems of grasses. As noted in my *Rhopalocera Africæ Australis*, Mr. D'Urban noted quite similar habits of the insect in British Kaffraria as long ago as 1861; while on the Bashee, Colonel Bowker observed that the butterfly was fond of settling on the ground or on small stones. The last-named gentleman took many specimens in Napoleon Valley, Zululand, in 1880.

Localities of *Alæna Amazoula*.

I. South Africa.
  B. Cape Colony.
    b. Eastern Districts.—Fort Murray, near King William's Town (*W. S. M. D'Urban*). Fort Warden, Kei River (*J. H. Bowker*).
    D. Kaffraria Proper.—Bashee River (*J. H. Bowker*).
  E. Natal.
    b. Upper Districts.—Great Noodsberg. Howick, near Pietermaritzburg. Estcourt (*J. M. Hutchinson*). Ladysmith (*J. H. Bowker*).
  F. Zululand.—Napoleon Valley (*J. H. Bowker*).

GENUS DELONEURA.

Deloneura, Trim., Trans. Ent. Soc. Lond., 1868, p. 81.

IMAGO.—*Head* wide, flattened anteriorly, scaly superiorly; *eyes* smooth; *palpi* of moderate length, without scales or hairs, ascendant, widely divergent, second joint much swollen, terminal joint slender, rather long, acuminate; *antennæ* of moderate length, stout, very gradually incrassate, extremity slightly curved outwardly, subacute at tip.

Thorax short, moderately stout, clothed with some short down anteriorly, and with scales and thin hair laterally and posteriorly; breast bare, except for some scattered groups of scales. *Fore-wings* large, broad; hind-margin entire, very convex; costa strongly arched from base to middle, thence nearly straight; apex pronounced; costal nervure very short, extending only one-third length of costa; subcostal nervure five-branched; first and second nervules emitted before extremity of discoidal cell; third and fourth very short, emitted (not far apart) nearer apex than cell, fourth ending at apex; discoidal cell short, rather narrow; upper disco-cellular nervule long, directed longitudinally as in *Pentila, middle one wanting, the radial nervules having a common origin at extremity of upper one,*—lower one long, oblique, joining third median nervule at a very pronounced angle not far beyond latter's origin; median nervules emitted near together at lower part of extremity of cell. *Hind-wings* almost ovate; hind-margin very convex, entire; costa moderately arched; costal nervure ending a little beyond middle; subcostal nervure branched a little before extremity of cell, first nervule terminating at apex; upper disco-cellular nervule short, oblique, lower one much longer, scarcely curved, joining median nervure at origin of second and third nervules; submedian nervure extending to anal angle; internal nervure unusually long, extending to considerably beyond middle of inner margin; discoidal cell short, rather narrow. *Legs* short, very thick, perfectly smooth; femora and tibiæ about equal in length; tibiæ without terminal spurs, but slightly spinulose beneath; tarsi rather long, spinulose beneath; *fore-legs* not differing in either sex (?) from the rest except in being rather smaller.

Abdomen rather short and thick, scaly above, downy beneath (especially at extremity).

This very isolated genus is difficult to place satisfactorily. In the marked feature of *the common origin of the radial nervules (so that the middle disco-cellular nervule is obsolete) in the fore-wing*, it stands alone in the Family, nor have I found the same arrangement in any other genus of butterflies.[1] When to this are added its wide head, very swollen second joint of palpi, perfectly smooth legs (with tarsus of front pair apparently fully developed in both sexes [2]), and large wings without spot or marking of any kind, it will be perceived that the sum of its characters does not warrant its close association with any of the other aberrant genera of *Lycænidæ*.

The only species, *D. immaculata*, Trim., from its general aspect and pale ochreous-yellow colour, might easily be mistaken for one of the smaller *Pierinæ*. It was one of Colonel Bowker's most valuable discoveries in Kaffraria; and so exceedingly rare and local does the insect appear to be, that during the past twenty-three years no addition has been made to the three examples originally secured.

[1] In *Hesperocharis*, a genus of South-American *Pierinæ*, there is some approach to this peculiarity, the middle disco-cellular nervule being very short, so that the two radials originate not far apart.

[2] As mentioned in my original notice of this genus (*loc. cit.*, p. 82, note), I believe that two of the only three specimens known to exist are males, judging from the smaller size and much more slender abdomen. The large example figured (*loc. cit.*, pl. v. f. 4) is undoubtedly a ♀. The fore-tarsi do not differ in these three individuals.

233. (1.) Deloneura immaculata, Trimen.

Deloneura immaculata, Trim., Trans. Ent. Soc. Lond., 1868, p. 83, pl. 5, f. 4.

Exp. al., 1 in. 5–9 lin.
Ochreous-yellow, without markings of any kind. UNDER SIDE.—Hind-wing and an ill-defined costal, apical, and hind-marginal border rather paler.

Head, with palpi and antennæ, dull-black,—the former with two spots on forehead, two on summit, and two behind eyes ochreous-yellow; antennæ tipped with ochreous-yellow. *Thorax* dull-black, with pale ochreous-yellow scales and short hairs superiorly, and four or five ochreous-yellow spots laterally. *Abdomen* ochreous-yellow, mixed with fuscous superiorly.

The three examples discovered by Colonel Bowker at the end of December 1863 at Fort Bowker, on the Bashee River, remain the only known representatives of this remarkable butterfly. The first specimen was captured on the 27th December, and the other two during the remaining days before the 1st January. Colonel Bowker described the insect as very rare, and only appearing for a few days; specimens were also most difficult to procure, owing to their habit of "whirling slowly with flapping wings round the tops of trees, rising and falling, sailing away and returning." He was struck with its resemblance to the "yellow tree-moth"—I believe a species of *Arva* (a day-flying Liparide form allied to *Orgyia*)—which abounds in the wooded parts of South Africa, and it is not impossible that *Deloneura* mimics these probably protected moths, and so may escape notice among the companies of the latter. It must be observed, however, that Colonel Bowker has in vain looked out for the butterfly during all his subsequent years of active search in the various forest-clad districts he has visited; and I think it unlikely to have escaped the notice of so practised a collector if it were really native to those tracts of country.

Locality of *Deloneura immaculata.*

I. South Africa.
   D. Kaffraria Proper.—Bashee River (*J. H. Bowker*).

GENUS ARRUGIA.

Arrugia, Wallengren, K. Vetensk.-Akad. Förhandl., 1872, p. 47.
? *Zeritis,* Trim., Rhop. Afr. Aust., ii. p. 278 (1866).

IMAGO.—*Head* small,—narrower in ♀; *palpi* long, with both second and terminal joints longer in ♀; *antennæ* short or very short, thick, blunt at tip,—in ♂ gradually incrassate from base, in ♀ of almost equal thickness from very near base.

Thorax very or exceedingly robust in ♂, and not much less so in ♀. *Wings* rather elongate, quite entire; *fore-wings* with costa nearly straight, a little deflected at apex; hind-margin slightly convex in ♂, decidedly so in ♀; subcostal nervure four-branched,—the first and second nervules originating at some distance apart, before extremity of discoidal cell,—the third and fourth together about midway between extremity of cell and apex,—the fourth terminating at apex; in the ♂, an ill-defined smooth patch over the median nervules at their origin; *hind-wings* with very short discoidal cell; subcostal nervure branched considerably before middle. *Legs* short, scaly, not hairy; tibiæ very much shorter than femora, and without terminal spurs; *fore-legs alike in both sexes*, those of ♂ *having the tarsi distinctly five-jointed and with a pair of terminal curved claws*.

Abdomen long and thick,—in ♀ bulky.

The three known species of this singular genus, which is confined to South Africa, bear much resemblance to the more robust members of the section of the genus *Zeritis* represented by *Thyra*, Linn., but the characters above given—especially the full development of the fore-tarsi in the ♂—amply serve to distinguish *Arrugia*. With the exception of *Lachnocnema*, and apparently also *Deloneura*, both endemic South-African genera, I do not know of any other form of *Lycænidæ* in which the fore-tarsi are equally developed in both sexes,—a feature indicative of approach to the *Pierinæ* in the family *Papilionidæ*.

Wallengren created the genus for the reception of his *Busula* and Linné's *Protumnus*. To these I have added a third, *A. brachycera*, which, to some extent intermediate between them, is distinguished from both by the extreme shortness of the antennæ. The colouring of all is very dull, and *A. brachycera* is particularly dingy, the paler specimens of *A. Protumnus* only presenting on the upper side a considerable space of ochre-yellow. On the under side the tint is mainly a hoary-grey varied with darker markings. The sexes of *Protumnus* and *Brachycera* are much alike, but those of *Busula* very different, owing principally to the exceptional size and mass of the thorax of the ♂ and the general suffused fuscous-ochreous of his wings, in contrast with the moderately robust thorax and darker conspicuously white-spotted wings of the ♀.

A. Protumnus is widely distributed over both the Eastern and Western Districts of Cape Colony, and has occurred in the Transvaal; while *A. Busula* has a large eastern range over Kaffraria, Natal, and the Transvaal, but is not known within the limits of the Cape Colony except in Basutoland. *A. brachycera* seems to be exceedingly local; it was numerous at Knysna on the southern coast of Cape Colony, but elsewhere I know only of its very rare occurrence, and that as a small dark variety, at Cape Town.

I have not met with *Busula* in life, but have frequently captured *Protumnus* and *Brachycera*. The latter both frequent the hottest and driest spots, resting on the bare ground after the manner of many species of *Zeritis*; they are with difficulty roused, and then only shift their position by a very short though rapid flight.

234. (1.) Arrugia Protumnus, (Linnæus).

Papilio Protumnus, Linn., Mus. Lud. Ulr. Reg., p. 340, n. 158 (1764);
    and Syst. Nat., i. 2, p. 794, n. 258 (1767).
♀ *Papilio Petalus*, Cram., Pap. Exot., iii. pl. ccxliii. ff. c, D. (1782).
Papilio Silvius, Fab., Mant. Ins., ii. p. 88, n. 800 (1787); and Ent. Syst.,
    iii. p. 342, n. 299 (1793).
Polyommatus Petalus, Godt., Enc. Meth., ix. p. 672, n. 171 (1819).
Papilio Protumnus, Donov., Nat. Repos., v. pl. 161 (1827).
♂ ♀ *Zeritis? Protumnus*, Trim. [part], Rhop. Afr. Aust., ii. p. 278, n. 173
    (1866).

Exp. al., (♂) 1 in. 3–6 lin.; (♀) 1 in. 5–10½ lin.

♂ *Dull fuscous-yellow-ochreous, with black spots, and dull blackish borders*. *Fore-wing*: a large, more or less defined, elongate spot at end of, and running into, cell; beyond it, from costa, a row of confluent spots, angulated inwardly on third median nervule and extending to middle of submedian nervure (the whole of the row below third median is often wanting or very faintly marked); costa rather narrowly bordered with greyish; hind-marginal blackish border broad, nearly even throughout; *a smooth greyish space covers median nervules at their origin*. *Hind-wing*: the broad blackish border leaves only an inner-marginal and discal ochreous space, not extending above discoidal nervule; crossing ochreous nearly to submedian nervure an ill-defined macular stripe. UNDER SIDE.—*Hind-wing and borders of fore-wing hoary-grey, the former crossed by two rows of faint-brownish confluent spots*. *Fore-wing: pale yellow-ochreous;* spots as above, but smaller and more distinct,—that closing cell divided into two, of which the inner is very much the smaller; below the inner of these two, outside cell, a small round spot; hind-marginal border faint-fuscous towards anal angle, sometimes marked interiorly by a row of very indistinct, small, dark spots. *Hind-wing:* markings variable and often very indistinct; macular row before middle very irregular and sometimes very much broken up,—a portion of it always marks end of cell, where it is often confluent with the more regular, broader, and better-defined discal row; two or three indistinct brownish spots near base. *Cilia* white or whitish, interrupted with dull-blackish on nervules.

♀ Similar, but *ochreous much clearer and yellower*, so that the black spots are more conspicuous; *borders brownish*. *Fore-wing*: spot at end of cell often divided, sometimes confluent with discal row on third median; no grey space at origin of median nervules. *Hind-wing:* a streak closing cell sometimes visible. UNDER SIDE.—As in ♂, but markings of *hind-wing* sometimes better marked.

This species varies much in both sexes, both as regards the extent and clearness of the ochreous-yellow of the upper side, and the definition, size, and shape of the spots on the under side of the hind-wing. Near Cape Town and at Stellenbosch, the dusky typical form (described by Linnæus and the ♀ figured by Cramer) prevails, and in these the discal black spots are almost always strongly developed on the upper side. At Triangle Station (elevation

3193 feet) in the Worcester District, a ♂ and two ♂s were taken by M. L. Péringuey, exhibiting much divergence from the typical form. The ♂ is of ordinary size, and, except for a rather marked yellow basal suffusion in the fore-wing, more obscure than usual on the upper side, the discal ochre-yellow being reduced to a row of small separate spots. The ♀s, on the contrary, are unusually large (exp. 1 in. 10½ lin.), with the yellow on the upper side much paler and largely developed from the bases outward. On the under side these three examples agree in having the markings of the hind-wing only faintly outlined, but in the ♂ the hoary tint still prevails, while in the ♀s a general faint-brownish tinge makes the markings even more indistinct. At Plettenberg Bay I met with two ♂s and a ♀ not unlike those just described, and agreeing with them in the remarkable detail of having the small submarginal blackish spots on the under side much less indistinct than in the typical form and sagittate in shape—in this particular resembling *A. brachycera*, Trim. The ♂s, however, are even duller than the Worcester ♂, having no basal yellow, and the pale space representing discal ochre-yellow, being very obscure dull-yellowish in both wings. On the under side these specimens have the markings much narrowed but very well defined; and in the hind-wing the terminal disco-cellular spot emits superiorly a long ray extending along upper part of discoidal cell almost to base. In these variations from the Western and Southern Districts of the Cape Colony, the ♂s are much more sombre-hued than the ♀s, but this does not appear to be the case farther to the North and East, for I have before me three ♂s from near Burghersdorp and one from the Transvaal which quite rival their ♀s (brighter in those interior tracts than elsewhere) in the extent and brightness of the upper-side ochre-yellow. Of this variation, ♀s, but not ♂s, have reached me from near Grahamstown, Griqualand West,[1] and the Carnarvon District of the Cape Colony; they (as well as the ♂s just mentioned) agree pretty closely with the typical form as regards the under-side markings, and all the ♀s have the lower spots of the discal row in the fore-wing obsolete on the upper side.

This curious butterfly only appears at the hottest time of the year, from the end of November to early in February. It is extremely local, occurring in spots of very limited extent, and is not by any means numerous in these favoured haunts. Its habits are quite those of the more sluggish species of *Zeritis*; it always frequents bare sandy spots, whether on low lands or mountain-sides, and almost invariably settles on the ground, seldom moving unless disturbed, and then taking but a very short though swift flight. Mr. Morant noted his Transvaal examples as captured on 21st and 23d October 1870.

Localities of *Arrugia Protumnus*.

I. South Africa.
 B. Cape Colony.
  a. Western Districts.—Cape Town. Stellenbosch. Triangle Station, Worcester District (*L. Péringuey*). Plettenberg Bay. Van Wyk's Vley, Carnarvon District (*E. G. Alston*). Garries and Springbokfontein, Little Namaqualand (*L. Péringuey*).
  b. Eastern Districts.—Grahamstown (*M. E. Barber*). New Year's River (*H. I. Atherstone*). Burghersdorp (*D. R. Kannemeyer* and *J. H. Bowker*).
  c. Griqualand West.—Vaal River (*J. H. Bowker*). Kimberley (*H. Grose Smith*).
 K. Transvaal. Kalkfontein (*W. Morant*).

---

[1] I have since seen a ♂ from Kimberley of this coloration belonging to Mr. H. Grose Smith; and another (November 1885) has reached me from the Carnarvon District.

235. (2.) Arrugia brachycera, Trimen.

PLATE IX. fig. 7 (♂).

Zeritis Protumnus, Trim. [part], Rhop. Afr. Aust., ii. p. 279 *obs.* (1866).
Arrugia brachycera, Trim., Trans. Ent. Soc. Lond., 1883, p. 353.

Exp. al., (♂) 1 in. 4-6 lin.; (♀) 1 in. 7-9 lin.

♂ *Dull fuscous-grey with a slight ochraceous tinge; in both wings a blackish terminal disco-cellular spot and row of discal spots (indistinct in hind-wing).* Fore-wing: hind-marginal area beyond discal spots darker than basal area; disco-cellular spot rather small, ill-defined inwardly; between it and discal spots a dull suffused space of pale dingy-grey tinged with ochraceous, radiating on the basal portion of the three median nervules; spots of discal row ill-defined, the first four confluent and forming a moderately broad costal bar as far as third median nervule,—the other four small, indistinct, separate, in a row inclining inwardly, between third median nervule and submedian nervure. *Hind-wing*: disco-cellular lunule, and median discal curved row of three or four spots indistinct or almost obsolete. *Cilia* dull-white, with rather narrow fuscous interruptions at extremities of nervules. UNDER SIDE.—*Hind-wing and apex of fore-wing dull houry-grey.* Fore-wing: basal area whitish-grey, scarcely separable from discal suffused space, which is dingy-whitish and much more extensive than on upper side, forming a band beyond the discal row of spots; this row and the disco-cellular spot are very distinctly defined on the pale ground-colour; near base, two less distinct small fuscous spots, one in cell, the other below it; outwardly edging discal dingy-whitish band, a row of somewhat ill-defined fuscous sagittate marks; hind-marginal border pale-brownish from a little below apex. *Hind-wing*: disco-cellular spot (irregularly reniform), and chain-like, almost regular discal row of spots, very pale brownish, with a thin dark-brown edging line (stronger on inner edge of row), relieved externally by a thin white line; in basal area, the traces of two highly irregular transverse rows of broken-up pale-brownish spots, of which the first and last spots of the outer row are least indistinct; parallel to and not far from hind-margin an almost obsolete row of minute black sub-sagittiform spots; hind-marginal border clouded with pale-brownish.

♀ *Rather paler and slightly more ochraceous.* Fore-wing: discal pale space less obscure, not radiating on median nervules, but *extending more or less distinctly beyond discal row of spots*. *Hind-wing*: spots not quite so indistinct. UNDER SIDE.—Hind-wing and apex of fore-wing less hoary, more brownish-grey. *Hind-wing*: spots of basal area (except first and last of outer row) altogether obsolete; sagittiform spots of submarginal row much larger and more distinct.

Variety ♂.—*Exp. al.* 1 in. 2-3½ lin.

Darker than type-form. *Fore-wing*: no paler discal space; blackish spots almost obsolete. *Hind-wing*: spots altogether obsolete. UNDER SIDE.—*Fore-wing*: much obscured; disc grey, the spots dusky and ill-defined.

Hab.—Cape Town.

The absence of yellow-ochreous colouring at once serves to distinguish this dingy Lycænide from *A. Protumnus* (Linn.), and tends to approximate it to *A. basuta*, Walgrn. From the latter, as far as the ♂ is concerned, *A. brachycera* may be known by its darker colouring, larger and more pronounced spots, and want of whitish on disc; while on the under side it is considerably darker, has the fore-wing spots much larger (with the marked exception of the spot near base below cell), and the discal row of hind-wing much broader and more regular. The ♀ is readily recognised by wanting the conspicuous white discal markings of the ♀ *A. basuta* on the upper side. Apart from pattern and colouring, however, *A. brachycera* exhibits a remarkable structural distinction in the extreme shortness of the antennæ. This is noticeable in both sexes, but especially in the ♀, whose antennæ are only about $2\frac{1}{2}$ lines in length,—shorter than in *Protumnus*, and much shorter than in *Basuta*.

The variety noted seems to be very rare. I have only met with three examples, all near Cape Town; and a fourth, in the collection of the South African Museum, was, I believe, captured by Mr. E. L. Layard in the same locality. It is, however, such a small obscure insect, that it would very readily escape notice. Two specimens I took occurred at a considerable elevation on the southern spur of Table Mountain, and the third at the base of the mountain itself.

With the exception of these specimens of the variety, all the *Brachycera* I have seen were taken at Knysna. Like *Protumnus*, they were strictly summer butterflies, appearing between the end of November and beginning of February. They settle on the bare ground, and I used often to find them sitting on the heaped-up dust of the waggon-roads, to which they would return after being roused by the passing passenger or vehicle. Their flight is weak and very short.

Localities of *Arrugia brachycera*.

I. South Africa.
    B. Cape Colony.
        a. Western Districts.—Cape Town [variety]. Knysna.

236. (3.) Arrugia Basuta, Wallengren.

PLATE IX. figs. 8 (♂), 8a (♀).

♂ ♀ *Zeritis? Basuta*, Wallengr., K. Sv. Vet.-Akad. Handl., 1857,—Lep. Rhop. Caffr., p. 46.
♂ ♀ *Zeritis Protumnus*, Trim., Var. A., Rhop. Afr. Aust., ii. p. 279 (1866).
Zeritis Basuta, Trim., Trans. Ent. Soc. Lond., 1870, p. 377.
Arrugia Basuta, Wallengr., K. Sv. Vet.-Akad. Förhandl., 1872, No. 3, p. 47.
♂ ♀ *Zeritis Zaraces*, Hewits., Trans. Ent. Soc. Lond., 1874, p. 354.

Exp. al., (♂) 1 in. 5–7 lin.; (♀) 1 in. 6–9 lin.

♂ Dull pale fuscous tinged with yellowish-ochreous; disco-cellular and discal blackish spots arranged as in Protumnus and Brachycera, but smaller, more separate, and less distinct (especially in hind-wing).

Fore-wing: immediately beyond terminal disco-cellular spot *a whitish spot*, and outwardly edging discal row of blackish spots *a series of six or seven more or less developed dull-whitish spots;* blackish spot in cell and another below its outside cell larger than in *Protumnus*, but suffused. *Hind-wing*: almost spotless, the paler and darker markings of the disc being extremely indistinct; in some specimens a broad deeper-fuscous shade over costal area from base to beyond middle. UNDER SIDE.—*Almost uniform whitish-grey, with a very faint yellowish tinge over disc of fore-wing; all the markings of hind-wing very thin and faint;* a common submarginal row of very small blackish spots, not *sagittate. Fore-wing:* all the blackish spots distinct, but separate and very small, except the spot below discoidal cell, which is enlarged, and confluent with a fuscous mark extending to base; a small additional spot in discoidal cell near base. *Hind-wing:* spots arranged as in *Protumnus*, but separate, very much smaller, and extremely indistinct; the discal row sharply interrupted on second subcostal nervule. *Cilia* greyish or greyish-white, very indistinctly varied with fuscous on nervules.

♀ *Darker, without yellow-ochreous tinge; all the black and whitish markings much enlarged. Fore-wing:* black spots strongly marked and more or less confluent, as in typical *Protumnus; the white spots very conspicuous,* and those of discal row occasionally confluent; usually a whitish spot between the terminal and interior disco-cellular black spots. *Hind-wing:* disco-cellular terminal spot and discal row of spots better marked than in ♂, sometimes quite distinct, and relieved by whitish suffused markings corresponding to the much more conspicuous white ones of fore-wing. UNDER SIDE.—*Hind-wing and border of fore-wing pale ashy-grey; disc of fore-wing much whiter. Fore-wing:* spots larger than in ♂, but still separate. *Hind-wing:* all the spots larger and much more distinct, being of a light-brownish tint (but smaller than in *Protumnus*). *Cilia* greyish mixed, but not regularly interrupted with fuscous.

The characters italicised afford ready distinctions between this species and its congeners. In its dingy colouring the ♂ is nearer to the ♂ *Brachycera;* but the ♀, though nearer to the ♀ *Protumnus* in the size and confluence of the black markings of the fore-wing, differs strikingly from both that species and the ♀ *Brachycera* in the conspicuous white spots of the same wing. The paleness and faint small spotting of the under side distinguish both sexes of *Basuta* from the other species; and the greater development of the thorax (which in the ♂ is most remarkable, and gives that sex the look of a Hesperide) is also a singular distinction.

I have not met with this interesting species in life. Colonel Bowker wrote to me that it frequented grassy spots at Fort Bowker, near the Bashee River; and Mr. W. Morant noted that he had found the species in the Orange Free State on stony hillsides during the month of December, and in the Transvaal on the 24th October 1870. Two of the three ♀s sent by Mr. T. Ayres from the Potchefstroom District have the white markings of the fore-wings better developed than in any other examples I have seen ; their general colouring is rather paler and their black spots are smaller than usual.

Localities of *Arrugia Basuta*.

I. South Africa.
   B. Cape Colony.
      d. Basutoland.—Maseru (*J. H. Bowker*).
   D. Kaffraria Proper.—Bashee River (*J. H. Bowker*).
   E. Natal.
      a. Coast Districts.—Pinetown (*W. Morant*).
      b. Upper Districts.—Estcourt (*J. M. Hutchinson*).
   K. Transvaal.—Potchefstroom District (*T. Ayres*). Locality not noted (*W. Morant*).

GENUS LACHNOCNEMA, *N.G.*[1]

Lucia [part], Hopffer, in Peter's Reise nach Mossamb.,—Ins., p. 412 (1862).
? *Lucia*, Trim., Rhop. Afr. Aust., ii. p. 280 (1866).

IMAGO.—*Head* small, roughly hairy (especially in front); *eyes* densely hairy; *palpi* long, ascendant, densely clothed beneath with bristly hair (especially on first and second joints),—terminal joint long, acute; *antennae* short, thick, very gradually and cylindrically clavate, blunt at tip.

Thorax short, moderately stout, or rather slender, very hairy throughout. *Fore-wings* rather long; hind-margin convex, entire; costa very slightly convex; apex pronounced; inner margin thinly ciliated; costal nervure short, ending about middle; subcostal nervure four-branched,—first and second nervules emitted before extremity of discoidal cell,—third and fourth at origins much nearer to extremity of cell than to apex (fourth ending at apex); discoidal cell of moderate width, rather short; upper radial nervule joined to subcostal nervure at extremity of cell ; disco-cellular nervules almost vertical,—lower one joining third median nervule at some distance from latter's origin. *Hind-wings* rather long, broad ; hind-margin very convex, entire ; inner margin with long thin cilia ; costal nervure extending to apex; subcostal nervure branched at extremity of discoidal cell ; disco-cellular nervules almost equal in length, rather oblique,—lower one joining third median nervule at an angle some way beyond latter's origin ; discoidal cell very short, rather broad ; submedian nervure unusually short; internal nervure very short. (Hind-marginal cilia in both wings dense and long.) *Legs* short, stout; femora and (especially) tibiae clothed with scales and with very long extremely dense woolly

[1] From λάχνη, wool, and κνήμη, tibia.

hair, hiding basal part of tarsi; tarsi scaly, and with a few short bristles, stout, spinose beneath: *fore-legs in both sexes like the rest*, except in being slightly smaller and perhaps even more hairy.

Abdomen of moderate length, arched dorsally, much compressed laterally.

As long ago as 1847, Boisduval (Appendix to Delegorgue's *Voyage dans l'Afrique Australe*, p. 588) observed that his *Lycœna Delegorguei* would doubtless constitute a new genus, and mentioned the characteristic features of its legs, antennæ, and palpi, but did not give it any generic name. Hopffer referred the butterfly to the genus *Lucia* of Swainson, and I, with the expression of much doubt and uncertainty, provisionally followed him in 1866. I find that *Delegorguei* (identified by Mr. A. G. Butler with *Bibulus*, Fab.) has no agreement with Swainson's type, the Australian *Lucia limbaria*, but is less remote from the Cingalese *Lucia Epius*, Westw., the type of Moore's genus *Spalgis* (*Proc. Zool. Soc. Lond.*, 1879, p. 137; and *Lep. Ceylon*, i. p. 70, 1881). From the latter, however, it is well distinguished by the densely hairy palpi, extraordinarily hairy tibiæ, the first subcostal nervule rising from the extremity of the discoidal cell of the hind-wings, and particularly by the completely articulated and two-clawed fore-tarsi of the male. This last character is, as far as I can learn, shown only in two other genera of the family, viz., *Arrugia* and *Deloneura*, and constitutes a distinction of great importance. Superficially, *Lachnocnema* bears considerable resemblance to the very hirsute Mediterranean and West Asian genus *Thestor*, Hübn. (especially to *T. Mauritanicus*, Lucas), but the latter genus is most remarkably distinguished by its extremely short and very massively spurred fore and middle tibiæ, while the fore-tarsi of the ♂ are of the ordinary unarticulated type with a single curved terminal claw.

I have separated from *L. Bibulus*, Fab., under the name of *L. D'Urbani*, the smaller and duller form alluded to at p. 281 of my earlier book, as I found that it was of constantly slenderer structure, and had a different station. I have received from the Limpopo River a very large ♀ *Lachnocnema*, which I believe represents a third species; it is very pale beneath, and has barely a trace of the characteristic steely dots.

In this genus the ♂ is of a plain uniform brown above, but the ♀ *Bibulus* has a more or less developed whitish or white disc in both wings, and the ♀ *D'Urbani* a discal suffusion of pale grey. On the under side the reddish-brown and brownish macular transverse bands, and the hind-margins are ornamented with glittering steely points or dots. *L. Bibulus* has a wide distribution over South-Eastern Africa, and *L. D'Urbani* almost as large a one, but the former only is recorded as a native of Mozambique, while the latter is not known from any place north-east of Natal.

237. (1.) **Lachnocnema Bibulus,** Fabricius.

♂ *Papilio Bibulus*, Fab., Ent. Syst., iii. 1, p. 307, n. 163 (1793).
♀ *Papilio Laches*, Fab., op. cit., p. 317, n. 199 (1793).
♂ *Papilio Bibulus*, Don., "Ins. Ind., pl. 46, f. 1 (1800)."
♂ ♀ *Lycæna Delegorguei*, Boisd., App. Voy. Deleg. l'Afr. Aust., p. 588 (1847).
♂ ♀ *Lucia Delegorguei*, Hopff., Peters' Reise Mossamb., Ins., p. 411 (1862).
♂ ♀ *Lucia? Delegorguei*, Trim., Rhop. Afr. Aust., ii. p. 280, n. 174 (1866).

Exp. al., (♂) 1 in.—1 in. 2 lin.; (♀) 10½ lin.—1 in. 2 lin.

♂ *Fuscous-brown, with a very slight purplish gloss. Fore-wing:* a faint indication of a darker spot at end of cell. UNDER SIDE.—Much paler. *Fore-wing:* a row of very brilliant steely dots along costa, larger and dark-edged beyond middle; on hind-margin a row of five larger steely spots, inwardly black-edged, between apex and first median nervule; spot at end of cell large, ill-defined, dark-brown; a little beyond it a similar larger, sub-quadrate mark on discoidal nervules, usually marked with some steely scales; costa near base and hind-marginal edge tinged with golden-ochreous. *Hind-wing:* a central straight transverse band of glossy ochrey-brown, of unequal width, marked on both its edges with steely points (of which there are also a few in middle of stripe) from costa beyond to inner margin before middle: a good-sized similarly hued spot in cell and two others on costa before middle; beyond middle a row of five contiguous, pale-edged, steely-centred ochrey spots, obsolete above second subcostal nervule; hind-marginal spots as in fore-wing, but more conspicuous and occupying the whole length—the three next anal angle larger than the rest, geminate, with black edges; inner marginal and discal region thickly irrorated with violaceous-white scales. *Cilia* greyish.

♀ *Bases with bluish-grey hairs and scales; in each wing a discal space of white outwardly ill-defined. Cilia* paler, nearly white in hind-wing, inconspicuously interrupted with fuscous. UNDER SIDE.—Markings similar to ♂, but *white occupies the ground*, except on hind-margins and on costa of fore-wing. *Fore-wing:* discal and cellular spots very conspicuous, the latter confluent with a brownish space on median nervure from base. *Hind-wing:* markings very conspicuous on white ground.

Mr. A. G. Butler first pointed out (*Cat. Fab. Diurn. Lep. Brit. Mus.*, 1869, p. 175) the close alliance between, if not the identity of, the *Delegorguei* of Boisduval and the *Bibulus* of Fabricius,—he having found both sexes well figured in Jones's *Icones*, quoted by Fabricius, op. cit. The description of *Bibulus* by this author is so inaccurate that no one without the aid of Jones's figure could have assigned it to *Delegorguei* ♂; but that of *Laches*, Fab., applies thoroughly to *Delegorguei* ♀.

Both sexes vary much in size; but as regards markings and colouring, while the ♂ seems remarkably constant, the ♀ exhibits much variation in the discal white on the upper side of both wings, ranging from a mere dull-whitish mark (in hind-wings scarcely distinguishable) to a broad space of pure-white occupying nearly all the disc except the upper part.

Bibulus has a wide range over Eastern South Africa, and is recorded also from Mozambique. In the Cape Colony I am not aware of its occurrence to the west of Port Elizabeth. The butterfly is by no means uncommon; it usually flies but little, keeping about the minor branches and twigs of shrubby plants. When at rest on these, it is with difficulty to be seen, the pattern and colouring of the under side according well with the bark and lichens, while the densely woolly legs form a solid base to the insect, disguising its real character and increasing its likeness to some vegetable growth on the twigs. The ♂ s appear at times to exhibit more activity, Colonel Bowker having in 1865 written from the Tsomo River in Kaffraria that the butterfly was "very numerous at the Ohita, Kreli's old kraal," and that "the ♂ s kept flying, often five or six together, about the tops of the trees, darting and whirling round and round like a lot of flies; while the ♀ s were quietly settled on the trees, feeding on the moisture from the bark or on the injured galls of the wild vine."

I have met with *Bibulus* pretty numerously in the height of summer, from the end of January to the end of March, and rarely in the winter, during June and August. Colonel Bowker has also captured it in May.

Localities of *Lachnocnema Bibulus*.

I. South Africa.
B. Cape Colony.
  b. Eastern Districts.—Port Elizabeth. Uitenhage. Grahamstown. Peddie District. Bathurst District (*M. E. Barber*). King William's Town (*W. S. M. D'Urban* and *J. H. Bowker*). Keiskamma (*M. E. Barber*).
D. Kaffraria Proper.—Butterworth, Tsomo River, and Bashee River (*J. H. Bowker*).
E. Natal.
  a. Coast Districts.—D'Urban. Verulam. Mapumulo. Mouth of Tugela River (*J. H. Bowker*).
  b. Upper Districts.—Maritzburg (*J. H. Bowker*). Estcourt and Weenen County (*J. M. Hutchinson*).
F. Zululand.—Napoleon Valley (*J. H. Bowker*).
H. Delagoa Bay.—Lourenço Marques (*Mrs. Monteiro*).

II. Other African Regions.
A. South Tropical.
  b. Eastern Coast.—" Querimba."—Hopffer.

238. (2.) Lachnocnema D'Urbani, *sp. nov.*

♂ ♀ *Lucia? Delegorguei*, Trim., part., Rhop. Afr. Aust., ii. p. 281 (1866).

Exp. al., (♂) 1 1½ lin.—1 in. 2 lin.; (♀) 10 lin.—1 in. 2 lin.

Very closely allied to *L. Bibulus* (Fab.)

♂ *Dull greyish-brown, much paler than in Bibulus; terminal discocellular darker spot in fore-wing more apparent.* UNDER SIDE.—*Very pale grey*, with a faint yellowish tinge over hind-wing and narrow costal border of fore-wing; steely spots as in *Bibulus*, but less brilliant; *the transverse band and discal row of spots much yellower.* Fore-wing: base dusky (except on inner margin) as far as end of discoidal cell;

quadrate mark on discoidal nervules much farther from extremity of cell. *Hind-wing*: central band much narrower at its beginning on costa, more irregular, being broken up into mostly separate unequal spots; discal row of spots also more irregular.

♀ *Very much paler and duller than in Bibulus, without dark costal borders and with only ill-defined dusky hind-marginal borders; no discal white patches, but a diffused very pale grey discal shade, inclining to whitish in hind-wing;* bluish-grey basal suffusion obsolete. *Fore-wing*: terminal disco-cellular fuscous spot rather conspicuous, isolated, much smaller. *Hind-wing*: a more or less indistinct sub-lunulate terminal disco-cellular fuscous spot. UNDER SIDE.—As in ♂; much duller than in *Bibulus*, the markings less distinct. *Fore-wing*: the discal quadrate marking often (in four out of eight examples) expanded into a series of dusky spots extending from near costa to first median nervule.

In addition to the differences above indicated, *D'Urbani* presents a slenderer body, less densely woolly legs, and considerably longer and narrower wings than *Bibulus*. Mr. W. S. M. D'Urban, who (as noted in my work above quoted) first brought specimens of this form to my notice, and was disposed to regard it as a distinct species, told me that in British Kaffraria it appeared earlier in the year than the typical *Bibulus*; and both Mrs. Barber and Colonel Bowker have noted its occurrence in a different station, viz., among long grass and low bushes. The former of these two observers wrote that the stronghold of this butterfly was along the coast of Bathurst District, where it occurred in great numbers; and the latter found it among long dry grass on the Bashee River, and noted that it kept on the wing for a long time together. I took a single ♀ flitting about some bushes on the margin of the New Year's River, in the Albany District, on 10th February 1870. The geographical range corresponds pretty closely with that of *Bibulus*, but I have not seen any examples from any part to the north-eastward of Natal.

I have great pleasure in naming this butterfly after Mr. W. S. M. D'Urban, F.L.S., Curator of the Exeter Museum, who added so much to our knowledge of the South-African *Rhopalocera* during his sojourn in British Kaffraria in 1860-61.

Localities of *Lachnocnema D'Urbani*.

I. South Africa.
   B. Cape Colony.
      b. Eastern Districts.—Grahamstown (*M. E. Barber*). New Year's River, Albany District. Between Kowie and Fish Rivers, Coast of Bathurst District (*M. E. Barber*). King William's Town (*W. D'Urban*).
   D. Kaffraria Proper.—Bashee River (*J. H. Bowker*).
   E. Natal.
      b. Upper Districts.—Estcourt (*J. M. Hutchinson*). Ladysmith, Biggarsberg, and Rorke's Drift (*J. H. Bowker*).

www.ingramcontent.com/pod-product-compliance
Lightning Source LLC
Chambersburg PA
CBHW031742230426
43669CB00007B/446